REFRACTORY METAL ALLOYS
Metallurgy and Technology

Editors

I. Machlin
Materials and Processes Branch
Naval Air Systems Command
Department of the Navy
Washington, D.C.

R. T. Begley
Astronuclear Laboratory
Westinghouse Electric Company
Pittsburgh, Pennsylvania

and

E. D. Weisert
Materials Research
Rocketdyne Division
North American Rockwell Company
Canoga Park, California

A Publication of The Metallurgical Society of AIME

REFRACTORY METAL ALLOYS
Metallurgy and Technology

Proceedings of a Symposium on Metallurgy and Technology of Refractory Metals held in Washington, D.C., April 25-26, 1968. Sponsored by the Refractory Metals Committee, Institute of Metals Division, The Metallurgical Society of AIME and the National Aeronautics and Space Administration, Washington, D.C.

℗ PLENUM PRESS • NEW YORK • 1968

Preface

This publication documents Proceedings of the Symposium on Metallurgy and Technology of Refractory Metal Alloys, held in Washington, D.C. at the Washington Hilton Hotel on April 25-26, 1968, under sponsorship of the Refractory Metals Committee, Institute of Metals Division, of the Metallurgical Society of AIME, and the National Aeronautics and Space Administration. The Symposium presented critical reviews of selected topics in refractory metal alloys, thereby contributing to an in-depth understanding of the state-of-the-art, and establishing a base line for further research, development, and application.

This Symposium is fifth in a series of conferences on refractory metals, sponsored by the Metallurgical Society of AIME. Publications issuing from the conferences are valuable technical and historical source books, tracing the evolution of refractory metals from early laboratory alloying studies to their present status as useful engineering materials.

Refractory metals are arbitrarily defined by melting point. A melting temperature of over 3500°F was selected as the minimum for this Symposium, thus excluding chromium and vanadium, which logically could be treated with other refractory metals in Groups VA and VIA of the periodic table. The Refractory Metals Committee is planning reviews of chromium and vanadium in subsequent conferences.

Of eleven metals with melting points above 3500°F, four have been advanced significantly in recent years, as bases for structural alloys, namely Cb, Mo, Ta, and W. The remaining seven (Re, Hf, Ir, Rh, Ru, Tc and Os) have received lesser development, a fact reflected in the Symposium program. Perhaps future Symposia will show a different emphasis.

At the time of the first High Temperature Materials Conference in 1957 (Cleveland, Ohio) only the Mo-0.5% Ti alloy was in commercial production. Currently (1968) several dozen useful Cb, Mo, Ta, and W alloys have evolved, with many obtainable in good quality as sheet, foil, tube, wire, bar, and forgings. Although complete information on aerospace, nuclear, and commercial applications was not available for presentation at the Symposium, it appears that refractory metals are serving essential structural functions, predominantly in prototype studies of propulsion, power generation, and hypersonic flight devices, with few requirements involving large quantities.

A major factor limiting the utility of refractory metals in high temperature processes continues to be the lack of alloys (coated or otherwise) reliably operable for long times (multi-hundreds of hours) in oxidizing environments. Nonetheless, there are several environments in which refractory metals are optimum materials, including vacua, inert gases, and liquid alkali metals. In oxidizing environments coated refractory metals can serve for limited periods, under appropriate conditions.

The U. S. capability in refractory metals for structural uses has
been attained with Government support, supplemented by industry, of programs ranging from basic research to fabrication of hardware. Much work
remains, if desired improvements in oxidation-resistance, strength,
ductility, weldability and fabricability are to be attained. However, it
appears likely that research efforts, in the future, will be selectively
directed towards solving those critical problems limiting the useful
potential of refractory metals. The invited papers in this publication
were intended to provide maximum information required in judging future
needs and research directions.

We are grateful to the National Aeronautics and Space Administration
for support enabling us to publish these proceedings. Much credit is due
to authors of papers, who have made significant professional contributions
by directing many hours of their leisure time to hard work. Our thanks
are also extended to session chairmen and to members of the Refractory
Metals Committee, who participated in organizing this Symposium.

Distribution of this book in a timely and economical manner has been
facilitated by photographic reproduction of authors' manuscripts, a procedure necessitating some relaxation of standards for editorial uniformity
and format. We trust that the advantage of timely publication will out-
weigh perfection of format. It is regretted that the following planned
papers were not available: Refractory Metals in Aerospace Structures, Use
of Refractory Metals in Aerospace Propulsion, and Molybdenum and Moly-
bdenum-Base-Alloys-Recent Developments.

April 1968   I. Machlin
                                    R. T. Begley
                                    E. D. Weisert

Contents

BASIC STRENGTHENING MECHANISMS
IN REFRACTORY METALS

B. A. Wilcox

ABSTRACT

Various mechanisms of strengthening refractory metals and
alloys are discussed. Special emphasis is placed on the microstructural
features which can be varied to produce strengthening, and relevant
illustrations are given. The general areas reviewed include: (a)
strengthening by second phase particles, (b) solid solution strengthening, (c) strain hardening and grain size refinement, (d) retaining
worked structures at high temperatures, (e) dynamic strengthening, and
(f) fiber reinforcement.

B. A. Wilcox is a Fellow in the Metal Science Group, Battelle Memorial
Institute, Columbus Laboratories, Columbus, Ohio.

INTRODUCTION

In order to capitalize fully on the potential of refractory metals in structural applications, it is necessary to strengthen the base metals. Thus, improving the strength of refractory metals has been one of the most important topics of research and development on these materials, and in previous AIME Refractory Metals Symposia this area has been the subject of major review papers (1,2).

This paper approaches strengthening in refractory metals by considering the various mechanisms that have been employed to increase the yield, ultimate and creep strengths. The subjects treated are: (a) strengthening by second phase particles, (b) solid solution strengthening, (c) strain hardening and grain size refinement, (d) retaining worked structures at high temperatures, (e) dynamic strengthening, and (f) fiber reinforcement. Radiation hardening is a special topic which is not considered here, since it is not a technique that materials designers intentionally use to strengthen refractory metals. Also, texture strengthening is not discussed, because in the major refractory metals (W, Mo, Cr, Ta, Nb, V) the BCC structure is not amenable to pronounced strengthening by preferred orientation, as compared to HCP metals.

The main emphasis here is placed on strengthening at room and elevated temperatures. Many of the strengthening mechanisms, e.g., dispersion hardening, solution strengthening, working and grain size refinement, etc., also are operative at low temperatures. However, at low temperatures additional factors, such as overcoming the Peierls barrier, are important. Conrad (3) has reviewed this area including numerous results on the BCC refractory metals. For practical purposes, ductility as well as strength is important to consider in refractory metal alloy development. In many cases those factors which influence the strength of refractory metals also affect the fracture behavior. However, this topic is not treated here, since it has been well reviewed in several recent papers (4,5).

A complete review of the literature on refractory metal strengthening is beyond the scope of this paper. Rather, the author has tried to give details of pertinent strengthening mechanisms, and to illustrate these with appropriate examples based on refractory metal alloys. Thus, it is the intention of this paper to provide the alloy developer with basic guidelines which will be of assistance in selecting methods to improve the strength of refractory metals.

It is often desirable in practice to employ more than one strengthening mechanism. This paper considers that the various strengthening mechanisms can supplement one another and they are often interrelated. Even though the strengthening effects may not be directly additive, it is convenient to formulate a phenomenological expression such that the strength, σ, of an alloy is written as:

$$\sigma = \sigma_o + \sum_i \sigma_i \quad . \tag{1}$$

Here σ_0 is defined as the lattice friction stress, or the stress required to move a dislocation through an otherwise perfect lattice of the pure base metal. The strength increments arising from each ith mechanism are then added to σ_0. A final expression is presented in the SUMMARY, which incorporates the important terms from each strengthening mechanism and serves to illustrate what structural features can be varied to produce strengthening.

STRENGTHENING BY SECOND PHASE PARTICLES

It is now apparent that strengthening by second phase particles can be both <u>direct</u> and <u>indirect</u>. This has been demonstrated for SAP-type alloys (Al + Al$_2$O$_3$) (6) and TD Nickel (Ni + 2 vol.% ThO$_2$) (7,8). <u>Direct</u> strengthening is caused by particles acting as barriers to dislocation motion during deformation. The strength increase can usually be ratio-nalized in terms of dislocation theories of dispersion hardening such as Orowan bowing between particles, cross-slip around particles, climb over particles, particle shearing, etc.[*] <u>Indirect</u> strengthening can arise when a dispersion-containing metal is thermomechanically processed so that the particles help to develop and stabilize a worked structure. Here, there is an additional strength increment due to the fine grain size and substructure, which can be beneficial at high temperatures if the structure is stable.

This section is devoted primarily to direct strengthening of refractory metals by second phase particles. Examples include cases where the particles were incorporated by: (a) precipitation during heat treatment, (b) inert oxide dispersoids added by powder metallurgy, and (c) internal oxidation and nitriding. Some examples of indirect strengthening by second-phase particles are given in the section on RETAINING WORKED STRUCTURES AT HIGH TEMPERATURES.

Precipitation by Heat Treatment

Perhaps the best documented refractory metal system which uti-lizes precipitation hardening as a means of improving the strength is based on Ti, Zr, and C additions to Mo, e.g., Mo-TZM and Mo-TZC. Chang (1,10-13) has studied this system extensively, and identified the carbide precipitates (Mo$_2$C, ZrC, and TiC), and the influence of heat treatment on structure and mechanical properties. This "TZC" type of precipitation hardening, which depends on the formation of interstitial compounds with reactive metals (Zr, Ti, Hf), also has been extended to Nb- (14,15), Ta- (16), and W-base (17) alloys.

---

[*] Excellent reviews of strengthening by second phase particles are given in References (6) and (9).

An example, where the direct strengthening effects of precipitates in refractory metals was rationalized in terms of a specific
mechanism, was reported by Wilcox and Gilbert (18). They produced fine
carbides in Mo-TZM by annealing for one hour at 2100°C and quenching into
molten tin at 250°C. The carbides, shown in Figure 1, had an average
particle diameter, $2\bar{r}_v$, of 560 Å, and a mean planar center-to-center
spacing, λ, of 3130 Å. The strengthening effect of these fine carbides
is illustrated in Figure 2. Here the strength and ductility of as-
quenched material are compared with furnace-cooled material, where the
carbides were very coarse and exerted no measurable strengthening effect.
Over the temperature range, 200-800°C, the as-quenched yield and ultimate
strengths are ~25-40,000 psi higher than those of the slowly cooled
specimens. At about 950°C there is a rapid drop in the strength of the
quenched specimens, which was attributed to coarsening of the carbides.

The yield strength increase caused by the fine carbide particles
in Figure 1 was explained by the Orowan mechanism of hardening by small
non-deforming particles. Here the shear yield strength, τ, of a metal
is given by

$$\tau = \tau_s + \tau_p .$$

(2)

The term τ_s is the shear yield strength of the matrix in the absence of
particles, and is comprised of the lattice friction stress, τ_0, and any
other terms which would add to the matrix strength in the particular
dispersion-hardened system being examined. The Orowan contribution,
τ_p, to the macroscopic yield strength was originally given by the
approximate relation

$$\tau_p \approx \frac{2Gb}{\lambda} ,$$

(3)

where G is the shear modulus of the matrix at the test temperature, b is
the Burgers vector, and λ is the mean planar center-to-center particle
spacing. More recent formulations of the Orowan stress increment have
been given by Kelly and Nicholson (9) and Ashby (19), the latter being
perhaps more rigorous. The Kelly-Nicholson relation gives τ_p as

$$\tau_p \approx \frac{Gb}{4\pi} \left(\frac{1}{\lambda/2-\bar{r}_s} \right) \varphi \, \ell n \left(\frac{\lambda/2-\bar{r}_s}{b} \right) .$$

(4)

The term $\bar{r}_s$ is the average planar particle radius and is defined, from
standard quantitative metallography, as $\bar{r}_s = \sqrt{2/3} \, \bar{r}_v$, where $\bar{r}_v$ is the
average true particle radius. $\varphi \approx 1.25$ is a factor which averages the
screw and edge character of a bowed dislocation loop. Ashby (19) derived
the Orowan stress for both screw and edge dislocations and found that

$$\tau_p(edge) = \left(\frac{1}{1.18} \right) \left(\frac{Gb}{2\pi} \right) \left(\frac{1}{\lambda-2\bar{r}_s} \right) \ell n \left(\frac{\bar{r}_s}{b} \right) ,$$

(5a)

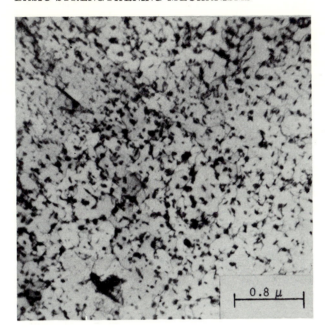

Fig. 1. Fine carbide
precipitates in Mo-
TZM annealed 1 hour
at 2100°C and quenched
into molten tin (after
Wilcox and Gilbert (18)).

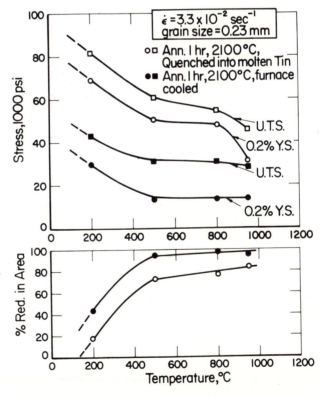

Fig. 2. Temperature de-
pendence of strength and
ductility of quenched and
slowly cooled Mo-TZM,
showing precipitation
strengthening effects in
quenched material (after
Wilcox and Gilbert (18)).

$$\tau_p(\text{screw}) = \left(\frac{1}{1-\nu}\right) \tau_p(\text{edge}) \quad , \tag{5b}$$

where ν is Poisson's ratio.

Wilcox and Gilbert (18) accounted for the strength increase due to the fine carbides in quenched Mo-TZM (Figure 1) by the Orowan mechanism, using Equation (4). The good agreement between experimental and calculated values of τ is shown in Table 1. The use of Equation (5) to calculate $(\tau_s + \tau_p)$ gave somewhat poorer agreement (18).

TABLE 1.   COMPARISON OF EXPERIMENTAL YIELD STRESSES OF
QUENCHED Mo-TZM WITH THOSE PREDICTED BY THE
OROWAN RELATION [AFTER WILCOX AND GILBERT (18)]

Test Temp., °C	τ_s, psi[a]	τ_p [from Eq.,(4)], psi	$\tau = \tau_s + \tau_p$, psi	Experimental τ, psi[a]
200	14,700	21,200	35,900	34,500
500	6,900	20,700	27,600	25,300
800	7,000	19,600	26,600	24,500
950	7,400	19,100	26,500	15,500(b)

(a) τ_s (for furnace-cooled specimens) and τ-experimental (for quenched specimens) were calculated taking the shear yield stress as 1/2 the 0.2% offset tensile yield stress in Figure 2.

(b)   Low value of τ is attributed to particle coarsening.

Other theories have been proposed to account for particle hardening, such as those based on stress-induced particle shearing (6) or prismatic cross-slip (20). However, these were not applicable to the fine-carbide particle hardening of Mo-TZM, and thus will not be discussed in detail here.

Hardening by Inert Oxides

The stability of oxide particles in metals, as compared to carbides and other precipitates which coarsen or go into solution at high temperatures, has led various investigators to attempt oxide-dispersion strengthening of refractory metals. As early as 1924, Jeffries (21) reported the effect of a ThO_2 dispersion on the mechanical properties and structural stability of W. From that time to the present, numerous other investigators have pursued research on the W-ThO_2 system (22-25). Oxide particles have also been added to Mo, and Jaffee (26) has

reported an extensive study where eleven oxides, in concentrations of
0.1 to 10 wt.%, were blended with Mo powder to produce dispersion-
hardened Mo alloys.

In earlier work, the oxide particles added to refractory
metals were relatively coarse, $2\bar{r}_v > 1\mu$, and optimum dispersion strengthening was not achieved. However recently, Sell, et al. (25), have achieved
very fine dispersions ($2\bar{r}_v \approx 400$ Å) of ThO_2 in W. Some of their results
are shown in Figure 3, where it is seen that the proportional limit of W
is markedly increased by the addition of 3.8 vol.% ThO_2. Similar effects
were found for the 0.2% yield strength and the ultimate strength, although
the increase in UTS was not as great since the unalloyed W work hardened
considerably. Also shown in Figure 3 is the dispersion-strengthening
increment predicted by the Orowan theory (Equation 5). Here the shear
stress, τ, was converted to tensile stress, σ, by the relation $\sigma = 2\tau$.
If it is assumed that the Orowan mechanism is operative here as the
<u>direct</u> strengthening mechanism, then over the temperature range ~ 400 -
1600°C the additional strength increment (above the Orowan curve) could
be due to <u>indirect</u> strengthening by the 3.8 vol.% ThO_2. That is, the
particles could have promoted a stable elongated structure during thermomechanical processing which provided an additional high temperature
strength increment. However, if the Orowan mechanism were operative
(as in the case of TD Nickel (8)), then the strength of the W-ThO_2 alloy
should not fall below the Orowan curves at $T \gtrsim 1700$-2000°C. This discrepancy could be accounted for if the average particle spacing, λ, were
actually larger than the reported value, i.e., the Orowan curves in
Figure 3 would then be lowered[*].

Internal Oxidation and Nitriding

Strengthening of metals by internal oxidation has been studied
in a number of systems. However, this technique has been exploited to
only a limited degree in the case of refractory metals. In Nb-1Zr and
Nb-1W-1Zr alloys, Tietz and co-workers (27,28) found that internal oxidation at ~800°C in low oxygen partial pressures (~10^{-5} to 10^{-6} torr in
a dynamic vacuum) produced significant increases in room temperature
tensile flow strength and high temperature (1200°C) creep strength. Some
typical values are given in Table 2, and it is noted that the strengthening
depends critically on the temperature and pressure of internal oxidation.
The strength increases here were of larger magnitude than the age hardening
effects associated with ZrO_2 precipitation produced by quenching and aging
a Nb-1Zr alloy (29,30).

---

[*] King, et al. (82), have, in fact, re-examined their results and included
additional experimental data. In their new analysis, the Orowan curve
lies below the experimental yield stress plot for the W-ThO_2 alloy at
$T \gtrsim 2100$°C and coincides with the Orowan plot over the range 2100°C <
$T < 2400$°C.

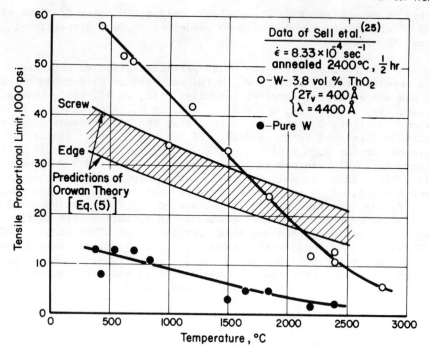

Fig. 3. The effect of dispersed ThO_2 particles on the strength of W.

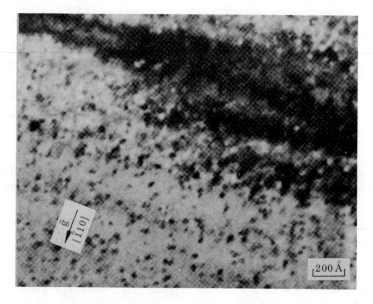

Fig. 4. Bright field transmission micrograph of Nb-1%Zr alloy internally oxidized at 800°C for 240 hours. Fine "precipitates" are probably Zr-O clusters (courtesy of D. J. Rowcliffe).

TABLE 2.   ROOM TEMPERATURE FLOW STRESSES OF A Nb-1% Zr
ALLOY INTERNALLY OXIDIZED FOR 20 HOURS (AFTER
BONESTEEL, ET AL. (28))

Internal Oxidation Treatment	Oxygen Content (ppm)	Flow Stress at $\epsilon=0.002$ (1000 psi)	Flow Stress at $\epsilon=0.08$ (1000 psi)
Untreated (As-Recrystallized)	80	24.5[a]	35.4
750°C at 2-6 x 10^{-6} torr	180	28.3[a]	41.1
800°C at 2-6 x 10^{-6} torr	230	37.1	55.1
800°C at 4-8 x 10^{-6} torr	380	45.6	63.3
800°C at 1-3 x 10^{-5} torr	290	42.1	61.1
850°C at 4-8 x 10^{-6} torr	210	25.6	42.0

(a)   Value represents upper yield strength, as specimen exhibited inhomo-
geneous deformation.

The internal oxidation treatments of the Nb-1Zr alloy (28)
resulted in an extremely fine dispersion of particles, as shown in the
transmission electron micrograph in Figure 4. Bonesteel, et al. (28),
concluded that the particles were coherent with the matrix and were
discs ~35 Å in diameter and ~10 Å in thickness. They suggested that the
particles were Zr-0 clusters. After annealing or creep testing at 1200°C
the clusters disappeared and only partially coherent or noncoherent
particles of α-ZrO$_2$ were observed. Although the exact strengthening
mechanism was not ascertained, the technique of internal oxidation
appears to be a promising way of strengthening Nb-base alloys containing
Zr. Similar effects might be expected if other reactive metal additions,
such as Ti or Hf, were made in place of Zr. These strengthening effects
might also be realized in other refractory metal alloys (e.g., Ta, V)
containing reactive metal additions.

A limited amount of work on internal nitriding of refractory
metals has been reported. The most extensive study was by Mukherjee and
Martin (31), who internally nitrided alloys of Mo-0.49 Ti, Mo-0.98 Ti,
and Mo-1.52 Zr by annealing in ammonia for 36 hours at 1500°C. The
resultant TiN and ZrN particles were about 0.2μ in diameter, and, compared
with the non-nitrided alloys, significantly improved the ultimate strength
from 20-900°C as well as the high temperature creep strength.

In a preliminary study, Wilcox (32) internally nitrided a low
carbon Mo-TZM alloy (Mo-0.5Ti-0.1Zr-0.001C) using the technique described
by Rudman (33). Here specimens were placed in a sealed evacuated Mo capsule containing mixed powders of Mo + Mo$_2$N, and were annealed for 6 hours
at 1400°C. The resultant structures are shown in the optical and electron
micrographs in Figure 5. Fine disc-shaped nitrides (either TiN or ZrN),

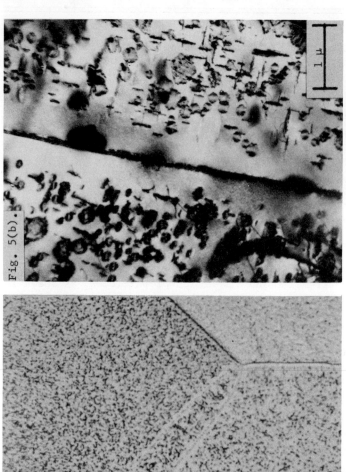

Fig. 5(b).

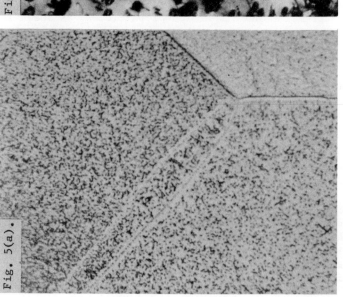

Fig. 5(a).

Fig. 5. Structure of internally nitrided low-carbon (10 ppm C) Mo-TZM (a) optical micrograph, 1500 X, (b) transmission electron micrograph showing disc-shaped nitrides and details of a denuded grain boundary.

about 0.1-0.5μ in diameter formed on $\{100\}$ planes within the grains.
However, most of the grain boundaries were denuded of these particles.
This treatment raised the hardness from 150 DPH (unnitrided) to 400-
450 DPH.

SOLID SOLUTION STRENGTHENING

Strengthening by solid solution additions has received extensive attention in refractory metal alloy development studies, and several
comprehensive review papers (1,34,35) have been devoted to solution
hardening of refractory metals exclusively. This section is limited to
solution strengthening by substitutional alloy additions, since interstitial solution hardening per se is not important in refractory metals
at high temperatures. The influence of interstitial elements on
strengthening is discussed later in the section on DYNAMIC STRENGTHENING.

Typical examples of solution strengthening in refractory
metals are shown in Figures 6, 7, and 8. Studies on binary single crystal alloys (Figure 6) revealed that in the Ta-Mo (36) and Nb-Mo (38)
systems, pronounced solution strengthening is evident, with the maximum
strength in Nb-Mo alloys occurring at the equiatomic composition.
However, in the Ta-Nb system (37) there is no significant solution
hardening. An extensive study of polycrystalline Nb-base alloys by
Bartlett, et al. (39), also showed that Ta was not an effective
strengthener of Nb (Figure 7). On the other hand, additions of W, Mo,
and V increased both the room temperature and elevated temperature tensile strength, with W being the most effective hardener per atomic
percent added.

Tungsten not only increases the tensile strength of Nb, but
also improves the high temperature creep properties. This is seen in
Figure 8, where increasing the W content in binary Nb-alloys lowers the
minimum creep rate and increases the rupture life at 1200°C. The influence of W, Mo, and V on the 1200°C creep rupture properties of binary
Nb-base alloys was studied by Bartlett, et al. (39). They, too, found
that W additions (and Mo to a somewhat lesser extent) increased the
rupture life. However additions of V to Nb had very little effect on
the rupture life, in contrast to the notable strengthening effect in
short-time tensile tests (see Figure 7). This lack of strengthening in
creep was attributed to the fact that V additions raise the high temperature self-diffusivity of Nb, as reported by Hartley, et al. (61). If the
creep process is diffusion controlled, then the creep rate, $\dot{\epsilon}$, and diffusivity, D, are related by $\dot{\epsilon} \propto D$. An increase in D leads to an increase
in $\dot{\epsilon}$, and this can be reflected by a decrease in the rupture life.
Similarly, Mo additions to Nb decrease the self-diffusivity (61) and
increase the creep-rupture life (39).

Traditionally, it has been considered that the chief cause of
substitutional solution strengthening arises from a difference in atomic
size between the solvent and solute elements. However, Fleischer (43-45)
has shown that differences in elastic moduli between solvent and solute

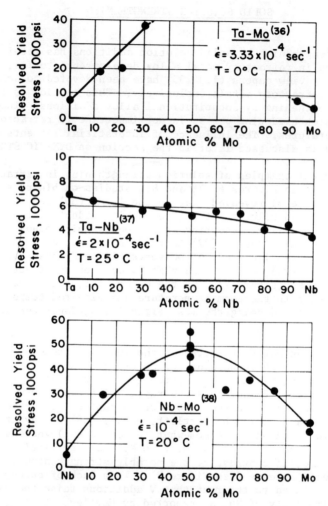

Fig. 6. Effect of composition on yield stress of binary single crystal
refractory metal alloys, resolved onto the most favorable
{011} <111> slip system.

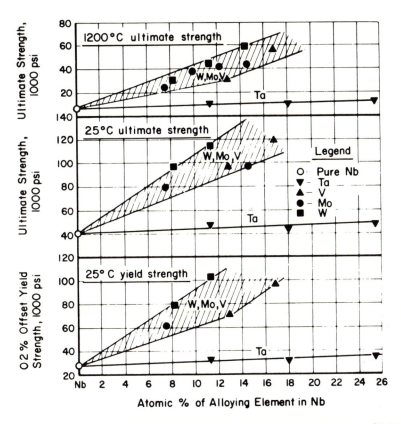

Fig. 7. The influence of substitutional alloying additions (W,Mo,V,Ta) on the strength of polycrystalline Nb at room temperature and 1200°C (after Bartlett, et al., 39).

can contribute strongly to solution hardening of FCC alloys, while Rudman (34) has considered that short-range ordering may be important in solution strengthening of refractory metals. The strengthening due to atomic size mismatch and modulus differences arises because of long-range elastic interactions between dislocations and solute atoms. Hardening by short-range ordering (SRO) occurs because the SRO state is a lower energy state than the random state. When a dislocation moves through an ordered region the order is destroyed and therefore additional energy must be supplied by an external force to cause the dislocation to move.

Following Rudman's (34) analysis, the strength, σ, of a solution-hardened alloy can be written as

$$\sigma = \sigma_o + \sigma_v + \sigma_G + \sigma_{SRO} \quad , \tag{6}$$

where σ_o is defined as the matrix strength in the absence of the solution strengthening alloy additions. The important terms in the strength contributions due to atomic size mismatch, σ_v and modulus differences, σ_G, are as follows: (43)

$$\sigma_v \propto \left| \frac{1}{a} \frac{da}{dc} \right| \quad , \tag{7a}$$

$$\sigma_G \propto \left| \frac{1}{G} \frac{dG}{dc} \right| \quad . \tag{8a}$$

In Equation (7a), a is the lattice parameter and da/dc is the rate of change of lattice parameter with solute concentration, c. Similarly, in Equation (8a), G is the shear modulus and dG/dc is the rate of change of modulus with composition. In the absence of data on da/dc and dG/dc, approximate estimates of the atomic size mismatch and modulus difference effects can be given by

$$\sigma_v \propto \left| \frac{V_B - V_A}{V_A} \right| \quad , \tag{7b}$$

$$\sigma_G \propto \left| \frac{G_B - G_A}{G_A} \right| \quad . \tag{8b}$$

Here A refers to the solvent and B the solute. V_A and V_B are the atomic volumes of the components and G_A and G_B are the shear moduli of the pure metals A and B.

It is more difficult to select the important terms in the short-range order contribution to strengthening, σ_{SRO}. Rudman (34) has given a relation describing σ_{SRO} for slip in BCC metals resolved on the

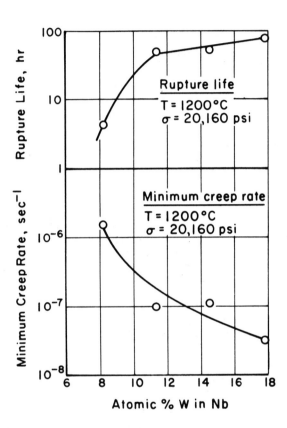

Fig. 8. The influence of W on the rupture life and creep rate of polycrystalline Nb-W alloys (after McAdam, 40).

$(1\bar{1}0)[111]$ system. The parameters, in functional form, are

$$\sigma_{SRO} \propto f(c,a,\alpha) \quad , \tag{9}$$

where c is the composition, a is the lattice parameter, and α is the short-range order coefficient obtained from X-ray diffuse scattering.

Unfortunately, there are very few SRO data for refractory metal alloys, and thus it is not possible to state how important the σ_{SRO} contribution is to solution strengthening. As noted in Table 3, it is generally accepted that atomic size mismatch is the most important contributor to solid solution strengthening of BCC refractory metals. In fact, Harris (41) was able to quantitatively separate the effects of atomic size mismatch from the modulus differences in binary Nb-base alloys containing W, V, Ta, and Zr additions of up to ~4 - 8 atomic percent. Zr has very nearly the same elastic modulus as Nb, but a different atomic size; Ta has the same atomic size as Nb, but a different modulus; and W and V differ from Nb in both atomic size and modulus.

Harris' results are replotted in Figure 9. In Figure 9a, the room temperature yield strength increase per atomic % of solute, $|(1/a) (da/dc)|$. Harris assumed that the only strengthening effects were those due to atomic size mismatch and modulus differences. Thus, in Figure 9a, a straight line is drawn between $d\sigma/dc = 0$ (pure Nb), and the data point for Zr, since there is essentially no modulus difference between Zr and Nb. The deviation from this line, δ, for Ta, V, and W is then taken as the strength increment due to elastic modulus differences. A good correlation is, in fact, obtained in Figure 9b when the modulus difference parameter, $\Delta E/E_{Nb}$ (E = Young's modulus), is plotted against δ. Thus in these binary Nb-base alloys it is concluded that the relative percentages of strengthening due to atomic size mismatch, σ_v, and modulus differences, σ_G, are as shown in Table 4.

WORKING AND GRAIN SIZE REFINEMENT

In this section, the strengthening effects arising from working and from refining the grain size and subgrain size are treated together since the two effects are interrelated. For example, one way of refining the "effective" grain size of a metal is to work it. A discussion of detailed work hardening mechanisms in refractory metals is beyond the scope of this paper, although some work has been reported in this area (46-48).

Cold or warm working metals and alloys to refine the structure has been perhaps the most widely used means of strengthening, and re-fractory metals are no exception. The beneficial effects of working on yield and ultimate strengths are illustrated in Figure 10. Here the strength and ductility of wrought-stress-relieved and recrystallized

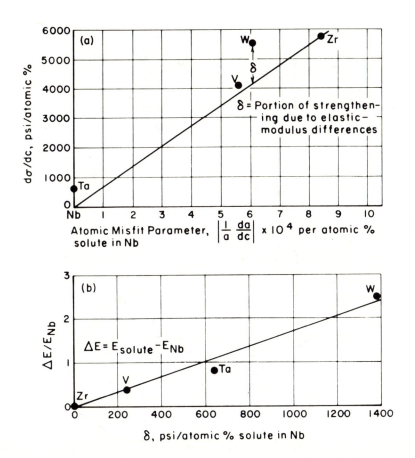

Fig. 9.   The influence of atomic size misfit $(1/a)(da/dc)$, and modulus
differences $(\Delta E/E_{Nb})$, on the room temperature yield strength
of polycrystalline Nb-base alloys containing W, Zr, V, and
Ta (after Harris, 41).

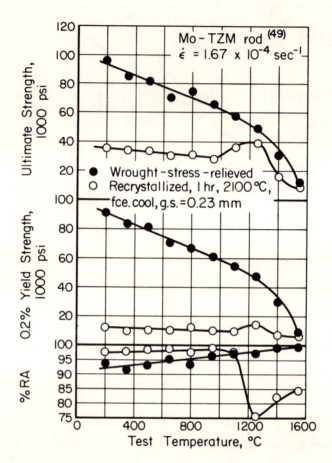

Fig. 10. A comparison of the strength and ductility of wrought-stress-
relieved and recrystallized Mo-TZM.

TABLE 3. VARIOUS WORKER'S OPINIONS AS TO THE RELATIVE IMPORTANCE OF (a) ATOMIC SIZE MISMATCH, (b) ELASTIC MODULUS DIFFERENCES, AND (c) SHORT RANGE ORDER EFFECTS IN SUBSTITUTIONAL SOLUTION STRENGTHENING OF REFRACTORY METALS

Author's opinion as to the most important solution strengthening contribution	System(s) Studied	Reference
Atomic size mismatch coupled with short range ordering	Numerous binary BCC refractory metal alloys	Rudman[34]
Elastic modulus differences	Various binary and ternary Nb-base alloys	McAdam[40]
Atomic size mismatch	Nb-Mo single crystals	Milne and Smallman[38]
Atomic size mismatch (generally)	Binary Nb-base alloys with Ta, V, W, and Zr	Harris[41]
Atomic size mismatch	Various Nb-base alloys	Tarasov, et al.[42]
Atomic size mismatch	Mo-, Nb-, and Ta-base alloys	Chang[1]

TABLE 4.   ATOMIC SIZE DIFFERENCE AND MODULUS DIFFERENCE CONTRIBUTIONS
TO YIELD STRENGTH OF BINARY NIOBIUM-BASE ALLOYS

Solute Addition to Nb	Percent of Solution Strengthening due to	
	Atomic Size Mismatch	Modulus Differences
W	75	25
Zr	100	0
Ta	0	100
V	93	7

Mo-TZM are compared over the temperature range 200-1550°C. A significant
increment in yield and ultimate strengths due to working is achieved up
to test temperatures of ~1200-1300°C. At higher temperatures the strength
of the wrought-stress-relieved material drops off rapidly because of
dynamic recovery and recrystallization, until at 1550°C there is relatively
little strength improvement over the recrystallized materials.

The strengthening due to working results from the fact that
this operation introduces more obstacles to dislocation motion, and thus
reduces the free slip distance. Two contributing factors, which have
been quantitatively assessed in the case of refractory metals, are an
increased dislocation density and an effective refinement of the grain
size. In numerous metals and alloys the strength increases as the square
root of the dislocation density, N, according to the relation

$$\tau = \tau_o + \alpha G b N^{1/2} \quad , \tag{10}$$

and this equation is obeyed for V[50] and Nb[51], as shown in Figure 11.
Here τ is the shear flow stress, G is the shear modulus, α is a constant,
and b is the Burgers vector.

Similarly, decreasing the grain size, d, causes an increase in
yield strength, σ_y, and flow strength according to the well known Hall-
Petch equation

$$\sigma_y = \sigma_o + k_y d^{-1/2} \quad . \tag{11}$$

The linear relation between σ_y and $d^{-1/2}$ holds for many recrystallized
refractory metals. However, in certain cases, particularly at large
grain sizes (small $d^{-1/2}$), the linearity breaks down, as has been shown
by Orava (52) for Mo (see Figure 12).

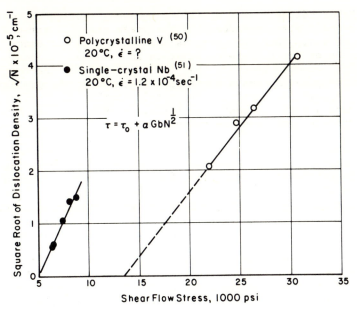

Fig. 11.   Relation between the shear flow stress and dislocation density
           for polycrystalline V (tension) and Nb single crystals
           (compression).

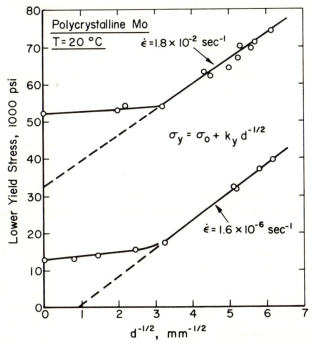

Fig. 12.   Room temperature grain size dependence of the lower yield
           stress of Mo (after Orava[52]).

Nevertheless, Equation (11) adequately represents most of the data on the grain size dependence of the yield and flow strength of refractory metals. There are various opinions as to the exact physical interpretation of the Hall-Petch equation in terms of dislocation models. There is little doubt, however, that the strengthening due to grain size refinement is caused by grain boundaries obstructing the propagation of slip from one grain to another. The total obstruction to slip propagation increases with increasing number of grain boundaries (finer grain size) and the strength level is raised.

Equation (11) has also been found to hold for worked metals, and the best examples reported have been for drawn wires, including various ferrous alloys (53-55), Mo-TZM (18), and W (56). The refinement of grain and subgrain size by drawing Mo-TZM is shown in Figure 13. The spacing between grain and subgrain walls decreases with increasing drawing strain (18), and the strength increases according to Equation (11). In the case of drawn wire the term, d, in Equation (11) is called the "barrier spacing", and is the distance between grain and/or subgrain walls measured perpendicular to the drawing direction. Figure 14 illustrates that the Hall-Petch relation is obeyed for drawn Mo-TZM wire. Also included in Figure 14 are Orava's (52) data on recrystallized Mo tested at approximately the same strain rate, and the straight line drawn through the wire data points is extrapolated through the points for recrystallized Mo.

RETAINING WORKED STRUCTURES AT HIGH TEMPERATURES

In the preceding section it was shown that working refractory metals increased the strength, both by refining the effective grain size and by increasing the dislocation density. Since refractory metals are often used in structural applications at high temperatures, it is desirable to retain the worked structure and the attendant strengthening at these temperatures. Several feasible means of accomplishing this have been demonstrated, and typical examples are shown in Figure 15. It is seen here that the recrystallization temperatures of refractory metals can be increased by (a) solid solution alloying, (b) "doping", which produces small gas bubbles, and (c) the presence of fine dispersoids and precipitates.

In Figure 15 A, it is noted that solid solution additions of W and Mo to Nb increase the annealing temperature for the start of recrystallization in one hour, whereas V and Ta additions have little effect on the recrystallization behavior. This can be rationalized, at least in part, by considering how the alloying additions affect the self-diffusivity. Since atom movements are involved in the processes of recovery and recrystallization, decreasing the self-diffusivity should slow down these processes. A rough correlation of the data of Bartlett, et al. (39), in Figure 15 A can be made with the diffusion results of Hartley, et al. (61) for binary Nb-base alloys containing V and Mo additions. Hartley, et al., found that V additions to Nb somewhat raised the chemical self-diffusivity at $1630^{\circ}C$, whereas additions of Mo to Nb decreased the

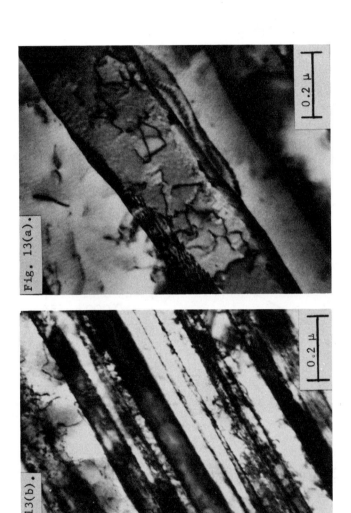

Fig. 13.   Fine structure of Mo-TZM wire (a) swaged from 0.25 inch dia.
bar to 0.080 inch dia rod (b) drawn from 0.080 inch dia. to
0.028 inch dia. wire (after Wilcox and Gilbert[18])

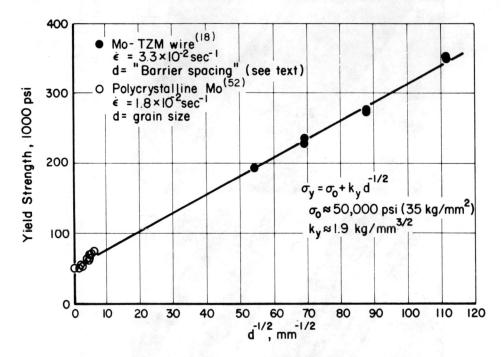

Fig. 14.   The influence of grain size and "barrier spacing" on the room
temperature yield strength of Mo and Mo-TZM.

diffusivity at 1800, 2000, and 2163°C. In Figure 15 A it is seen that V additions have little effect on the recrystallization temperature of Nb-V alloys, but Mo markedly increases the recrystallization temperature. In the absence of high temperature diffusivity data, a crude estimate of the effect of alloying on recrystallization temperature can be made by considering how the solute addition affects the alloy melting temperature. Often, if the solute addition increases the alloy melting point, it will decrease the self-diffusivity and raise the recrystallization temperature.

Another means of raising the recrystallization temperature of refractory metals is by adding second phase particles such as inert oxide dispersoids, precipitates, or small gas bubbles or voids. Here the particles can help to retard dislocation motion during recovery and inhibit grain boundary migration during recrystallization and grain growth (69). According to the terminology used in the section on STRENGTHENING BY SECOND PHASE PARTICLES, this would be an indirect strengthening effect of particles. Small gas bubbles are involved in this discussion, since in certain cases they can behave as second phase particles (70).

The influence of gas bubbles is considered first, and this is a special case relating to doped or "non-sag" tungsten lamp filaments. The term doping refers to small additions of K, Al, and Si as oxides. This technique of raising the recrystallization temperature and altering the recrystallized grain morphology of heavily drawn W wire has been used for many years in the lamp industry. The recrystallized grains in doped W wire and sheet are elongated in the direction of working, as seen in Figure 16. An example of how the recrystallization temperature is raised is shown in Figure 15 B. At a given annealing temperature, a longer time is required to initiate recrystallization in silicate doped W sheet than in undoped W.

Until recently, there was no generally accepted explanation for the doping effect, and even today various workers are not in complete agreement. Several investigators (62-66) have shown recently that during annealing of doped W, bubbles or small particles ($\sim$100-1000 Å diameter) form in rows parallel to the working direction. Walter (63,66) concluded that these were particles of mullite ($Al_6Si_2O_{13}$). However, the electron diffraction contrast experiments of Koo and Moon (62,65) and Das and Radcliffe (64) have shown conclusively that the "particles" are gas bubbles which probably form by the volatilization of the dopant during annealing. Examples of these bubbles are shown in Figure 17. Koo and Moon (65) have proposed that the bubbles retard the recrystallization process by inhibiting the rate of nucleation of new grains. The stringer-like distribution of the bubbles could inhibit the lateral movement of the new (recrystallized) grain boundaries giving rise to the elongated interlocking grain structure (63).

Dispersed oxide particles and precipitates raise the recrystallization temperature of refractory metals in the same manner as do the bubbles in doped W. Figure 15 C illustrates that the 1-hour recrystallization temperature of powder metallurgy W is $\sim$1800°C, whereas addition of

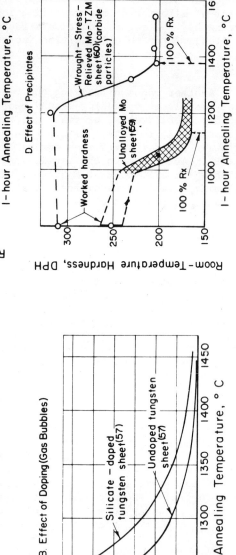

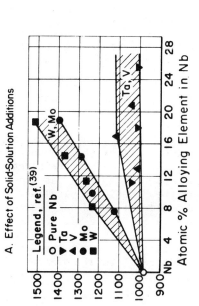

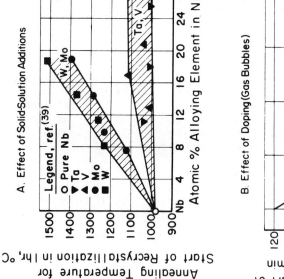

Fig. 15. The influence of: A. Solid solution additions, B. Gas bubbles (doping), C. Dispersed particles, and D. Precipitates, on raising the recrystallization temperatures of refractory metals.

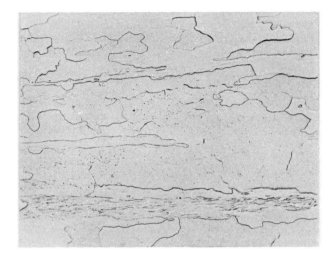

Fig. 16. Micrograph of doped W sheet (218 alloy), recrystallized by
          annealing 1 hour at 1600°C, 250 X. (courtesy of D. J. Maykuth).

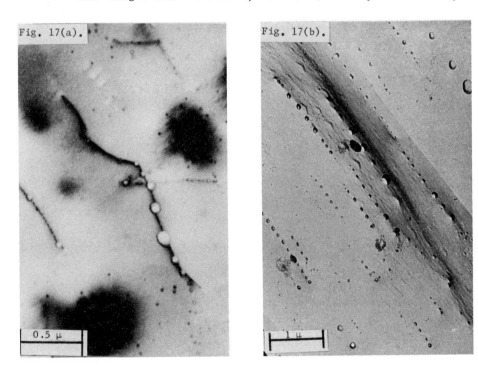

Fig. 17. Gas bubbles in doped W that has been annealed for 1/2 hour at
          2500°C. (a) Transmission electron micrograph of 0.005 inch
          thick ribbon, (b) Replica electron fractograph of 0.035 inch
          diameter wire (courtesy of R. C. Koo and D. M. Moon).

3.8 vol.% ThO_2 raises the recrystallization temperature to ~3000°C.
Similarly, the carbide particles in Mo-TZM (Figure 15 D) help to promote
a higher recrystallization temperature than that of unalloyed Mo.

DYNAMIC STRENGTHENING

In this paper dynamic strengthening refers to strengthening
arising during deformation. In the case of refractory metals, interstitial elements play a dominant role in dynamic strengthening. This
can occur in two ways: (1) Interstitial elements can interact with dislocations to produce classical dynamic strain aging. These effects are
most pronounced in the Group VA metals, Ta, Nb, and V, which have higher
solubilities for C, O, N, and H than do the Group VIA metals, Mo, W,
and Cr. (2) Interstitial elements can interact with reactive metal
solutes, such as Zr, Ti, and Hf, to cause precipitation during deformation.
This has also been referred to by Chang (13) as "strain-induced precipitation".

The most common observation of dynamic strengthening is the
occurrence of peaks in plots of strength versus test temperature, and an
example of this is seen in Figure 10 for recrystallized Mo-TZM. The
peak in ultimate strength in the temperature range ~1000-1400°C was
believed to be caused by dynamic precipitation of carbides (TiC, ZrC, and
possibly Mo_2C) during tensile deformation (13,49). Other phenomena
commonly associated with dynamic strengthening are serrated deformation
curves and negative strain rate sensitivity (71). Dynamic strengthening
can also be beneficial under creep conditions, and this topic has been
reviewed by Cottrell (72).

Dynamic strain aging in Nb, Ta, and V due to the interaction of
O, N, and C, with dislocations, generally occurs over the temperature
range ~200-500°C. This strengthening effect is dissipated at higher
temperatures because of the increased mobility of interstitial atoms with
increasing temperature. However, it is possible to shift the dynamic
strengthening effects to higher temperatures by reactive metal additions,
and this is illustrated in Figure 18. Here it is seen that the dynamic
strengthening peak temperature is ~300°C for Nb-10W and unalloyed Nb.
However, additions of Zr cause the peak temperature to shift to ~800°C
for Nb-10W-2.5Zr and Nb-1Zr alloys. Wilcox and Allen (77) discussed this
behavior, and tentatively concluded that the influence of Zr was to interact with oxygen, such that precipitation or clustering (Zr-0) during
deformation impeded dynamic recovery.

Irrespective of the exact mechanisms of dynamic strengthening,
it should be possible, in principle, to design an alloy system such that
more than one mechanism is operative. Figure 19 is a schematic plot of
tensile flow stress versus test temperature in a system containing three
hypothetical dynamic strengthening mechanisms, each operating over a
different temperature range. If such an alloy system could be prepared,
then it is seen in Figure 19 that the average flow strength could be

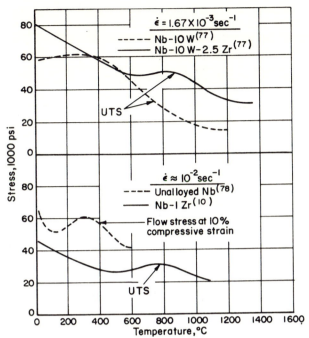

Fig. 18.   The influence of Zr on the temperature ranges where dynamic
           strengthening occurs in Nb alloys.

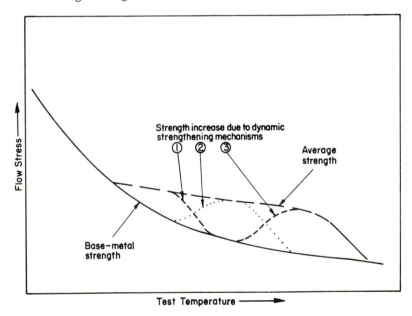

Fig. 19.   Schematic plot showing how a sequence of dynamic strengthening
           mechanisms can raise the average flow stress over a wide
           temperature range.

increased over a considerable temperature range. This concept is presently being exploited in manganese steels containing nitrogen (73,74,79), and has been found to be particularly useful in raising the intermediate temperature ($\sim$400-600°C) creep strength.

One feature associated with dynamic strengthening, which is apparently not well known, is the observation that during serrated yielding there is a greater increase in dislocation density per unit strain than if serrated yielding were not present. This phenomenon has been discussed by Wilcox and Rosenfield (71) and is schematically illustrated in Figure 20. It was seen earlier that $\sigma \propto N^{1/2}$, where N is the dislocation density. Thus warm working refractory metals in the dynamic strengthening temperature range should produce structures with an increased dislocation density, which would result in strengthening during subsequent deformation. If the dynamic strengthening process involved precipitation on dislocations and sub-boundaries during deformation, an additional benefit might be derived since dynamic recovery would be inhibited. This could lead to a finer grain or subgrain size, d, and as shown earlier, $\sigma \propto d^{-1/2}$.

FIBER REINFORCEMENT

In this paper strengthening by fiber reinforcement is treated only briefly, since there has been very little work in this area where the matrix has been a refractory metal. Most of the applications of refractory metals in fiber reinforced composites have been as the reinforcing filament, e.g., W and Mo wires in nonrefractory matrix metals like copper or nickel alloys.

The basic equations relating the ultimate tensile strength, σ_c, of a fiber reinforced composite to the properties of the components have been reviewed by Kelly and Davies (75) and Kelly and Tyson (76). For continuous fibers, σ_c is given by

$$\sigma_c = \sigma_f V_f + \sigma_m' (1-V_f) \quad , \tag{12}$$

and for discontinuous fibers,

$$\sigma_c = \sigma_f V_f \left(1 - \frac{1}{2\alpha}\right) + \sigma_m' (1-V_f) \quad , \tag{13}$$

where $\alpha = \ell/\ell c$ is a measure of the ratio of the actual fiber length, ℓ, to the critical load transfer length, ℓ_c. The other terms in Equations (12) and (13) are:

σ_f = fiber ultimate tensile strength
V_f = volume fraction of the fiber
σ_m' = the stress in the matrix at the composite failure strain.

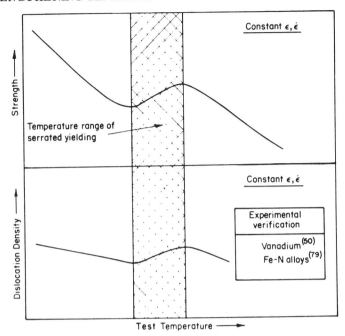

Fig. 20.   Schematic plots showing how the dislocation density per unit
           strain increases in the temperature range where dynamic
           strengthening and serrated yielding occur.

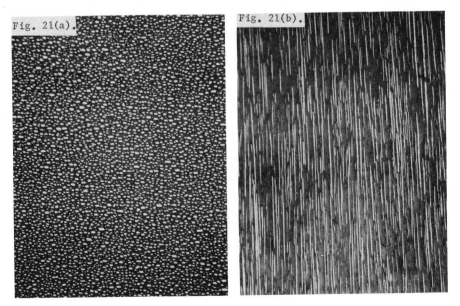

Fig. 21.   Microstructure of unidirectionally solidified Ta-Ta$_2$C
           eutectic (a) Transverse section, 200 X, (b) Longitudinal
           section, 200 X (courtesy of M. J. Salkind).

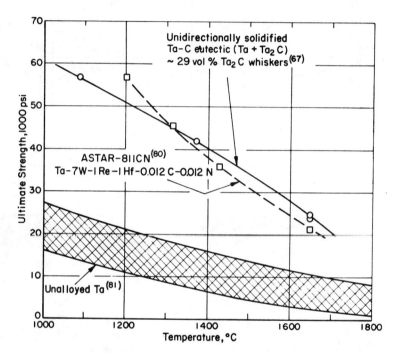

Fig. 22. A comparison of the high temperature ultimate strength of a
eutectic Ta + Ta$_2$C composite with those of unalloyed Ta and
a strong Ta-base alloy.

Two composite systems, where the matrix was a refractory metal, have been studied by Lemkey, et al. (67,68). Using unidirectional solidification of Nb-C and Ta-C eutectic alloys, they grew composites with discontinuous fibers of Nb_2C and Ta_2C. Examples of the type of microstructures produced are seen in Figure 21 for the Ta-Ta_2C system. The carbide fibers were very stable at high temperatures, and the strengthening effects could be rationalized on a law of mixtures basis by Equation (13).

The ultimate strength of the Ta + Ta_2C composite is compared in Figure 22 with the strengths of unalloyed Ta and a strong Ta-base alloy, Ta-7W-1Re-1Hf-0.012C-0.012N. Over the temperature range ~1200-1650°C, the strength of the eutectic composite compares very favorably with that of the high strength Ta-base alloy.

SUMMARY

This paper has treated the various strengthening mechanisms which have been used in producing strong refractory metal alloys. The examples used to illustrate the various mechanisms were, by and large, based on simple systems. It is obvious, of course, that implementing more than one mechanism can produce combined strengthening effects, and this procedure has been common practice in refractory metal alloy development. An example of this is seen in the case of the strong Ta-base alloy, Ta-7W-1Re-1Hf-0.012C-0.012N (80). W and Re are added as solid-solution strengtheners. A portion of the Hf is in solution and the remainder combines with the C and N to form the precipitate Hf(C,N). This precipitate, together with Ta_2C particles, contributes to the strengthening. If the alloy is tensile tested in the wrought-stress-relieved condition the yield and ultimate strengths are higher than in the recrystallized condition. This follows because of the higher dis-location density and smaller effective grain size in the wrought material.

It is now appropriate to summarize by presenting a phenomeno-logical equation which contains, in functional form, the important para-meters that can be varied to increase the strength of refractory metals. Equation (1) is now written as:

$$\sigma = \sigma_0 + f\left(\frac{Gb}{\lambda}\right) + f\left(\left|\frac{1}{a}\frac{da}{dc}\right|\right) + f\left(\left|\frac{1}{G}\frac{dG}{dc}\right|\right) + \sigma_{SRO} + f\left(GbN^{1/2}\right)$$

$$+ f\left(d^{-1/2}\right) + f\left(\sigma_f V_f\right) \quad . \tag{14}$$

Increasing the strength, σ, can thus be accomplished by:

(a) Incorporating small second phase particles spaced closely together (small λ). Particles (and small voids or bubbles) can also contribute indirectly to strengthening by helping to develop and/or pin a stable elongated grain and subgrain structure.

(b) Solid solution alloying to (1) increase the alloy shear modulus, G, (2) increase the absolute magnitude of the atomic misfit parameter $(1/a)(da/dc)$, (3) increase the absolute magnitude of the modulus difference parameter $(1/G)(dG/dc)$, (4) promote short range ordering, and (5) lower the high temperature alloy self-diffusivity, which in turn retards recovery and recrystallization.

(c) Mechanical working to increase the dislocation density, N, or decrease the effective grain size, d.

(d) Adding strong fibers (high σ_f) in conveniently high volume fractions, V_f

ACKNOWLEDGMENTS

The author is grateful to his colleagues R. I. Jaffee, P. C. Gehlen, and A. R. Rosenfield for their helpful comments and criticisms during the preparation of this manuscript. Appreciation is also expressed to D. J. Rowcliffe, D. J. Maykuth, R. C. Koo, D. M. Moon, and M. J. Salkind for providing original micrographs which were used in this paper.

REFERENCES

(1)   W. H. Chang, <u>Refractory Metals and Alloys</u>, Vol. 11, AIME Met. Soc.
      Conf., M. Semchyshen and J. J. Harwood, eds., Interscience,
      New York, 1961, p. 83.

(2)   R. W. Armstrong, J. H. Bechtold, and R. T. Begley, <u>Refractory</u>
      <u>Metals and Alloys</u>, Vol. 17, AIME Met. Soc. Conf., M. Semchyshen
      and I. Perlmutter, eds., Interscience, New York, 1963, p. 159.

(3)   H. Conrad, <u>The Relation Between the Structure and Mechanical</u>
      <u>Properties of Metals</u>, NPL Symposium No. 15, Her Majesty's Stationery
      Service, London, 1963, p. 476.

(4)   H. Conrad, <u>High Temperature Refractory Metals</u>, Vol. 34, AIME Met.
      Soc. Conf., R. W. Fountain, J. Maltz, and L. S. Richardson, eds.,
      Gordon and Breach, New York, 1966, p. 113.

(5)   A. R. Rosenfield, E. Votava, and G. T. Hahn, "Dislocations and the
      Ductile-Brittle Transition", paper presented at ASM Seminar on
      Ductility-Limitations and Utilization , Cleveland, October 14, 15,
      1967 (to be published by Am. Soc. Met.).

(6)   G. S. Ansell, "The Mechanism of Dispersion-Strengthening: A
      Review", paper presented at the Oxide Dispersion Strengthening
      Conference, Bolton Landing, New York, June 27-29, 1966 (to be
      published by Gordon and Breach).

(7)   B. A. Wilcox and A. H. Clauer, "High Temperature Creep of Ni-ThO$_2$
      Alloys", ibid.

(8)   B. A. Wilcox and R. I. Jaffee, "Direct and Indirect Strengthening
      Effects of ThO$_2$ Particles in Dispersion Hardened Nickel", paper
      presented at the International Conference on the Strength of Metals
      and Alloys, Tokyo, September 4-8, 1967 (to be published by Japan
      Inst. Met.).

(9)   A. Kelly and R. B. Nicholson, <u>Precipitation Hardening</u>, Progress in
      Mat. Sci., Vol. 10, No. 3, MacMillan, New York, 1963.

(10) W. H. Chang, "A Study of the Influence of Heat Treatment on Micro-
      structure and Properties of Refractory Alloys", ASD-TR-62-211,
      April, 1962.

(11) W. H. Chang, Trans. ASM, <u>56</u>, 107 (1963).

(12) W. H. Chang, Trans. ASM, <u>57</u>, 527 (1964).

(13) W. H. Chang, ibid., p. 565.

(14) F. Ostermann and F. Bollenrath, "Investigation of Precipitates in
      Two Carbon-Containing Columbium-Base Alloys", AFML-TR-66-259,
      December, 1966.

(15) R. T. Begley, J. L. Godshall, and D. L. Harrod, "Development of
      Dispersion Hardened Creep Resistant Columbium Alloys", paper
      presented at the Fourth Symposium on Refractory Metals, French
      Lick, Indiana, October 3-5, 1965 (to be published by Gordon and
      Breach).

(16) W. H. Chang, "Effect of Carbide Dispersion in Tantalum-Base Alloys",
      ibid.

(17) P. L. Raffo and W. D. Klopp, "Solid Solution and Carbide Strength-
      ened Arc Melted Tungsten Alloys", ibid.

(18) B. A. Wilcox and A. Gilbert, Acta Met., 15, 601 (1967).

(19) M. F. Ashby, "A Theory of the Critical Shear Stress and Work
      Hardening in Dispersion-Hardened Crystals", paper presented at the
      Oxide Dispersion Strengthening Conference, Bolton Landing, New York,
      June 27-29, 1966 (to be published by Gordon and Breach).

(20) H. Gleiter, Acta Met., 15, 1213 (1967).

(21) Z. Jeffries, Trans. AIME, 70, 303 (1924).

(22) R. I. Jaffee, B. C. Allen, and D. J. Maykuth, Plansee Proceedings-
      Powder Metallurgy in the Nuclear Age, F. Benesovsky, ed., Springer-
      Verlag, Vienna, 1962, p. 770.

(23) A. Gilbert, J. L. Ratliff, and W. R. Warke, Trans. ASM, 58, 142
      (1965).

(24) J. L. Ratliff, D. J. Maykuth, H. R. Ogden, and R. I. Jaffee, Trans.
      AIME, 230, 490 (1964).

(25) H. G. Sell, W. R. Morcom, and G. W. King, "Development of Dispersion
      Strengthened Tungsten Base Alloys", AFML-TR-65-407, Part I,
      November, 1965, Part II, November, 1966.

(26) R. I. Jaffee, The Metal Molybdenum, J. J. Harwood, ed., Am. Soc.
      Met., Cleveland, 1958, p. 330.

(27) D. J. Rowcliffe, R. M. Bonesteel, and T. E. Tietz, "Strengthening
      of Niobium-Zirconium Alloys by Internal Oxidation", paper presented
      at the Oxide Dispersion Strengthening Conference, Bolton Landing,
      New York, June 27-29, 1966 (to be published by Gordon and Breach).

(28) R. M. Bonesteel, D. J. Rowcliffe, and T. E. Tietz, "Mechanical
      Properties and Structure of Internally Oxidized Niobium-1% Zirconium
      Alloy", paper presented at the International Conference on the
      Strength of Metals and Alloys, Tokyo, September 4-8, 1967 (to be
      published by the Japan Inst. Met.).

(29) J. R. Stewart, W. Lieberman, and G. H. Rowe, Columbium Metallurgy,
      D. L. Douglas and F. W. Kunz, eds., Interscience, N. Y., 1961, p. 407.

(30) D. O. Hobson, <u>High Temperature Materials II</u>, Interscience, New York, 1961, p. 325.

(31) A. K. Mukherjee and J. W. Martin, J. Less Common Met., <u>2</u>, 392 (1962). See also <u>3</u>, 216 (1961); <u>5</u>, 117 (1963); and <u>5</u>, 403 (1963).

(32) B. A. Wilcox, unpublished work.

(33) P. S. Rudman, Trans. AIME, <u>239</u>, 1949 (1967).

(34) P. S. Rudman, "Solution Strengthening in Refractory Metal Alloys", paper presented at The Fourth Symposium on Refractory Metals, French Lick, Indiana, October 3-5, 1965 (to be published by Gordon and Breach).

(35) L. L. Seigle, <u>The Science and Technology of Tungsten, Tantalum, Molybdenum, Niobium, and Their Alloys</u>, N. E. Promisel, ed., Pergamon Press, Oxford, 1964, p. 63.

(36) L. I. Van Torne and G. Thomas, Acta Met., <u>14</u>, 621 (1966).

(37) B. C. Peters and A. A. Hendrickson, Acta Met., <u>14</u>, 1121 (1966).

(38) I. Milne and R. E. Smallman, Trans. AIME, <u>242</u>, 120 (1968).

(39) E. S. Bartlett, D. N. Williams, H. R. Ogden, R. I. Jaffee and E. F. Bradley, Trans. AIME, <u>227</u>, 459 (1963).

(40) G. D. McAdam, J. Inst. Met., <u>93</u>, 559 (1964-65).

(41) B. Harris, Phys. Stat. Sol., <u>18</u>, 715 (1966).

(42) N. D. Tarasov, R. A. Ulyanov, and Ya. D. Mikhailov, Fiz. Metallov i Metallovedenie, <u>18</u>, 740 (1964).

(43) R. L. Fleischer, Acta Met., <u>9</u>, 996 (1961).

(44) R. L. Fleischer, Acta Met., <u>11</u>, 203 (1963).

(45) R. L. Fleischer and W. R. Hibbard, Jr., <u>The Relation Between the Structure and Mechanical Properties of Metals</u>, NPL Symposium, No. 15, Her Majesty's Stationery Office, London, 1963, p. 261.

(46) R. A. Foxall, M. S. Duesbery, and P. B. Hirsch, Canadian J. Phys., <u>45</u>, 607 (1967).

(47) D. K. Bowen, J. W. Christian, and G. Taylor, ibid., p. 903.

(48) W. A. Spitzig and T. E. Mitchell, Acta Met., <u>14</u>, 1311 (1966).

(49) B. A. Wilcox, A. Gilbert, and B. C. Allen, "Intermediate Temperature Ductility and Strength of Tungsten and Molybdenum TZM", AFML-TR-66-89, April, 1966.

(50) J. W. Edington and R. E. Smallman, Acta Met., 12, 1313 (1964).

(51) J. W. Edington, "Effect of Strain Rate on the Dislocation Substruc-
      ture in Deformed Niobium Single Crystals", paper presented at
      Symposium on the Mechanical Behavior of Materials Under Dynamic
      Loads, San Antonio, Texas, September 6-8, 1967 (to be published by
      Springer Verlag).

(52) R. N. Orava, Trans. AIME, 230, 1614 (1964).

(53) J. D. Embury and R. M. Fisher, Acta Met., 14, 147 (1966).

(54) J. D. Embury, A. S. Keh, and R. M. Fisher, Trans. AIME, 236, 1252
      (1966).

(55) V. K. Chandhok, A. Kasak, and J. P. Hirth, Trans. ASM, 59, 288
      (1966).

(56) E. S. Meieran and D. A. Thomas, Trans. AIME, 233, 937 (1965).

(57) E. Nachtigall, Plansee Proceedings-Sintered High Temperature and
      Corrosion Resistance Materials, F. Benesovksy, ed., Pergamon Press,
      London, 1956, p. 313

(58) B. C. Allen, D. J. Maykuth, and R. I. Jaffee, J. Inst. Met., 90,
      120 (1961-62).

(59) M. Semchyshen and R. Q. Barr, ASTM Spec. Tech. Pub. No. 272, 1959.

(60) C. W. Neff, R. G. Frank, and L. Luft, "Refractory Metals Structural
      Development Program", ASD-TR-61-392, Part II, October, 1961.

(61) C. S. Hartley, J. E. Steedly, and L. D. Parsons, Diffusion in Body-
      Centered-Cubic Metals, Amer. Soc. Met., Metals Park, Ohio, 1965,
      p. 51.

(62) R. C. Koo, Trans. AIME, 239, 1996 (1967).

(63) J. L. Walter, Trans. AIME, 239, 272 (1967).

(64) G. Das and S. V. Radcliffe, "Internal Void Formation in Tungsten",
      paper presented at 97th AIME Annual Meeting, New York, February
      26-39, 1968.

(65) R. C. Koo and D. M. Moon, "Nucleation and Growth of Bubbles in Doped
      Tungsten", ibid.

(66) E. F. Kock and J. L. Walter, Trans. AIME, 242, 157 (1968).

(67) F. D. Lemkey, B. J. Bayles, and M. J. Salkind, "Research Investiga-
      tion of Phase-Reinforced High Temperature Alloys Produced Directly
      From the Melt", Tech. Rept. No. AMRA CR-64-05/4, July 31, 1965.

(68) F. D. Lemkey and M. J. Salkind, _Crystal Growth_, H. S. Peiser, ed., Pergamon Press, New York, 1967, p. 171.

(69) M. F. Ashby and J. Lewis, "On the Interaction of Inclusions with Migrating Grain Boundaries, Tech. Rept. No. 547, Contract Nonr-1866(27), November, 1967.

(70) P. Coulomb, Acta Met., _7_, 556 (1959).

(71) B. A. Wilcox and A. R. Rosenfield, Mat. Sci. and Engr., _1_, 201 (1966).

(72) A. H. Cottrell, _Creep and Fracture of Metals at High Temperatures_, Proc. of NPL Symposium, Her Majesty's Stationery Office, London, 1956, p. 141.

(73) L.M.T. Hopkin, J. Iron and Steel Inst., _203_, 583 (1965).

(74) J. Glen, J. Iron and Steel Inst., _190_, 114 (1958).

(75) A. Kelly and G. J. Davies, Met. Rev., _10_, 1 (1965).

(76) A. Kelly and W. Tyson, _High Strength Materials_, V. F. Zackay, ed., Wiley, New York, 1965, p. 578.

(77) B. A. Wilcox and B. C. Allen, J. Less Common Met., _13_, 186 (1967).

(78) E. T. Wessel, L. L. France, and R. T. Begley, _Columbium Metallurgy_, D. L. Douglas and F. W. Kunz, eds., Interscience Pub., New York, 1961, p. 459.

(79) J. D. Baird and C. R. MacKenzie. J. Iron Steel Inst., _202_, 427 (1964).

(80) R. W. Buckman and R. C. Goodspeed, "Development of Dispersion Strengthened Tantalum Base Alloy", NASA-CR-72316, 1967.

(81) F. F. Schmidt, "Tantalum and Tantalum Alloys", DMIC Rept. 133, July 25, 1960.

(82) G. W. King, H. G. Sell, and W. R. Morcom, "An Investigation of the Yield Strength of a Dispersion Hardened W-3.8 v/o ThO_2 Alloy" (to be published).

....

HIGH TEMPERATURE CREEP AND FRACTURE BEHAVIOR
OF THE REFRACTORY METALS

By

R. T. Begley, D. L. Harrod and R. E. Gold

The creep properties of the pure, polycrystalline refractory metals are reviewed and summarized. The creep behavior of all the pure metals is correlated in terms of their diffusivities and elastic moduli following the Sherby-type analysis. Experimental values of the activation energy and stress dependence of creep are also summarized for the pure metals, and the structural features of deformation and fracture are described. The general effects of alloying are delineated in terms of solid solution, dispersed phase, and strain hardening contributions to high temperature creep strength.

Westinghouse Astronuclear Laboratory
Pittsburgh, Pennsylvania 15236

41

I. INTRODUCTION

The technological importance of the refractory metals arises primarily as a result of their capability of performing useful engineering functions at temperatures far beyond those possible with other metals. Consequently, the high temperature mechanical behavior of refractory metals and alloys is a subject of considerable interest from both theoretical and engineering viewpoints. In recent years increasing emphasis has been directed to the time-dependent deformation behavior of these materials, reflecting their growing importance in high temperature applications where creep deformation is a limiting design parameter.

The purpose of this paper is to review the high temperature creep and fracture behavior of the refractory metals, the high melting transition metals in Groups VA (V,Cb,Ta) and VIA(Cr,Mo,W) of the periodic table. No attempt is made to describe in detail all aspects of the creep behavior: rather, this review is concerned with summarizing experimental data on the creep behavior of the pure, polycrystalline refractory metals, and delineating the general effects of alloying and structure on creep and fracture behavior.

II. CREEP OF PURE METALS

General Behavior

The creep curves of the pure metals are normal in the sense that primary, steady-state and tertiary stages of creep are observed, with the presence and extent of the different stages being functions of the temperature and stress. Some of the general features common to all of the pure metals are illustrated in Figure 1.

Most high temperature tests have been of relatively short duration, ranging from tensile tests at an imposed strain rate of $\sim 10^{-3} sec^{-1}$ and lasting for less than 1 hour to 100-hour stress-rupture tests. As a rough guide, stresses which produce rupture in 100 hours will produce several percent or more primary creep, a steady-state creep rate on the order of $10^{-6} sec^{-1}$, and up to a third or more of the test life will be in the tertiary stage. Creep ductility is quite high, generally 50-100% elongation. Since primary creep is dependent mainly upon stress, very low creep rates on the order of $10^{-9} sec^{-1}$, will necessarily require stresses so low that there will be little if any measurable primary creep strain. These general comments pertain to material prepared by melting processes; as shown later, powder metallurgy tungsten and molybdenum are considerably stronger than the melted material and the ductility is less.

The stress and temperature dependence of creep deformation is summarized for tungsten and molybdenum in Figure 2 and for tantalum and columbium in Figure 3. The sources of the data used in making these plots are listed in Table 1 (1-15). In these figures the stress, on a log scale, required to give the indicated deformation is plotted as a function of homologous temperature, T/Tm (absolute degrees). Clearly,

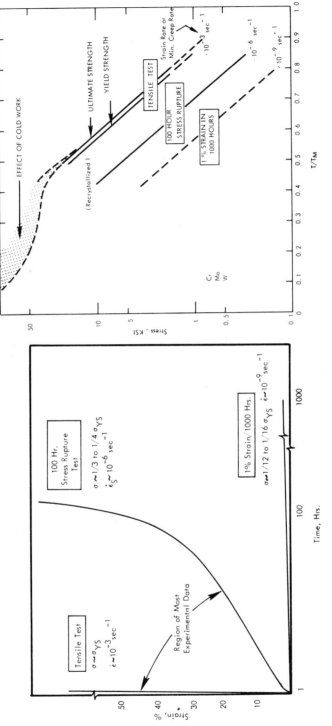

FIGURE 2 – Summary of High-Temperature Deformation of the Group VI$_A$ Refractory Metals (Melted Material Only)

FIGURE 1 – General Features of Time-Dependent Deformation of Pure Refractory Metals Above ~0.5 T$_M$ (Melted Material Only)

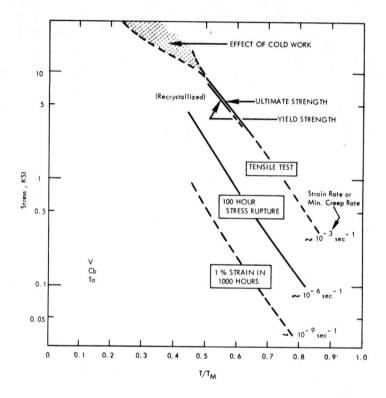

FIGURE 3 – Summary of High-Temperature Deformation of the Group V_A Refractory Metals (Melted Material Only)

TABLE 1 – Sources of Data for Pure Metals

Metal	Interstitial Content, ppm			Starting Material Condition		Test Conditions	Grain Size (mm)	Reference
	C	N	O					
V	460	260	900	Arc melted	1 hr. 1652°F in 10^{-6} torr	Tensile creep in 10^{-6} torr	---	1
V	600	400	1000	Arc melted	1 hr. 1652°F in 10^{-5} torr	Tensile creep in 10^{-5} torr	---	2
Cb	40	50	70-120	Zone refined	1 hr. 2100°F in 10^{-5} torr	Compression creep in 10^{-6} torr	0.20	3
Cb	<50	<60	100	Zone refined	1 hr. 2100°F in 10^{-5} torr	Tensile creep in 10^{-6} torr	0.13/0.35	4
Ta	60-170	15	60-80	EBM	8 hrs. 3270°F in 10^{-5} torr	Tensile creep in 10^{-5} torr(or He)	3.0	5
Ta	25	20	46	EBM	Wrought	Tensile creep in 10^{-7} torr	(a)	6
Cr	---	10	200	Arc melted	Wrought	Compression creep, "rough" argon	0.022	7
Cr	100	500	410	Arc melted	1 hr. 2900°F in hydrogen	Tensile tests in argon	0.50	8
Mo	68	16	184	Arc melted	2 hrs. test temp. in H₂	Tensile creep in hydrogen	0.50(a)	9
Mo	130-200	110-180	30-35	Powder metallurgy	30 min. test temp.	Tensile creep in 2 psig helium	0.07	10
Mo	50-200	5	10	Arc melted	1 hr. 2730°F in 10^{-5} torr	Compression creep in 10^{-6} torr	0.05/0.08	11
W	6	10	5	Arc melted & EBM	15 min. test temp.	Tensile creep in 10^{-6} torr	0.40	12
W	30	70-110	---	Powder metallurgy	30 min. test temp.	Tensile creep in 2 psig helium	0.05	13
W	---	---	---	Arc melted	2 hrs. test temp. in H₂	Tensile creep in hydrogen	0.3/5.0	9
W	22	18	22 τ	Powder metallurgy	1 hr. 3630°F in <10^{-6} torr	Tensile creep in 10^{-6} to 10^{-8} torr	0.02	14
W	<10	<5	29	Powder metallurgy	5 min. 2020°F	Tensile creep in 10^{-7} torr	(a)	15

(a) Fully recrystallized and/or grain growth during tests noted.

the implication is that the metals within each Periodic Group have similar creep properties at equal fractions of their melting points, and within the variability of the experimental data the log σ vs. T/Tm plots are linear.

Group V_A Versus Group VI_A Metals

The creep behavior of the Group V_A and VI_A metals are compared in Figure 4 where the stress to produce a steady–state creep rate of $10^{-6}sec^{-1}$ is plotted against T/Tm. With the exception of the limited data on chromium, the data for the metals in a given Group fall on a common curve. At equal fractions of their melting points the Group VI_A metals are significantly stronger than the Group V_A metals, by a factor of about three. The same general trend is apparent regardless of the creep parameter used in making the comparison. For example, Bechtold, Wessel, and France (16) showed the same trend in terms of stress to produce rupture in 1 and 100 hours.

Only data on melted material was used in preparing Figure 4. As shown in Figure 5, powder metallurgy tungsten and molybdenum are decidedly stronger than arc or electron beam melted material. The most obvious differences in the structures of the two types of materials are the much finer grain size and the grain boundary impurities and precipitates in the powder material. The data on melted chromium was for very fine grained and quite impure material, and it is noted that the chromium data more nearly fit the curve for powder material than they do for the melted material.

Role of Modulus and Diffusivity

Compensation for differences in melting point by means of homologous temperature provides by itself good correlation of the creep properties of the metals within a given Periodic Group, but there still remains the marked difference between the two Groups. In fact, at a given temperature the strength of the pure metals does not increase in the same order as their melting points. Thus, the creep strength of molybdenum at any temperature sufficiently below its melting point is higher than that of tantalum even though tantalum has a much higher melting point.

The two temperature dependent properties of pure metals which are commonly thought to influence creep behavior are elastic modulus E and self–diffusivity D. Sherby (17) proposed that the steady–state creep rate $\dot{\epsilon}$ of pure metals could be correlated by means of the expression:

$$\dot{\epsilon} = S \left(\frac{\sigma}{E}\right)^n D \tag{1}$$

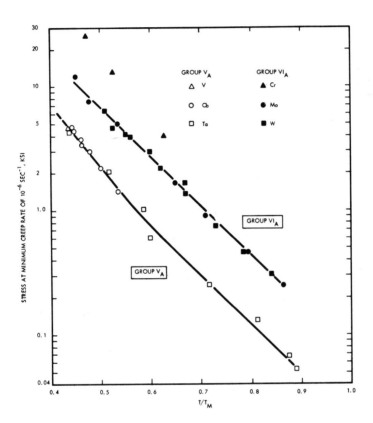

FIGURE 4 – Stress at $\dot{\epsilon} = 10^{-6}\,\text{sec}^{-1}$ vs. T/T_M for Group V_A and
Group VI_A Metals (Melted Material Only)

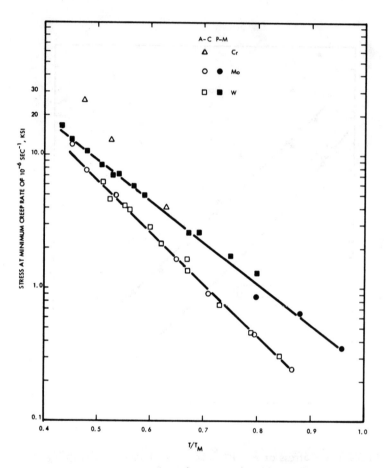

FIGURE 5 – Stress at $\dot{\epsilon} = 10^{-6}\,sec^{-1}$ vs. T/T_M for Arc–Cast and Powder Metallurgy Group VI_A Metals

The structure term S (grain morphology, dislocation density and distribution, etc.) and the stress exponent n are assumed to be constants. For comparison with Figures 2-5, Eq. (1) can be rewritten in terms of log σ to give:

$$\log \sigma = \frac{1}{n} \log \frac{\dot{\epsilon}}{S} + \log E - \frac{1}{n} \log D \qquad (2)$$

The temperature dependence of E is shown in Figure 6 and that of D in Figure 7. On a T/Tm basis, the modulus of the Group VI_A metals is higher than that of the Group V_A metals, on the average by a factor of about two, and the diffusivity is lower, on the average by about one order of magnitude. Hence, the Group VI_A metals are expected, according to Eq. (2), to be stronger by a factor of about three than the Group V_A metals at equal T/Tm, as indeed they are. With these data for E and D it can be shown that the sum of the last two terms in Eq. (2) decreases approximately linearly with increasing T/Tm, hence the linearity of the plots in Figures 2 to 5.

All of the available creep data on the pure metals prepared from the melt are plotted as $\dot{\epsilon}/D$ versus σ/E in Figure 8, following Sherby. The heavy lines are drawn to have a slope which gives n = 5, and since $S = \dot{\epsilon}/D$ when σ = E, extrapolation gives $S \approx 10^{27}$-10^{29} cm^{-2}. The values near the top and to the left of the band are for low temperature, high stress tests on tungsten for which the stress dependence appears to be exponential. Sherby and Burke (26) have shown for a wide range of metals that the transition from a power law to an exponential stress dependence of creep rate occurs at $\dot{\epsilon}/D \approx 10^9$ cm^{-2}. This also seems to be true for the refractory metals. Quantitative agreement is clearly sensitive to the precise values of not only D and E, but also n and S. Even so, the good fit obtained in the Sherby plot suggests that the creep properties of the pure metals can be correlated qualitatively in terms of their moduli and diffusivities.

The appropriate modulus to use in the Sherby equation is the unrelaxed modulus. Even though the moduli shown in Figure 6 were determined by a dynamic method, the drop at high temperatures probably reflects relaxation processes. In the absence of data to the contrary, the temperature dependence of the unrelaxed modulus is generally assumed to be linear, and might possibly be estimated by extrapolation of the approximately linear portion of the low temperature data. However, the moduli used in preparing the Sherby plot were the actual values shown in Figure 6.

Activation Energy and Stress Dependence

Experimental values of the activation energy and stress dependence of creep are tabulated in Table 2.

Only very limited data have been reported for vanadium and chromium. The activation energy for chromium was estimated by using the data of Landau et al (7) on the temperature dependence of the steady-state creep rate.

R. T. BEGLEY, D. L. HARROD, AND R. E. GOLD

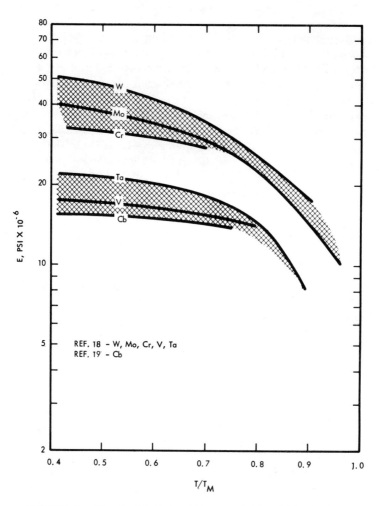

FIGURE 6 – Elastic Modulus (Dynamic) Vs. T/T_M for Group V_A and Group VI_A Refractory Metals

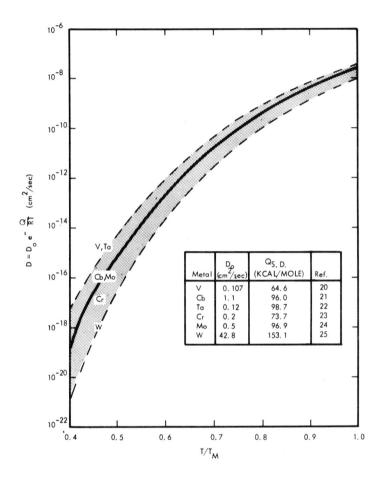

FIGURE 7 - Diffusivity Vs. T/T_M for Group V_A and Group VI_A Refractory Metals

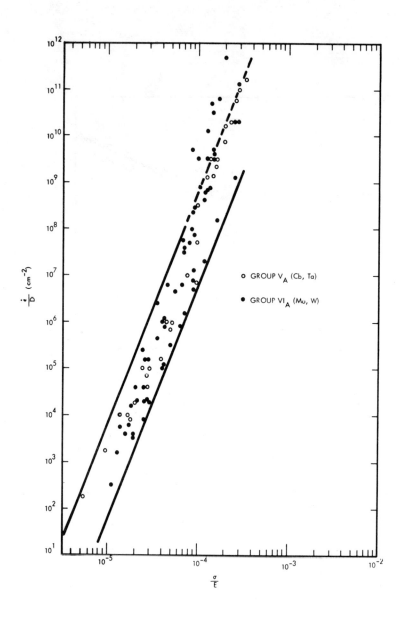

FIGURE 8 – Sherby Plot for the Refractory Metals

TABLE 2 – Summary of Activation Energy and Stress Dependence of Creep of the Pure Refractory Metals

Metal	Apparent Activation Energy for Creep (Kcal/Mole) / Temperature Range T/T_M	Method	$Q_{Self-Diff.}$ (Kcal/Mole) (a)	Strain Rate Range (sec^{-1})	Stress Range (ksi)	Stress Dependence of $\dot{\epsilon}$	Material	Reference
V	65	Isothermal	64.6	10^{-8} to 10^{-7}	11.4		Melted	2
Cb	33 to 77	Isothermal	96.0	10^{-8} to 10^{-6}	1 to 6	$\sigma^{4-4.5}$	Melted	3
Cb	95	ΔT		10^{-8} to 10^{-4}	1.5 to 7		Melted	1P
Cb	106	Isothermal		10^{-8} to 10^{-4}	1.5 to 7		Melted	27
Cb & Cb-O	98, 92	ΔT		10^{-8} to 10^{-6}	1.5 to 7.5		Melted	4
Cb-C	92/130	Isothermal		10^{-9} to 10^{-6}	1.5 to 3	σ^5	Melted	4
Cb	23	ΔT		10^{-9} to 10^{-7}	4 to 8		Melted	28
Ta	75, 114	ΔT	98.7	10^{-7} to 10^{-2}	0.05 to 3	σ^4	Melted	5
Ta	185	Isothermal		(Isothermal Analysis of Data from Reference 5)				29
Cr	86	Isothermal	73.7	10^{-9} to 10^{-7}	6.7 to 20.2		Melted	7
Mo	62-88	Isothermal	96.9	10^{-8} to 10^{-3}	3 to 9	$\sigma^{4.5}$	Powder	10
Mo	54-76	Isothermal		10^{-8} to 10^{-5}	3.5 to 10	σ^{5-6}	Melted	11
Mo	114	ΔT		10^{-8} to 10^{-5}	3.5 to 10	σ^{5-6}	Melted	11
Mo	132-165	ΔT		10^{-7} to 10^{-5}	0.2 to 3	$\sigma^{5.1}$	Melted	9
W	160	Isothermal	153.1	10^{-6} to 10^{-3}	1.5 to 7	$\sigma^{6.3}$	Powder	13
W	141	Isothermal		10^{-7} to 10^{-3}	2 to 10	$\sigma^{5.8}$	Melted	12
W	106	Isothermal		10^{-7} to 10^{-3}	2 to 10		Melted	12
W	105	Isothermal		10^{-8} to 10^{-3}	7 to 25	σ^7	Powder	14
W	72-135 (Tensile Tests)	Isothermal		10^{-3} to 10^{-1}	25 to 40	σ^{5-7}	Powder	30
W	160	Isothermal		10^{-6} to 10^{-4}	15 to 30	$\dot{\epsilon}^{0.7/1.7}\sigma$	Powder	31 via 14
W	78-88	Isothermal & ΔT		10^{-7} to 10^{-5}	0.4 to 1.7	$\sigma^{4.3}$	Melted	9

(a) Activation energies for self-diffusion from references in Figure 7.

Brinson and Argent (3) studied the creep of pure Cb over a relatively narrow temperature range. The isothermal data were complicated by grain growth during some of the tests. At the lower temperatures where no grain growth occurred, a value of Q_C = 33-44 Kcal/mole was obtained, and since this value is low relative to Q_{SD} it was concluded that climb was not rate controlling. At slightly higher temperatures where grain coarsening occurred the activation energy was found to be Q_C = 50-77 Kcal/mole. On the assumption that grain coarsening initially led to a decreased creep rate (eventually followed by accelerated creep with further coarsening) it was proposed that self-diffusion could have been rate controlling and that the low apparent values of Q_C were due to the grain size effect.

Rawson and Argent (4) studied the creep of pure Cb and Cb-O and Cb-C alloys. Using the differential-type test (designated by ΔT in Table 2) the activation energy for creep of pure Cb and Cb-O alloys was found to be Q_C = 98 Kcal/mole, in agreement with Q_{SD}. The sudden increase in temperature in these differential type tests resulted in a period of transient creep. The activation energy was calculated using the initial creep rate following the temperature change and not the eventual new steady-state rate, along with the steady-state creep rate that existed prior to the temperature change. The presence of the transient response following the temperature change was taken to mean that the structural parameters were not constant independent of temperature and therefore the activation energies previously determined (Brinson and Argent) by the isothermal method were not valid. For the Cb-C alloys the carbides present apparently stabilized the dislocation structure since no transients were observed in differential-type tests. This suggested that the isothermal method would give valid results and the value obtained by the isothermal method for the activation energy was Q_C = 92 Kcal/mole, again in agreement with Q_{SD}. These results by Brinson and Argent and by Rawson and Argent therefore suggest that the activation energy for high temperature diffusion controlled creep should not be affected by interstitial impurities, and that the test method employed influences the experimental values of the activation energy.

Gregory and Rowe (28), using the differential method, observed a change in the activation energy for creep of columbium with temperature from Q_C = 23 Kcal/mole below ~0.45 Tm to Q_C = 92-130 Kcal/mole at higher temperatures. This transition was interpreted in terms of a change from cross-slip to climb controlled deformation.

Green (5) studied the creep of tantalum from 0.6-0.9 Tm and from three tests using the differential method obtained a value of Q_C = 114 Kcal/mole, in general agreement with Q_{SD}. Flinn and Gilbert (29) used Green's data to determine Q_C by the isothermal method and obtained Q_C = 75 Kcal/mole below ~0.8 Tm and Q_C = 185 Kcal/mole at the higher temperatures, which they interpreted in terms of two discrete (but unspecified) processes. As with columbium, these data on tantalum illustrate the marked difference that can occur in the results between the two test methods.

The values of the activation energy for creep of molybdenum determined by Green et al (10) and by Conway and Flagella (9) at very high temperatures are as high or higher than Q_{SD}, and are considerably higher than the value determined by Carvalhinhos and Argent (11) at ~0.5 Tm. The latter authors did not find any transient effects by the differential method and the results by the differential and isothermal methods were in good agreement. However, by both methods the activation energy was found to be stress dependent, with the higher stresses giving the higher activation energies! This was interpreted in terms of the stress dependence of the structural parameters along the lines of the Barrett-Nix model of creep controlled by the non-conservative motion of jogged screw dislocations (32).

For tungsten, as for molybdenum, the activation energy for creep at very high temperatures is as high or higher than Q_{SD} and is considerably higher than the values obtained at lower temperatures. Klopp et al (12) suggested that transient and steady-state creep are controlled by different mechanisms at the lower temperatures, perhaps cross slip for transient creep (Q_C=106 Kcal/mole) and climb for steady-state creep (Q_C= 141 Kcal/mole). On the other hand, Gilbert et al (14) interpret their value of Q_C= 105 Kcal/mole in terms of the non-conservative motion of jogs on screw dislocations, with the low apparent value of Q_C being due to enhanced diffusion along short-circuit paths, e. g. , grain boundaries and dislocations. Further, on the basis of estimates of the activation volume, they favor the jog mechanism to the climb of edge dislocations even at the higher test temperatures.

Inspection of the activation energies for creep listed in Table 2 reveals that Q_C is often considerably less than Q_{SD} at temperatures below ~0.6 Tm, and often much higher at higher temperatures. The high values might reflect a contribution due to the temperature dependence of the unrelaxed modulus. Sherby and Burke (26) have shown that, in Eq. (1), $Q_C = Q_{SD} - nR \dfrac{T^2}{E} \dfrac{dE}{dT}$ where E is the unrelaxed, isotropic modulus. Thus Q_C is equal to or greater than Q_{SD} and an estimate for tungsten indicates that the modulus correction could amount to more than 50 Kcal/ mole at very high temperatures.

The stress dependence of steady-state creep has generally been expressed in terms of a power function, i. e. , $\dot{\varepsilon} \propto \sigma^n$. The values of n have ranged from about 4 to 7.

Structural Features and Fracture

Structural features associated with creep deformation have been described for Cb (3,27), Ta (5), Mo (10,11), and W (9,12,13,31). Similar features have been observed in all of these metals and, presumably, would be the same for V and Cr.

Coarse, wavy slip, on single and intersecting slip systems, occurs in which the slip band spacing is inversely proportional to the applied stress. This is taken as evidence of extensive cross slip and to be consistent with a high stacking fault energy and the absence of partial dislocations. In very coarse grained (~3 mm) tantalum, Green (5) found that no coarse slip was observed at stresses below 400 psi. This is consistent with the observation by Brinson and Argent (3) that although coarse slip was observed in fine grained (~0.2 mm) columbium below 0.5 Tm and at high stresses, there was very little evidence of coarse slip at higher temperatures where the stresses were much lower. Thus, as with other pure metals, coarse slip is more pronounced at high stresses, which generally means lower temperatures, and the critical stress below which coarse slip is not observed (slip band spacing approaching the grain size) is higher the smaller the grain size.

Rather pronounced subgrain structures form in all of the pure metals and apparently at all stresses and temperatures but with the subgrain size being smaller at high stresses and low temperatures. The subgrains may be approximately equiaxed or they may be banded. Subgrain formation seems to be particularly pronounced near grain boundaries, and their interaction with the grain boundaries may cause the grain boundaries to be serrated. Based on the angles at which the subgrains interact with the grain boundaries, it is deduced that the misorientation across the well developed subgrains must be rather high, i. e., the subgrains are rather high angle, hence high energy, boundaries.

There seems to have been no quantitative measurements made of the contribution of grain boundary sliding to creep strain in any of the pure metals, although qualitative evidence of grain boundary sliding and migration has been obtained for all of them, and seems to be particularly prominent in Cb and Ta (Group V_A). For coarse grained tantalum, Green (5) deduced from the small offsets at grain boundaries at the surface that grain boundary sliding made only a small contribution to the creep strain. He estimated that about half of the strain was due to coarse slip and most of the remaining half was due to fine slip or so-called slipless flow. Brinson and Argent (3) observed that grain boundary sliding and migration were more pronounced in 0.2 mm grain size columbium the higher the temperature and the lower the stress. They also observed considerable evidence of fold-formation at triple points in columbium.

Although grain boundary sliding is quite prominent in columbium and tantalum, there is no evidence of void formation and intercrystalline cracking. The ease with which fold formation and grain boundary migration occurs seems to preclude intercrystalline cracking in these metals. Hence creep fracture in columbium and tantalum is transcrystalline and the fracture elongation is quite high, 50 to 100% and more.

Powder metallurgy molybdenum and tungsten fail in creep by intercrystalline fracture, with the ductility being generally 5 to 25% elongation. When prepared from the melt, molybdenum and tungsten, like the Group V_A metals, exhibit transcrystalline creep fracture with quite high ductility.

Most of the creep tests on the pure metals have been of relatively short duration and with rather high creep rates. Since the pure metals are weak, the stresses have also been rather low. These conditions would tend to favor transcrystalline fracture. However, in tests on arc melted tungsten, Conway and Flagella (9) found that the material failed by transcrystalline fracture with 70% elongation after 4000 hours and with a steady-state creep rate of $\sim 10^{-8} \text{sec}^{-1}$, which occupied about half of the life.   Thus, the ease with which grain boundary migration and fold formation occurs in the pure metals would probably lead to transcrystalline fracture under any creep conditions.   The reasons for intercrystalline fracture in powder metallurgy Group VI$_A$ metals are not known, but no doubt involve the influence of grain boundary impurities and precipitates on the structural features of deformation.

Cold Working, Grain Size, Impurities and Environment

Variability in creep test data is often attributed to microstructural and compositional differences.   The difference in creep properties between powder and melted material is an extreme example.   Except for the powder material, microstructural and compositional differences were not taken into account in the previous figures, hence the trends established exist in spite of these differences.   The apparent and implied lack of scatter is due in part to the manner of presenting the data.   From Eq. (1), for example, variations in $\dot{\epsilon}$ are directly proportional to variations in S, but the stress is proportional to the nth-root of S.

Cold Working.   The pure metals, after being heavily cold worked, all recrystallize in 1 hour or less at $\sim 0.5 \text{Tm}$.   Hence cold working is not expected to be an effective mechanism of strengthening in high temperature creep.   In short time tensile tests some effect of cold working may persist up to $\sim 0.6 \text{Tm}$, at least in the Group VI$_A$ metals, but in longer time tests or at higher temperatures history effects from prior strain hardening should be eliminated.   The substructure present following complete recrystallization can be influenced by prior mechanical-thermal processing, and this could influence creep, particularly the primary stage.   Even so, there seems to be no direct evidence of an influence of mechanical-thermal processing on creep properties above 0.6 Tm.

Grain Size.   Brinson and Argent (3) tested columbium in the range 0.45 to 0.54 Tm.   The initial grain size was about 0.2 mm.   However, grain coarsening and recrystallization occurred during testing at the higher temperatures and stresses.   Low values for the apparent activation energy were interpreted as being due to enhanced creep strength as the grains initially coarsened, the implication being that the coarse grained material was stronger.   When grains grew (presumably via secondary recrystallization) to a very large size (comparable to the specimen diameter) an inflection towards accelerated creep was often but not always observed in the creep curve. The implication here is that the coarse grained material was weaker.   These observations and interpretations suggest that the creep rate first decreases with increase in grain size, passes through a minimum, and then increases with further increase in grain size.

Stoop and Shahinian (27) observed sudden bursts of accelerated creep in columbium-oxygen alloys.   Their interpretation was that the accelerated creep was not due to recrystallization nor to strain aging but was due to accelerated recovery by dislocation climb, with the creep rate eventually leveling back off as the substructure formed.

Klopp et al (12) have performed perhaps the only extensive study of grain size effects in creep of the pure refractory metals. Their work was on arc-and EB-melted tungsten. In the temperature range 0. 45 to 0. 67 Tm, their results were: ultimate tensile strength $\propto L^{-0.12}$, yield strength $\propto L^{-0.25}$ and stress to produce a minimum creep rate of $10^{-6}sec^{-1} \propto L^{-0.074}$, where L = grain size. Thus, the fine grained material is stronger and the effect of grain size is more pronounced the shorter the test time, which is equivalent to less complete recovery or more effective strain hardening.

For arc-melted molybdenum tested at 0. 65 Tm, Conway and Flagella (9) found that the steady-state creep rate increased by a factor of three as the pre-test anneal-ing temperature increased from 0. 65 to 0. 93 Tm. This suggested that the finer grained material was stronger.

Green et al (10) tested powder metallurgy molybdenum and observed grain growth during testing, presumably due to secondary recrystallization. Below 0. 68 Tm, grain growth led to increased creep rate and above 0. 79 Tm grain growth led to decreased creep rate. This suggested that the fine grained material was stronger at low temperatures and the coarse grained material was stronger at high temperatures.

Impurities. Rawson and Argent (4) studied the effect of oxygen and carbon additions to the melt on the creep strength of columbium. Oxygen additions up to 1200 ppm had little influence on creep at temperatures above 0. 48 Tm, but decreased the creep rate somewhat at lower temperatures. Transmission electron micrographs showed precipitates, presumably oxides, on dislocation networks after creep testing the alloys whereas no such precipitates were observed in the pure columbium. Since the precipitates probably formed upon cooling, the strengthening due to oxygen at the lower temperatures was attributed to an oxygen-clustering mechanism. Carbon additions up to 600 ppm, which yielded carbides, led to decreased creep rates. The highest test temperature was 0. 54 Tm. With 600 ppm carbon the creep rate was about one-tenth that of pure columbium.

Sheely (33) found that oxygen additions up to 6400 ppm had no effect on tensile properties ($\dot{\epsilon} \sim 10^{-3}sec^{-1}$) above $\sim$0. 36 Tm. Stoop and Shahinian (27), how-ever, found that up to 3500 ppm oxygen additions to columbium produced an order of magnitude strengthening in terms of both minimum creep rate and rupture time at temperatures in the range 0. 55 to 0. 62 Tm.

Schmidt et al (34) studied tantalum containing, separately, additions of up to 560 ppm O, 225 ppm N and 995 ppm C. In terms of 100 hour rupture life at 0. 45 Tm, oxygen produced no strengthening, whereas the increase in strength due to nitrogen was 13% and that due to carbon was 200%. The strengthening due to carbon was found to be greater in the 100 hour rupture tests than in 0. 1 hour rupture tests, the latter being increased by 108%. The carbon content was considerably in excess of the solubility limit at the test temperature and was thus present as a $Ta_2 C$ precipitate.

Conway and Flagella (9) obtained the same creep results on arc cast molybdenum sheets obtained from different vendors, implying no effect due to differences in impurities if such differences existed.   Klopp et al (12) found that arc melted tungsten was about 15% stronger than EB melted tungsten of the same grain size in tensile tests at temperatures up to 0. 7 Tm, which they attributed to impurities.   However, there was no difference in the results obtained in longer time creep tests, $\dot{\epsilon}=10^{-6}sec^{-1}$.

Environment.   The marked reactivity of the pure metals with interstitials suggests the necessity to consider the possible effects of the test environment on creep properties.   Systematic studies are lacking but a few specific effects have been reported.

McCoy and Douglas (35) tested columbium at 0. 46 Tm.   With tests in argon as a reference, they found a slight strengthening effect due to the addition of oxygen, a rather pronounced strengthening due to nitrogen, and a very pronounced weakening due to both wet and dry hydrogen.

Conway and Flagella (9) have run tests on tungsten and molybdenum in vacuum, argon and hydrogen.   Short time tests (~10 hours) on arc-cast molybdenum at 0. 65 Tm gave about the same results for vacuum and hydrogen, although the specimens tested in vacuum were slightly stronger.   On powder metallurgy tungsten, tests in argon and hydrogen gave the same results at 0. 67 Tm but the tests in hydrogen were slightly stronger at 0. 78 Tm.   This latter effect was interpreted as possibly being due to additional sintering occurring during testing in hydrogen.

III. ALLOYING EFFECTS

Alloy strengthening of the refractory metals has been a topic of extensive study in recent years, although the emphasis has been on improvement of short time strength rather than optimization of creep properties.   Alloy strengthening mechanisms in the refractory metals have been described in several review papers (36,37), in addition to the comprehensive review by Wilcox which appears in this volume.   Thus we will confine our attention in this paper to a phenomenological description of the more significant effects of alloying on creep behavior, rather than treating in detail specific strengthening mechanisms.

To obtain a perspective on the trend of alloy development it is useful to examine the refractory metal alloys which exhibit the most attractive high tempera-ture properties.   The great majority of these alloys have compositions of the general type:

$$M + M_{S.S.} + M_{R.E.} + M_I = Alloy \tag{3}$$

where: M – matrix element
$M_{S.S.}$ – substitutional solute(s)
$M_{R.E.}$ – reactive element solute(s)
M_I – interstitial solute(s)

The notation $M_{R.E.}$ refers to a solute which has a higher negative free energy of formation for carbides and/or nitrides and oxides than the matrix element. Obviously some specific elements can serve as both substitutional solute strengtheners and as reactive solute additions, for example Hf in tungsten. Eq. (3) implicitly defines the two major means of strengthening which have been most widely and effectively utilized in the refractory metals; solid solution strengthening and strengthening by dispersed second phase particles. The interaction of a reactive element with a residual or intentionally added interstitial element to provide precipitation hardening has been shown to be a very effective means of strengthening all of the Group V_A and VI_A metals, and is typified by the Mo-TZM and Mo-TZC alloys which are strengthened by the precipitation of titanium and zirconium carbides.

The improvement in creep strength which can be realized from solid solution and dispersed phase strengthening is illustrated in Figure 9, which compares the creep strength of pure columbium, a Cb-11 a/oW-1 a/o Hf solid solution alloy, and an identical alloy containing an additional 1.3 a/o C. All materials were in the recrystallized condition, and had equivalent grain sizes. At 0.5 Tm the substitutional solutes increase the stress required for a creep rate of $10^{-6}sec^{-1}$ by a factor of about 12 compared to unalloyed columbium. In the carbon containing alloy, the precipitation of a columbium-hafnium carbide significantly enhances the creep resistance of the solid solution matrix. However, the pronounced temperature dependence of the dispersed phase strengthening increment reflects the major limitation of refractory metal alloys strengthened by the precipitation of carbide, oxide, and nitride phases. Since strength retention is strongly influenced by precipitate stability, these phases in general became rather ineffective creep strengtheners at temperatures of >0.6 Tm because of precipitate agglomeration and solutioning effects. Nevertheless, the pronounced improvement in properties which can be achieved at intermediate (homologous) temperatures with carbide and nitride precipitates has stimulated considerable research aimed at controlling precipitate size, distribution and morphology through suitable thermal or thermal-mechanical treatments in order to optimize mechanical properties.

Solute Strengthening

As shown previously, differences in the creep strength of the pure refractory metals can be rationalized quite well in terms of differences in their elastic moduli and self-diffusion rates. Thus, as has been indicated by Sherby (17), solutes which increase elastic modulus and lower atomic mobility should be effective in improving high temperature creep resistance. Rigorous demonstration of the validity of this concept in the refractory metals is difficult because of the lack of data for specific solute elements, particularly with respect to the composition and temperature range evaluated. However, the general trend exists. Based upon the results of a number of studies, the alloy additions presently used in creep resistant refractory metal alloys are summarized in Table 3.

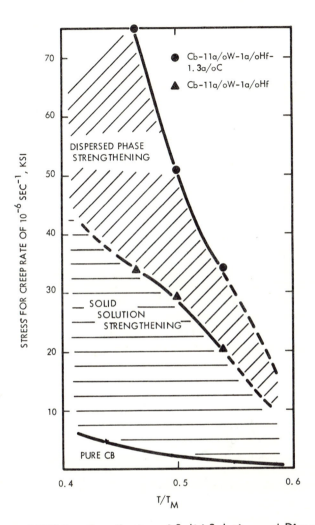

FIGURE 9 - Contribution of Solid Solution and Dispersed
Phase Strengthening in a Columbium Alloy

TABLE 3 - Summary of Alloy Additions Utilized in Creep Resistant
Refractory Metal Alloys

Alloy Base	Substitutional Solutes	Reactive Elements	Interstitial Elements
V	Cr, Cb, Ta, Mo, Fe	Zr, Ti	C, N
Cb	W, Mo, Ta	Zr, Hf	C, N
Ta	W, Re, Mo	Zr, Hf	C, N
Cr	W, Mo, Cb, Ta	Ta, Cb, Ti, Zr, Hf	C
Mo	W, Re	Ti, Zr, Cb	C, N
W	Re, Hf, Cb, Ta	Hf	C

In vanadium, Pollack et al (1) found that Cr, Cb, Ta, Mo, and Fe improved the creep properties of vanadium at 700 and $800^{\circ}C$. However, the relative effectiveness of the specific solutes cannot be unambiguously established from this study since the alloys contained additions of zirconium and carbon which provided a dispersed phase strengthening contribution. Böhm and Schirra (2) observed effective strengthening of vanadium by columbium and molybdenum, in the temperature range of 650 to $800^{\circ}C$ (0. 43 to 0. 5 Tm).

An extensive study of the effect of various substitutional solutes on the creep strength of columbium was carried out by McAdam (38), whose results are summarized in Figure 10. The strengthening contribution of the individual elements are expressed as a percentage of the effect of tungsten, the most potent solute. The data are for a temperature of $1200^{\circ}C$ (0. 54 Tm) at a single stress and are derived from values of the log rupture time, which correlated well with secondary creep rate. McAdam showed good correlation between creep strength and calculated rigidity modulus and melting point values, with the melting point showing the most marked effect. To a first approximation increase in melting point may reasonably be related to a decrease in atomic mobility, although data on the effect of solute additions on D are very limited. Bartlett et al (39) also show tungsten to be a very effective creep strengthener of columbium at $1200^{\circ}C$, as is molybdenum to a lesser degree.

McAdam (Figure 10) observed that additions of vanadium, as well as Hf, Zr, and Ti, make a negative contribution to the creep strength of columbium alloys, even though vanadium has a potent effect in increasing the short time elevated temperature tensile strength of columbium (56). Based upon the elastic modulus data shown in Figure 6, vanadium additions should slightly increase the elastic modulus of columbium; however, Hartley et al (40) have shown that vanadium also increases the self-diffusivity of columbium. An increase in D could overshadow the slight favorable effect of vanadium on the elastic modulus and result in a decrease in creep resistance. The weakening of columbium by additions of titanium has also been observed by Chang (41) in a complex columbium alloy. Titanium has a low elastic modulus (16×10^6 psi at room temperature) and has been reported to decrease the activation energy for the self-diffusion of columbium (42,43). In contrast,

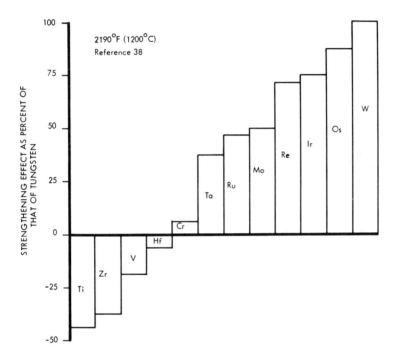

FIGURE 10 – Strengthening Effect of Alloying Elements in Columbium

molybdenum significantly increases the elastic modulus of columbium and decreases the self-diffusion rate (44,45) and is an effective strengthener (38).

Degradation in creep strength of vanadium alloys resulting from large additions of titanium has also been reported (2), as shown in Figure 11. In V-Ti binary alloys, the 650°C (0.43 Tm) ultimate tensile strength increased significantly with increasing titanium content, but the 1000 hour stress-rupture strength was drastically reduced at titanium concentrations over approximately 5 w/o. Additions of titanium to a V-20 w/o Cb alloy also resulted in a marked decrease in stress rupture strength with increasing titanium level although the tensile strength increased slightly. It is interesting to note however, that the 650°C, 1000 hour stress rupture properties of a V-2.5 w/o Ti alloy was reported by Böhm and Schirra (2) to be approximately 50,000 psi, while the ultimate tensile strength of pure vanadium at this temperature is only 8500 psi. It seems apparent that this pronounced increase in strength arising from a relatively small addition of titanium is not primarily a solution strengthening effect, but is rather a manifestation of dispersed phase strengthening as a result of the interaction of the titanium with the interstitial elements present in the vanadium.

The strengthening of tantalum by substitutional solutes generally parallels the trend in columbium, although the data are more limited. Buckman and Goodspeed (46) and Bartlett et al (47) found tungsten, and to a lesser extent molybdenum to be effective strengtheners. Small additions of rhenium, on the order of 1 a/o, were observed by Buckman and Goodspeed to significantly enhance the creep strength of Ta-W-Hf alloys, while Hf additions had a weakening effect, similar to that observed in columbium by McAdam (38).

Data on the effect of individual solutes on the creep properties of the Group VIA elements are quite limited. In chromium, Greenaway (48) showed effective creep strengthening by W, Mo, Ta, Cb, and Zr in tests conducted at 950°C. Clark and Wukusick (49) compared the effect of additions of V, Mo, and W on the elevated temperature tensile properties of chromium at 1095 and 1205°C, showing that W was twice as effective as Mo at 1205°C, with vanadium being ineffective. Both tungsten and molybdenum significantly improve the creep strength of carbide strengthened chromium alloys (50).

Tungsten and rhenium additions improve the creep strength of molybdenum at temperatures as high as 2200°C (9). At 1095°C (0.47 Tm) Semchyshen (51) found that very small additions of Co, V, Al, and W, on the order of 1 a/o or less, resulted in modest improvements in the 100 hour rupture stress. However, these effects were far overshadowed by the improvement achieved with small additions of Ti, Zr, Hf, and Cb which are primarily associated with interstitial-solute interactions. Furthermore, Semchyshen found that while a Mo-1 a/o Ti alloy had a 100 hour rupture stress of 20,000 psi, increasing the titanium content to 4 a/o reduced the 100 hour rupture stress by almost a factor of 2, again illustrating the weakening effect of larger additions of titanium.

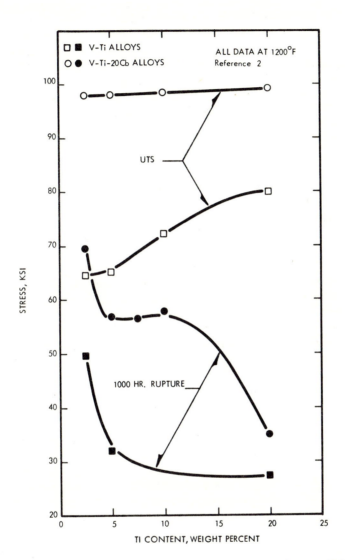

FIGURE 11 – Effect of Ti Content on the UTS and 1000
Hour Stress-to-Rupture at 1200° F

Conway and Flagella (9) performed creep-rupture tests on W and W-25w/oRe from 1600 to 2600°C. On the basis of rupture life and secondary creep rate they found W-25w/o Re to be stronger than W at temperatures below 2000°C(0.62 Tm) but at higher temperatures W-25w/o Re was inferior to pure tungsten. Studies by Klopp and co-workers (52,53,54)(Figure 12) indicate that hafnium significantly improves the creep strength of tungsten at 3500°F (0.6 Tm). Columbium and rhenium additions were found to be less effective. The potency of the solute additions in improving creep properties, expressed in terms of stress for a minimum creep rate of $10^{-6}sec^{-1}$, was in the same order as their effect on short-time tensile strength. The effectiveness of hafnium and columbium at such high temperatures suggests an interstitial coupling effect with these reactive solutes. However, the residual interstitial levels of the tungsten used by Klopp et al were quite low, on the order of 0.05a/o. To determine if an interstitial-solute interaction contributes to the strength of the alloys, specimens were annealed at 4000°F for 1 hour and both slow cooled and helium quenched. Metallographic examination showed that the slow cooled sample had a globular precipiate at the grain boundaries, while the quenched sample was single phase. Creep tests at 3500°F showed the creep strength of the slowly cooled sample to be 4800 psi versus 5400 psi for the quenched material. Since these two heat treatments had relatively little effect on creep strength, it may be inferred that interstitial-solute interaction made a relatively insignificant contribution to strength. Solutes can increase strength by segregation to a jog, and may be strongly bound to vacancies, thus reducing jog and vacancy mobility. Raffo and Klopp (54) pointed out that the binding energy between solutes and jogs or vacancies could be expected to depend on the difference in atomic size between solute and solvent, and showed a reasonable correlation between solute atom size and strength for their alloys. However, Seigle's (55) review of solute strengthening in refractory metals indicated that atomic size is of secondary importance, as did McAdam (38) in his studies of Cb alloys.

A review of the solutes utilized in the most creep resistant alloys shown in Table 3 indicates that the most effective strengtheners are generally those which increase the elastic modulus and/or melting temperature of the matrix, although there are several exceptions to this trend. The converse is also generally observed in that elements such as Ti which decrease melting temperature and activation energy for self-diffusion degrade the creep properties of alloys. While other factors are undoubtedly of importance, it seems apparent that the role of substitutional solutes on elastic modulus and diffusivity may largely dictate their effectiveness as creep strengtheners.

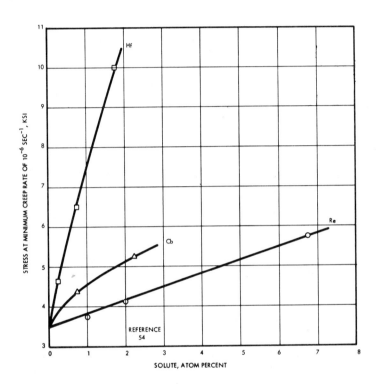

FIGURE 12 - Influence of Alloying on the 3500°F Creep
Strength of Binary Tungsten Alloys

Dispersed Phase Strengthening
_____

Carbides. Creep strengthening effects arising from the precipitation of reactive metal carbides have been demonstrated in V (1), Cb (41,57,58), Ta (46), Cr (49,59), and W (54) base alloys analogous to those in the TZC type molybdenum alloys. The strength of these alloys is controllable by heat treatment and is critically dependent upon precipitate stability. For example, studies by Raffo and Klopp (54), subsequently extended by Rubenstein (60), showed that W-Hf-C alloys containing a fine HfC precipitate were superior to solid solution W-Hf alloys at temperature as high as 3500°F (0.6 Tm) as shown in Figure 13. However Rubenstein's studies suggested that precipitate coarsening could be fairly rapid at this temperature, indicating that the strengthening contribution of the HfC precipitate may be considerably less in longer time creep tests.

Strengthening by boride phases has been explored to a limited extent in Cb alloys (58), but the strengthening effects were not as pronounced as those observed with carbides.

Nitrides. Precipitation strengthening by nitrides has also been shown in V (1), Cb (61), Ta (46), and Mo (62) alloys. Tantalum base alloys strengthened by the precipitation of hafnium and zirconium nitrides have creep properties superior to any other tantalum alloys at 1315°C (0.49 Tm). A Ta-W-Hf-Re-Zr-N alloy tested at 20,000 psi at 1315°C exhibited less than 0.2% strain in 200 hours (70). These nitride strengthened alloys overage rapidly at 1300-1350°C but even in the over-aged condition have attractive mechanical properties. Begley et al (61) reported potent strengthening in Cb-Hf-N alloys resulting from the precipitation of HfN, but here too, rapid overaging occurred at relatively low temperatures, approximately 1200°C.

Oxides. Attempts to strengthen refractory metals by precipitation of stable oxide phases have not had the same success as with carbides and nitrides. Oxygen solubility in the Group VI_A metals is very low, and oxides tend to precipitate at grain boundaries. Although the oxygen solubility is high in the Group V_A elements, the addition of strong oxide formers such as Zr and Hf drastically reduces oxygen solubility. Barber and Morton (63) have shown the oxygen solubility in a Cb-1w/o Zr alloy to be less than 100 ppm at 1000°C compared to a solubility of 3800 ppm in unalloyed columbium. Moderate aging response in Cb-1w/oZr was observed by Hobson (64) and by Stewart et al (65) at 925°C which was attributed to the precipitation of ZrO_2. At higher zirconium and oxygen concentrations than those investigated by Hobson, massive oxides formed which could not be redistributed by subsequent heat treatment and no improvement in mechanical properties could be realized (66).

Significant strengthening by oxides has been achieved, however, when Cb and Ta alloys containing a reactive solute have been internally oxidized. Bonesteel et al (67) internally oxidized Cb-1w/oZr and Cb-1w/oW-1w/oZr alloys by exposure in a dynamic vacuum of approximately 10^{-5} torr at temperatures of 800 and 900°C.

Subsequent creep testing at 1200°C showed improvement in creep strength compared to non-oxidized material. Electron microscopy studies of the internally oxidized samples showed a high density of coherent precipitates less than 100 Å in size. After creep testing at 1200°C only large noncoherent precipitates of monoclinic ZrO_2 were observed.

Buckman (68) achieved oxide strengthening in a Ta-8w/oW-2w/oHf alloy by oxidizing in a partial pressure of oxygen at 538°C, and subsequently diffusing the oxygen into the matrix by a high vacuum anneal for 50 hours at 982°C. This material was creep tested at 1200°C (0.45 Tm) at 21,000 psi and initially exhibited strength superior to that of the uncontaminated material. The secondary creep rate of the internally oxidized material was 0.0052% hr.$^{-1}$ versus 0.044% hr.$^{-1}$ for the un-contaminated alloy. However, after approximately 75 hours the creep rate of the internally oxidized alloy increased to 0.03% hr.$^{-1}$ indicating loss of coherency of the oxide precipitate. Transmission electron microscopy confirmed the presence of very fine HfO_2 precipitates in the oxygen contaminated material although precipitate growth occurred during creep at 1200°C, as shown by the electron micrograph in Figure 14.

The results of Bonesteel and Buckman indicate that while internal oxidation can significantly improve creep strength, the effectiveness of the strengthening achieved appears to be limited to fairly low temperatures because of precipitate instability. However, these data illustrate another feature of considerable importance in the evaluation of the creep behavior of refractory metal alloys. Contamination with oxygen (as well as carbon and nitrogen) during testing can lead to spurious creep results unless particular care is taken to eliminate all sources of contamination.

Appreciable differences in the creep strength of a Ta-8w/oW-2w/oHf alloy were observed by Buckman (69) depending upon test environment. Testing at 10^{-6} torr in a liquid nitrogen trapped, oil diffusion pumped, vacuum system resulted in higher creep strength than when testing was carried out at pressures of 10^{-8} torr or less in a sputter-ion pumped system. Contamination by oxygen and carbon from backdiffusion of diffusion pump oil in the 10^{-6} torr tests resulted in the formation of fine HfO_2 and (Ta,Hf)C precipitates which provided strengthening. Even at temperatures at which overaging occurs rapidly, the presence of a constant source of contamination can provide strengthening until all of the reactive element in the sample is saturated. Consequently, the use of ultra-high vacuum equipment operating at pressures below 10^{-8} torr is becoming increasingly utilized for long time creep testing of refractory metals (71,72). However, the review by Inouye in this volume shows that degassing, decarburization and sublimation effects can also occur under high vacuum conditions depending upon the individual constituents of the alloys. Hence considerable care must be exercised to achieve a test environment which does not interact with the test specimen.

Although much attention has been directed to the development and under-standing of precipitation strengthened refractory metal alloys, the remarkable high temperature properties of S.A.P. and TD-nickel has focused increasing attention on

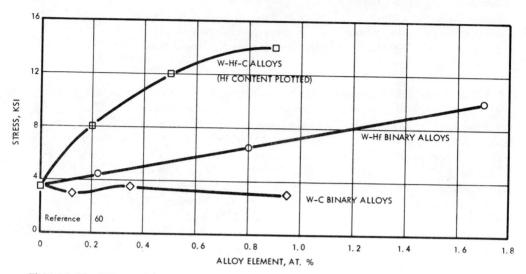

FIGURE 13 - Effect of C and Hf Content on Stress for $\dot{\epsilon} = 10^{-6}$ sec^{-1} at 3500° F. Data from Step-Load Creep Tests of Recrystallized W Alloys

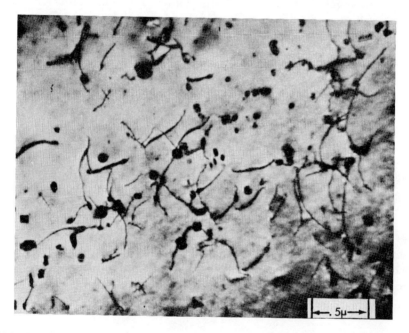

FIGURE 14 - Transmission Micrograph of Internally Oxidized Ta-Base Alloy (T-111, 165 ppm O_2) After Testing at 2200°F (Ref. 68)

refractory metal systems strengthened by fine, insoluble particles, since these systems do not undergo precipitate agglomeration and solution effects which limit the useful temperature range of precipitation hardened alloys. Indeed, studies of tungsten containing a ThO_2 dispersion have been carried on for many years and predate the successful development of S. A. P. However, the early work on thoriated tungsten did not result in a true dispersion strengthened structure. By careful control of processing to prevent particles agglomeration, Sell, Morcom, and King (75) recently demonstrated remarkable tensile strength in a W-3. 8v/oThO_2 alloy at temperatures up to 2800°C. Although creep data are very limited, preliminary results indicate high creep strength at temperatures up to 2200°C (0. 67 Tm). Studies of the stability of the structure indicate that creep strength can be realized for long time, very high temperature applications.

Structural Features

Alloy development studies have quite naturally placed heavy emphasis on compositional effects per se, but it is also widely recognized that the attendant microstructures and dislocation substructures lie at the heart of strengthening mechanisms. The carbides, nitrides, and oxides used in the refractory metal alloys exhibit equilibrium phase relationships with the matrix and consequently are subject to microstructural control by heat treatment. The specific mechanisms whereby precipitation reactions confer creep strength are presently only poorly understood. However, a great variety of microstructural features associated with precipitate dispersion and morphology have been observed and correlated with properties, as described in detail in the references in the preceding section.

Phase diagrams have now been determined over major portions of the useful composition range of a number of refractory metal ternary and higher order systems (77). In many specific alloys, the composition and crystal structure of the precipitating phases have been determined. In some cases the creep properties of a series of alloys in a given system have been correlated with the different phase fields of the phase diagram (78).

In the absence of phase diagrams, the interstitial/reactive metal atom ratio is often taken as a guide in selecting the concentrations of these elements to be used. It has frequently been observed, in carbide strengthened alloys, that optimum strengthening is achieved with carbon/reactive metal atom ratios of approximately unity. This was observed in W-Hf-C alloys by Rubenstein (60). In Cb alloys McAdam (58) found the best results at a carbon/reactive metal atom ratio of~0. 8. Begley and Cornie (74), also in Cb alloys, found the optimum ratio varied somewhat with temperature, as shown in Figure 15. At 2000°F the creep strength, as indicated by 100 hour rupture data, increased with carbon/reactive metal atom ratio up to a ratio of at least 1. 15. At 2400°F the maximum creep strength was achieved at a ratio of~0. 8. Similar data are lacking to determine if an optimum interstitial/reactive metal atom ratio exists in oxide and nitride strengthened alloys.

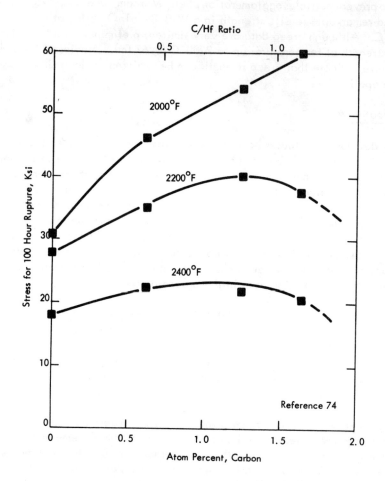

FIGURE 15 – Effect of Carbon and Hafnium Content on the Stress
for 100 Hour Rupture of Cb-22W-2Hf Alloy

In addition to solid solution and dispersed phase strengthening, strain hardening is the other major means for strengthening refractory metals. This is discussed in some detail in papers by Wilcox and by Perkins in this volume. Whereas the pure metals recrystallize in 1 hour at temperatures below 0.5 Tm, precipitate stabilization of deformation substructures may raise the 1 hour recrystallization temperature to 0.6 Tm and perhaps even up to 0.7 Tm. Alloys of the Group VI_A metals in particular are often used in the strain hardened condition. Stress-rupture data for Cb-TZM reported by Perkins (73) shows that for short times at 1315°C (0.55 Tm) strain hardened material stabilized by platelet CbC precipitates had much superior rupture properties compared to unworked material, Figure 16. However, for times greater than about 20 hours the benefits of strain hardening were lost due to structural instability. The effectiveness of strain hardening Group VI_A metals to enhance short time creep strength is therefore presently limited to below ~0.6 Tm, and to considerably lower temperatures for very long time applications.

In the Group V_A metals strain hardening appears to be considerably less effective, as shown by the data of Figure 17 for a strain hardened and solution annealed Cb-W-Hf-C alloy (74). At 1200°C (0.5 Tm) the warm worked plus stress relieved condition resulted in a marked decrease in rupture life and an increase in creep rate compared to the solution annealed condition. Similar effects have been observed in carbide strengthened Ta alloys.

The stress dependence of the steady-state creep rate is often quite high in the dispersed phase alloys as compared to the pure metals and solid solution alloys. Rubenstein (60) showed stress exponents ($\dot{\varepsilon} \propto \sigma^n$) for recrystallized and solution annealed plus aged W-Hf-C alloys varying from 10 to 13 compared to about 5 for pure tungsten and W-Hf binary alloys. Begley, Cornie, and Goodspeed (57) observed values of about 8 in a carbide strengthened Cb alloy at 1100 and 1200°C, but only 3.4 at 1315°C where the carbide precipitate tended to change from an acicular shape to more regular polyhedra. Studies by Perkins (73) indicated a stress exponent of 7 in a strain hardened Cb-TZM alloy containing a platelet CbC precipitate, but values of only 4 for strain hardened material stabilized by spherical CbC precipitates. Recrystallized Cb-TZM having a platelet CbC precipitate had a stress dependence of 40. Thus, precipitate morphology appears to have a pronounced effect on the stress dependence of creep.

The activation energy for creep of dispersed phase alloys is also often observed to be quite high. For example, Cornie (unpublished data) has measured activation energies for creep in Cb-W-Hf-C alloys upwards of 300 Kcal/mole by the ΔT technique. Mukherjee and Martin (62) observed an activation energy of 206 Kcal/mole on a nitrided Mo-Ti alloy by the isothermal technique. Such high values for the apparent activation energy probably reflect structural instabilities occurring during deformation.

Fracture

It has been shown that the pure metals fail in creep by transcrystalline fracture

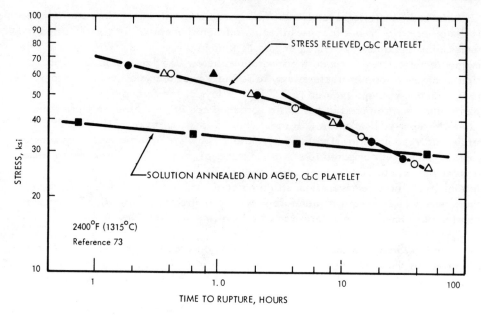

FIGURE 16 – Effect of Structure on the Stress Rupture Properties of Cb-TZM

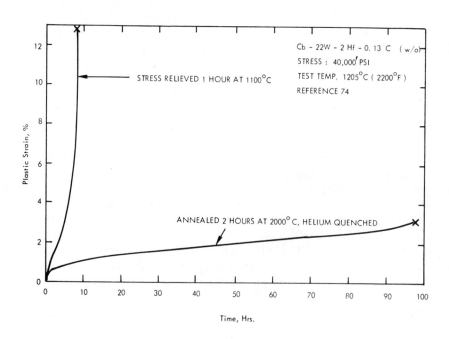

FIGURE 17 – Effect of Thermal-Mechanical Treatment on the Creep
Behavior of a Cb-W-Hf-C Alloy

with quite high ductility. This stems from pronounced grain boundary migration and fold formation at triple points which mitigate the stress concentrations arising from grain boundary sliding. Alloying by both solid solutions and dispersed phases often promotes intercrystalline fracture which in turn frequently leads to low creep ductility. Alloying of course strengthens the matrix. Disparity in the shear strengths of the matrix and grain boundary would tend to promote intercrystalline cracking if the strains due to grain boundary sliding cannot be relaxed by deformation of the stronger matrix. Precipitate denuded grain boundary zones would promote such a situation. Precipitates might restrain grain boundary migration, and microcracks might form at matrix-grain boundary interfaces due to low interfacial adhesion. However, grain boundary precipitates can exert a favorable influence by decreasing the free lengths of grain boundary, which would reduce stress concentrations. Although the general features of intercrystalline cracking are known (79,80), and although intercrystalline cracking is common in refractory metal alloys, few systematic studies of fracture in these alloys have been conducted.

Begley and Godshall (76) studied grain boundary sliding and fracture in Cb-1w/oZr and Cb-10w/oW-5w/oV-1w/oZr alloys. Grain boundary sliding, as measured by offsets in fiducial markings, was observed in both alloys. However, grain boundary migration and fold formation prevented intergranular fracture in the relatively weak Cb-1w/oZr alloy, whereas intergranular cracking was observed in the stronger alloy after only several percent creep strain. In a still more highly alloyed Cb base alloy, Begley, Cornie and Goodspeed (57) observed that grain size and morphology had a pronounced effect on creep fracture, as illustrated in Figure 18. Both material conditions gave similar creep curves through the steady-state stage, and both exhibited intergranular cracking. However, the grain boundary cracks could not propagate far in the fine grained and fibered (duplex) structure, with the result that a long third stage of creep occurred which led to much higher ductility and more than a doubling of the rupture life.

Mukherjee and Martin (62) observed enhanced intergranular cracking in nitrided Mo-Ti and Mo-Zr alloys which they suggested may have been due to inter-granular nitrides inducing microcracks, either by fracturing of the precipitates or by adhesive failure at the matrix-precipitate interface.

Wilcox, Gilbert and Allen (81) observed a tensile ductility minimum at inter-mediate temperatures ($\sim$1100-1500°C) in recrystallized Mo-TZM. The ductility minimum was due to intercrystalline cracking and occurred in the same temperature range in which dynamic strengthening was prominent. This apparently is an example of a disparity in shear strength between the matrix and grain boundary region.

IV. SUMMARY

The degree of strengthening achieved in the Group V_A and VI_A refractory metals by alloying is summarized in Figure 19. The alloys shown were picked as being representative of the stronger alloys for each metal; a number of other alloys of equal strength might have been chosen.

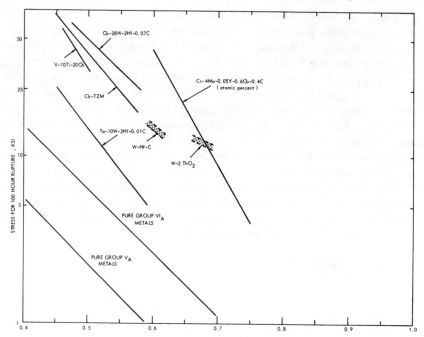

FIGURE 18 – Effect of Microstructure on Creep-Rupture Characteristics of Cb-28W-2Hf-0.067C

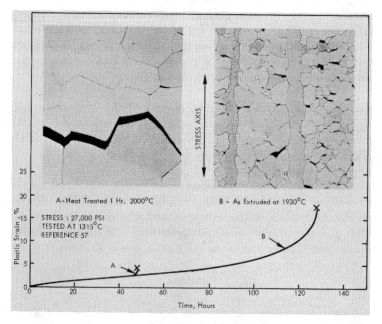

FIGURE 19 – Summary of Alloy Strengthening Effects in Refractory Metals

It is apparent that remarkable success has been realized in alloying of the pure metals for high strength at high temperatures. Solid solution, dispersed phase, and strain hardening mechanisms have all been used, the first two particularly in the Group V_A metals and the latter two mostly with the Group VI_A metals. The Group V_A alloys may contain as much as 10 to 20 a/o solute, as compared with usually less than 2 a/o solute in the Group VI_A alloys. Thus, the Group V_A metals are highly alloyed whereas the Group VI_A alloys are relatively lean.

The success achieved in raising the strength-temperature curves up the stress axis at temperatures near 0.5 Tm has been far greater than that realized in moving the curves out the temperature axis. This is an inherent consequence of the nature of the precipitation reactions utilized in the refractory metal alloys. Thus, in the range 0.4 to 0.6 Tm, dispersed phase strengthening offers great flexibility through thermal-mechanical processing. But, unfortunately, the phase relationships and high atomic mobility lead to gross structural instability at temperatures above 0.6 Tm. The relatively inert dispersoids such as ThO_2 in W offer some promise, but whether or not the carbides and nitrides can be complexed by further alloying to give the necessary stability at the very high temperatures is problematical. These instability problems will become even more critical with continued emphasis on very long time creep applications.

V. REFERENCES

1. W. Pollack, R. W. Buckman, R. T. Begley, K. C. Thomas and E. C. Bishop, "Development of High Strength Vanadium Alloys", Final Report, AEC Contract AT(30-1) - 3487, Westinghouse Electric Corp., June, 1967

2. H. Bohm and M. Schirra, "Investigations of the Stress-Rupture and Creep Behavior of Binary and Ternary Vanadium Alloys", J. Less Common Metals, 12, 1967, pp. 280-293

3. G. Brinson and B. B. Argent, "The Creep of Niobium", J. Inst. of Metals, 91, 1962-63, pp. 293-298

4. J. D. W. Rawson and B. B. Argent, "The Effect of Oxygen and Carbon on the Creep Strength of Niobium", J. Inst. of Metals, 95, 1967, pp. 212-216

5. W. V. Green, "High-Temperature Creep of Tantalum", Trans AIME, October, 1965, pp. 1818-1825

6. R. L. Stephenson, "Creep-Rupture Properties of Unalloyed Tantalum, Ta-10%W and T-111 Alloys", ORNL-TM-1994, December, 1967

7. C. S. Landau, H. T. Greenaway and A. R. Edwards, "Properties of Chromium and Chromium-Tungsten Alloys", J. Inst. of Metals, 89, 1960-61, pp. 97-101

8. J. W. Pugh, "The Tensile and Stress-Rupture Properties of Chromium", Trans ASM, 50, 1958, pp. 1072-1080

9. J. B. Conway and P. N. Flagella (and others), Progress Reports on AEC Fuels and Materials Development Program, GE-NMPO, Cincinnati, Ohio, 1961-1967

10. W. V. Green, M. C. Smith and D. M. Olson, "Short-Time Creep-Rupture Behavior of Molybdenum at High Temperatures", Trans AIME, 215, 1959, pp. 1061-1066

11. H. Carvalhinhos and B. B. Argent, "The Creep of Molybdenum", J. Inst. of Metals, 95, 1967, pp. 364-368

12. W. D. Klopp, W. R. Witzke and P. L. Raffo, "Effects of Grain Size on Tensile and Creep Properties of Arc-Melted and Electron-Beam-Melted Tungsten at 2250° to 4140°F", Trans AIME, 233, 1966, pp. 1860-1866

13. W. V. Green, "Short-Time Creep-Rupture Behavior of Tungsten at 2250° to 2800°C", Trans AIME, 215, 1959, pp. 1057-1060

14. E. R. Gilbert, J. E. Flinn and F. L. Yaggee, "Multimechanism Behavior in the Creep of Tungsten", Paper Presented at the Fourth Symposium on Refractory Metals, French Lick, Indiana, October 3-5, 1965, to be published

15. H. E. McCoy, "Creep-Rupture Properties of W and W-Base Alloys", ORNL-3992, August, 1966

16. J. H. Bechtold, E. T. Wessel and L. L. France, "Mechanical Behavior of the Refractory Metals", Refractory Metals and Alloys, AIME Metall. Soc. Conferences, Volume 11, 1960, pp. 25-81

17. O. D. Sherby," Factors Affecting the High Temperature Strength of Polycrystalline Solids", Acta Met, 10, 1962, pp. 135-147

18. P. E. Armstrong and H. L. Brown," Dynamic Young's Modulus Measurements Above 1000°C on Some Pure Polycrystalline Metals and Commercial Graphites", Trans AIME, 230, 1964, pp. 962-966

19. P. E. Armstrong and H. L. Brown," Anomalous Temperature Dependence of Elastic Moduli of Niobium Metal", Trans ASM, 58, 1965, pp. 30-37

20. R. P. Agarwala, S. P. Murarka and M. S. Anand, " Diffusion of Vanadium in Niobium, Zirconium and Vanadium", Acta Met, 16, 1968, pp. 61-67

21. T. S. Lundy, F. R. Winslow, R. E. Pawel and C. J. McHargue," Diffusion of Nb-95 and Ta-182 in Niobium (Columbium) ", Trans AIME, 233, 1965, pp. 1533-1539

22. R. E. Pawel and T. S. Lundy," The Diffusion of Nb^{95} and Ta^{182} in Tantalum", J. Phys Chem Solids, 26, 1965, pp. 937-942

23. J. Askill," Self-Diffusion in Chromium", Diffusion in Body-Centered Cubic Metals, Chapter 18, ASM, Metals Park, Ohio, 1965

24. J. Askill and D. H. Tomlin," Self-Diffusion in Molybdenum", Phil Mag, 8, 1963, pp. 997-1001

25. R. L. Andelin, J. D. Knight and M. Kahn," Diffusion of Tungsten and Rhenium Tracers in Tungsten", Trans AIME, 233, 1965, pp. 19-24

26. O. D. Sherby and P. M. Burke," Mechanical Behavior of Crystalline Solids at Elevated Temperature", First Technical Report to NASA (SC-NGR-05-020-084) August 1, 1967

27. J. Stoop and P. Shahinian," Effect of Oxygen on Creep-Rupture of Niobium", High Temperature Refractory Metals, Part 2, AIME Metallurgical Society Conferences, Volume 34, 1966, pp. 407-432

28. D. P. Gregory and G. H. Rowe," Mechanisms of Creep in Columbium and Columbium-1% Zirconium Alloy", Columbium Metallurgy, AIME Metallurgical Society Conferences, Volume 10, 1960, pp. 309-341

29. J. E. Flinn and E. R. Gilbert," Discussion of High Temperature Creep of Tantalum", Trans AIME, 236, 1966, pp. 1512-1513

30. G. W. King and H. G. Sell," The Effect of Thoria on the Elevated-Temperature Tensile Properties of Recrystallized High-Purity Tungsten", Trans AIME, 233, 1965, pp. 1104-1113

31. J. W. Pugh," Tensile and Creep Properties of Tungsten at Elevated Temperatures", Proc ASTM, 57, 1957, pp. 906-915

32. C. R. Barrett and W. D. Nix," A Model for Steady State Creep Based on the Motion of Jogged Screw Dislocations", Acta Met, 13, 1965, pp. 1247-1258

33. W. F. Sheely," Mechanical Properties of Niobium–Oxygen Alloys", J. Less Common Metals, 4, 1962, pp. 487–495

34. F. F. Schmidt, W. D. Klopp, W. M. Albrecht, F. C. Holden, H. R. Ogden and R. I. Jaffee," Investigation of the Properties of Tantalum and Its Alloys", WADD Technical Report 59-13, March, 1960

35. H. E. McCoy and D. A. Douglas," Effect of Various Gaseous Contaminants on the Strength and Formability of Columbium", Columbium Metallurgy, AIME Metallurgical Society Conferences, Volume 10, 1960, pp. 85–118

36. R. W. Armstrong, J. H. Bechtold and R. T. Begley," Mechanisms of Alloy Strengthening in Refractory Metals", Refractory Metals and Alloys II, AIME Metallurgical Society Conferences, Volume 17, 1963, pp. 159–190

37. W. H. Chang," Strengthening of Refractory Metals", Refractory Metals and Alloys , AIME Metallurgical Society Conferences, Volume 11, 1961, pp. 83–117

38. G. D. McAdam, "Substitutional Niobium Alloys of High Creep Strength", J. Inst. of Metals, 93, 1964–65, pp. 559–564

39. E. S. Bartlett, D. N. Williams, H. R. Ogden, R. I. Jaffee and E. F. Bradley," High–Temperature Solid–Solution–Strengthened Columbium Alloys", Trans AIME, 227, 1963, pp. 459–467

40. C. S. Hartley, J. E. Steedly and L. D. Parson," Binary Interdiffusion in Body–Centered Cubic Transition Metal Systems", Diffusion in Body–Centered Cubic Metals, Chapter 4, ASM, Metals Park, Ohio, 1965

41. W. H. Chang," Influence of Heat Treatment on Microstructure and Properties of Columbium–Base and Chromium–Base Alloys", ASD-TDR-62-211, Part IV, March, 1966

42. R. F. Peart and D. H. Tomlin," Diffusion of Solute Elements in Beta–Titanium", Acta Met, 10, 1962, pp. 123–134

43. G. B. Gibbs, D. Graham and D. H. Tomlin," Diffusion in Titanium and Titanium–Niobium Alloys", Phil Mag, 8, 1963, pp. 1269–1282

44. V. P. Lubinov, P. V. Geld and G. P. Shveykin, Isv. Akad. Nauk SSSR, Met i Gorn, Delo 5, 137, 1964

45. P. V. Geld, Z. Ya. Velmozhnyi, V. D. Lubinov and G. P. Shveykin, Isv. Vysshikh Uchehn. Zavedenic Tsvetn, Met. 2, 135, 1966

46. R. W. Buckman, Jr. and R. C. Goodspeed,"Considerations in the Development of Tantalum Base Alloys," This Volume

47. E. S. Bartlett, F. F. Schmidt and H. R. Ogden," Properties of Ta–W–Mo Alloys", High Temperature Refractory Metals, Part 2, AIME Metallurgical Society Conferences, Volume 34, 1966, pp. 326–345

48. H. T. Greenaway," Creep Testing of Chromium Alloys", Report ARL/Met. 55, Aeronautical Research Laboratories, Melbourne, Australia, October, 1964

49. J. W. Clark and C. S. Wukusick, Preliminary Information Reported by General Electric Co., Evendale, Ohio, Under a NASA Contract (cited in Ref. 50, p. 12)

50. D. J. Maykuth and A. Gilbert," Chromium and Chromium Alloys", DMIC Report 234, October 1, 1966

51. M. Semchyshen," Development and Properties of Arc-Cast Molybdenum-Base Alloys", The Metal Molybdenum, (Edited by J. J. Harwood), Chapter 14, ASM, Cleveland, Ohio, 1958

52. P. L. Raffo, W. D. Klopp and W. R. Witzke," Mechanical Properties of Arc-Melted and Electron-Beam-Melted Tungsten-Base Alloys", NASA TN D-2561, January, 1965

53. W. D. Klopp, W. R. Witzke and P. L. Raffo," Mechanical Properties of Dilute Tungsten-Rhenium Alloys", NASA TN D-3483, September, 1966

54. P. L. Raffo and W. D. Klopp," Mechanical Properties of Solid-Solution and Carbide-Strengthened Arc-Melted Tungsten Alloys", NASA TN D-3248, February, 1966

55. L. L. Seigle," Structural Considerations in Developing Refractory Metal Alloys", The Science and Technology of Selected Refractory Metals, (Edited by N. E. Promisel) AGARD Conference on Refractory Metals held in Oslo, Norway, June 23-26, 1963, pp. 63-93

56. R. T. Begley and J. H. Bechtold," Effect of Alloying on the Mechanical Properties of Niobium", J. Less Common Metals, 3, 1961, pp. 1-12

57. R. T. Begley, J. A. Cornie and R. C. Goodspeed," Development of Columbium Base Alloys", AFML-TR-67-116, November, 1967

58. G. D. McAdam," The Influence of Carbide and Boride Additions on the Creep Strength of Niobium Alloys", J. Inst. of Metals, 96, 1968, pp. 13-16

59. N. E. Ryan," The Formation, Stability and Influence of Carbide Dispersions in Chromium", J. Less Common Metals, 11, 1966, pp. 221-248

60. L. S. Rubenstein," Effects of Composition and Heat Treatment on High-Temperature Strength of Arc-Melted Tungsten-Hafnium-Carbon Alloys, NASA TN D-4379, February, 1968

61. R. T. Begley, J. L. Godshall and R. Stickler," Precipitation Hardening Columbium-Hafnium-Nitrogen Alloys", Fifth Plansee Seminar, June 22-26, 1964, Reutte/Tyrol, pp. 401-420

62. A. K. Mukherjee and J. W. Martin," The Effect of Nitriding upon the Creep Properties of Some Molybdenum Alloys", J. Less Common Metals, 5, 1963, pp. 403-410

63. A. C. Barber and P. H. Morton," A Study of the Niobium-Zirconium-Carbon and Niobium-Zirconium-Oxygen Systems", High Temperature Refractory Metals, Part 2, AIME Metallurgical Society Conferences, Volume 34, 1966, pp. 391-406

64. D. O. Hobson," Aging Phenomena in Columbium–Base Alloys", <u>High Temperature Materials II</u>, AIME Metallurgical Society Conferences, Volume 18, 1963, pp. 325–334

65. J. R. Stewart, W. Lieberman and G. H. Rowe," Recovery and Recrystallization of Columbium–1% Zirconium Alloy", <u>Columbium Metallurgy</u>, AIME Metallurgical Society Conferences, Volume 10, 1960, pp. 407–434

66. R. T. Begley, R. L. Ammon and R. Stickler," Development of Niobium Base Alloys", WADC TR 57–344, Part VI, February, 1963

67. R. M. Bonesteel, J. L. Lytton, D. J. Rowcliffe and T. E. Tietz, " Recovery and Internal Oxidation of Columbium and Columbium Alloys", AFML-TR-66-253, August, 1966

68. R. W. Buckman, Jr. , Unpublished Research on Studies of Internally Oxidized T-111 Alloy, Westinghouse Astronuclear Laboratory

69. R. W. Buckman, Jr. , " Operation of Ultra High Vacuum Creep Testing Laboratory", <u>Transactions Vacuum Metallurgy Conference, 1966,</u> American Vacuum Society, 1967, pp. 25–37

70. R. W. Buckman, Jr. and R. C. Goodspeed, Unpublished Research, Westinghouse Astronuclear Laboratory

71. R. H. Titran and R. W. Hall," Ultrahigh-Vacuum Creep Behavior of Columbium and Tantalum Alloys at 2000° and 2200°F for Times Greater Than 1000 Hours", NASA TN D-3222, January, 1966

72. TRW, Inc. ," Generation of Long Time Creep Data on Refractory Alloys at Elevated Temperatures", Final Report on Contract NAS 3-2545, June 6, 1967

73. R. A. Perkins," Effect of Processing Variables on the Structure and Properties of Refractory Metals", AFML-TR-65-234, Part II, May, 1967

74. R. T. Begley and J. A. Cornie, Unpublished Research, Westinghouse Astronuclear Laboratory

75. H. G. Sell, W. R. Morcom and G. W. King," Development of Dispersion Strengthened Tungsten Base Alloys", AFML-TR-65-407, Part II, November, 1966

76. R. T. Begley and J. L. Godshall," Some Observations on the Role of Grain Boundaries in High Temperature Deformation and Fracture of Refractory Metals", Paper Presented at the Fourth Symposium on Refractory Metals, French Lick, Indiana, October 3-5, 1965

77. E. Rudy and others, Aerojet-General Corporation," Ternary Phase Equilibria in Transition Metal-Boron-Carbon-Silicon Systems", AFML-TR-65-2, 1965-66

78. R. W. Buckman, Jr. and R. C. Goodspeed," Development of Dispersion Strengthened Tantalum Base Alloy", Sixth Quarterly Report on Contract NAS 3-2542, NAS CR-54658

79. R. N. Stevens," Grain Boundary Sliding in Metals", Metallurgical Reviews, 11, 1966, pp. 129-142

80. R. C. Gifkins, Fracture, (Edited by B. L. Averback, D. S. Felbeck, G. T. Hahn and D. A. Thomas), Technology Press and J. Wiley and Sons, New York, 1959

81. B. A. Wilcox, A. Gilbert and B. C. Allen," Intermediate-Temperature Ductility and Strength of Tungsten and Molybdenum TZM", AFML-TR-66-89, April, 1966

THE EFFECT OF THERMAL-MECHANICAL TREATMENTS ON THE
STRUCTURE AND PROPERTIES OF REFRACTORY METALS

Roger A. Perkins

Abstract

The control of properties by thermal-mechanical treatments is based on
mechanical deformation and annealing according to a schedule that will
produce a unique and stable combination of grain structure, dislocation
substructure, texture, and dispersion of precipitated phases. The formation and stabilization of deformation substructures in Cb, Ta, Mo, W,
and their alloys is reviewed. The effect of thermal-mechanical treatments on grain structure and texture also is considered. Examples are
given to illustrate how the ductile-to-brittle transition behavior, low
temperature strength and ductility, high temperature strength, creep
resistance, and formability are affected by the structural changes that
can be produced. Most of the results obtained to date indicate considerable promise for major advances in refractory metal technology by the
application of thermal-mechanical treatments. The challenge for the
future is to relate the basic concepts and understanding of the control of
structure and properties to the metal working processes required to
produce metals in appropriate shapes or forms.

R. A. Perkins is a Senior Member of the Research Laboratories —
Mechanical Metallurgy at the Lockheed Palo Alto Research Laboratory,
Lockheed Missiles & Space Company, Palo Alto, California.

Introduction

With few exceptions, the strength of a metal, particularly at high temperatures, is not uniquely determined by its composition and heat treatment. Twenty years ago Dorn, Goldberg, and Tietz (1) demonstrated that prior thermal-mechanical history had a significant effect on the flow-stress and deformation behavior of metals. This classic study disproved the validity of mechanical equations of state and proved that the flow-stress for any instantaneous values of strain, strain rate, and temperature was governed by the entire thermal-mechanical history of the material. In the following year, Tietz, Anderson, and Dorn (2) proved that the strain hardened state of a metal depends on the temperature and rate of straining as well as on the total strain. Variation of the strain hardening temperature in a range well below that for recovery or recrystallization also was observed to influence the rate of strain hardening and the subsequent recovery behavior. Studies such as these conducted by Professor Dorn and his coworkers at the University of California during the late 40's laid the foundation for a new approach to strengthening of metals: controlled thermal-mechanical processing.

The approach to strengthening by thermal-mechanical treatments is based on mechanical deformation and annealing according to a schedule that will produce a unique combination of grain structure, low angle boundary substructure, and dispersion of precipitated phases (if any). The primary objective in processing is the regulation of micro-structure and properties. Unlike normal practices for the working of metals, shape change, product yield, and quality are of secondary importance.

In recent years, considerable interest has been generated in the application of thermal-mechanical treatments to a wide variety of materials. A Sagamore conference in 1962 reviewed the various treatments and their effects on metals (3). Most of the interest and activity in recent years has centered on the significant increases in strength achieved by the introduction of plastic deformation into the heat treatment cycle of alloy steels. Kula (4) presented a comprehensive review of the application of thermomechanical treatments for strengthening steel in 1966. Thomas, Zackay, and Parker (5) pointed out that most metals in their current state of development exhibit only a fraction of their theoretical yield strength. Iron has been strengthened to about 40% of its maximum theoretical yield by heavy cold working. Refractory metals have been strengthened to only 15% of their theoretical yield. It is evident that considerable room for improvement exists and current trends indicate that hope for significant advancement lies in the possibility of controlling microstructure (5).

The development of high strength refractory metal alloys by and large has followed the traditional approach based on modification of composition and heat treatment. In most cases, the mechanical working of these materials has been regarded as a means for refining grain structure, changing shape, and controlling quality. Until recently, comparatively little effort has been devoted to the use of thermal-mechanical treatments to enhance the properties of these materials. Rostoker (6) reviewed the primary working of refractory metals at the 9th Sagamore conference in 1962 and concluded that the future of refractory metals was dependent on improving our understanding of the fundamentals of deformation processing. Ingram and Ogden (7) prepared a comprehensive review on the effect of fabrication history on the structure and properties of refractory metals in 1963. Houck and Ogden (8) in 1964 prepared a summary of knowledge on thermal-mechanical variables affecting the properties of refractory metal alloys. This was part of a survey on metal deformation processing conducted by the Materials Advisory Board in support of its Metalworking Process and Equipment Program. Significantly, these authors pointed out that thermal mechanical treatments by far had the greatest influence on properties. They raised the hope for future control of properties with great precision by such treatments.

The purpose of this paper is to review some of the more recent developments in thermal-mechanical processing of refractory metals and to assess the potential of this approach to improving properties. There are many illustrations of the control of structure and properties by thermal-mechanical treatments but only a few of the more timely examples will be discussed. The intent is to demonstrate the significance and overall importance of controlled deformation processing for refractory metals and the paper does not represent a comprehensive state-of-the-art survey.

Control of Structure

A major requirement for metals and alloys in load-bearing applications is resistance to deformation. Since plastic flow occurs by the movement of dislocations, metals are strengthened by obstacles to the generation and motion of dislocations. These obstacles are commonly referred to as lattice defects and include solute atoms, precipitates, grain boundaries, subboundaries, stacking faults, dislocations, and vacancies. The mechanical properties of an alloy are governed by the type, density, and distribution of various defects and these factors in turn can be manipulated by suitable thermal-mechanical treatments.

An improved foundation for strengthening metals logically should be based on the role of various defects and the manner in which they are influenced by processing. The condition or state of a material often is defined solely in terms of one or more aspects of process history, i.e., recrystallized, wrought, stress relieved, etc. Microstructural considerations such as dislocation substructures, dispersed phases, grain size and shape, and crystallographic orientations (texturing) all too often are ignored. It is these latter factors that are important, however, in describing the state of a material and in assessing the influence of thermal-mechanical treatments.

Dislocation Substructures

The dislocation substructure has a controlling effect on the strength of metals and alloys. It is now known that the theoretical strength of metals can be approached from either of two directions: (1) by creating highly perfect (dislocation free) structures or (2) by creating highly imperfect (dislocation saturated) structures. Whisker technology is an example of the first approach while strain hardening is an example of the second. Substructure as discussed here consists of the line defects and imperfections that exist within any given grain of a polycrystalline metal or alloy. These defects include a variety of dislocation arrays such as tangles, forests, cells, networks, and low angle subboundaries.

All metals have a dislocation substructure, even in the cast or fully annealed state. Substructures are modified by plastic deformation and heat treatment. The exact nature of the substructure at any time depends upon a complex interaction of internal factors such as composition, grain structure, and dispersed phases with external factors such as type and amount of strain, strain rate, and temperature. These latter factors are referred to as the thermal-mechanical history of the material.

The importance of substructure has long been recognized but it is only in recent years that the tools required to study it in detail have become available.

The development of transmission electron microscopy has shed new light on the basic features of substructure associated with deformation and annealing processes. Most of these studies have been made with high purity metals, notably iron, nickel, and aluminum. The behavior of all metals is strikingly similar; however, some differences are noted, particularly for body centered cubic metals. In 1959, the Air Force Materials Laboratory initiated a comprehensive study on substructure in refractory metals and its relation to mechanical behavior. The results are presented in three summary technical reports. (9, 10, 11) In 1962, some of the more significant findings from this program and other studies were reviewed in an Air Force Materials Laboratory Symposium on substructure and the mechanical behavior of metals. (12) The reader is referred to these documents for more detailed descriptions of behavior than those presented in this review.

As a metal is deformed, dislocations are freed from pinning obstacles and new dislocations are created by a complex variety of processes. Cross slip is the most common source of dislocation multiplication in BCC metals. Moving dislocations interact with each other. Local stress fields are reduced and lower energy configurations are produced. As a result, the dislocations cease to move and become tangled. The tangles in turn act as barriers to other dislocations, resulting in the formation of networks of tangled dislocations throughout each individual grain. A typical array of tangled dislocations developed in a Cb-1W-1Zr alloy strained 10% in tension at room temperature is shown in Figure 1A.

Tangled dislocations are effective barriers to dislocation motion. Keh (13) established experimentally a direct correlation between dislocation density and the flow properties of iron:

$$\sigma_f = \sigma_o + 0.17 \text{ Gb } \sqrt{N}*$$

where

σ_f = flow stress
σ_o = lattice frictional stress
G = shear modulus
b = Burgers vector
N* = density of dislocations within tangled regions

Most theories of work hardening today predict a dependency of flow stress on the square root of dislocation density for FCC and BCC metals. Strength is increased by providing a high density of uniformly distributed dislocations within each grain. The maximum density of dislocations observed in heavily cold worked metals is about $10^{12}/cm^2$. (5)

The refractory metals tend to obey this general square root relationship. Van Torne and Thomas (14) found that the flow stress of Cb was proportional to the square root of the dislocation density. A high purity grade of Cb, however, had a lower dependency of flow stress on density of dislocations ($N^{1/7}$). They concluded that purity influenced the relationship. Owen (11) demonstrated that the flow stress of tantalum also was a linear function of the square root of dislocation density. The data for a number of samples could reasonably be represented by a modification of the basic Keh equation.

Substructure Formation

Many studies have been made to characterize substructure formation in refractory metals. Stephens (15) found that the dislocation density in tungsten and tungsten-rhenium alloys was a linear function of strain to 5%. Iron also shows a linear dependency out to at least 7% strain. Work hardening of the tungsten alloys appeared to correlate with dislocation density. A W-9% Re alloy work hardened much more rapidly than pure tungsten. The dislocation density in the alloy was found to be four times that in the unalloyed material after 2% strain. (15) The dislocations in the alloy were arranged in a well defined cellular array after 2% strain. This behavior is quite unusual since much higher strains are needed to produce cellular arrays in most BCC metals.

The work of Berghezan and Fourdeux (16) on high purity columbium is more typical of refractory metal behavior. The development of a dislocation substructure on straining in tension is heterogeneous, varying from grain to grain depending on orientation. In the early stages (2 to 3% strain) long dislocation lines and loops are formed. At slightly higher strains, (4 to 5%) definite tangles are formed in some grains and the structure is less heterogeneous. At strains from 5 to 7%, long skeins of tangled dislocations are formed and crude cells are observed in some grains. At still higher deformation (over 8%) well defined cellular networks will form. The cells are highly regular, with a square or rectangular shape and an average size of 0.5 by 0.5 to 0.5 by 1.0 microns. The structure is very similar to that found in other metals.

The amount of strain required to produce a cellular array of dislocations varies for different materials. Cellular structures are developed in a Cb-1W-1Zr alloy strained 10% in tension at room temperature. (17). (Figure 1A) Van Torne and Thomas (14) also found well defined cells in unalloyed Cb (impure) after 12% strain at room temperature. In high purity Cb, however, these investigators found no cellular formation at this strain level. Over 22% strain was needed to create cellular arrays in the high purity metal. This behavior is similar to that found for high purity tantalum by Jewett and Weisert (18). Although the dislocation density increased with strain, strains of 25% or more were needed to promote the formation of regular cells.

High purity refractory metals are characterized by rapid dislocation multiplication on yielding followed by low rates of multiplication on further straining. Large strains appear to be required for cell formation. It is known that internal factors such as alloy composition, grain structure, and precipitates do affect the substructure. As shown by Stephens (15), solid solution alloy additions in some cases enhance cell formation at low strains. In a study of high purity Fe-Si alloys, Walter (19) also found that solid solution additions decreased the average cell size. Stephens (15), on the other hand, states that some solid solution additions, i.e., 3%Ta in W depresses cell formation on straining. The effects of solid solution additions are not clear and more detailed studies are needed to develop an understanding of alloy behavior.

It has long been known that the strength of metals is influenced by grain size. Grain boundaries can be a source of weakness by acting to generate dislocations. On the other hand, they can act to impede dislocation motion and be a source of strength. Jewett and Weisert (18) found that at low strains, the grain boundaries in tantalum generate dislocations while at high strain, they impede dislocation motions. Stephens (15) found that the dislocation density in W and W-Re alloys increased with decreased grain size. Thomas, et al, (5) report that grain size affects the cell size in iron. Grain boundaries generate dislocations and fine grains tend to give fine cells.

Structural factors within each grain also have an important effect on the formation of dislocation substructures. The distribution and density of dislocations in tantalum after a given strain was shown to be dependent on the initial arrangement of dislocations (20). If the initial distribution was random, the rate

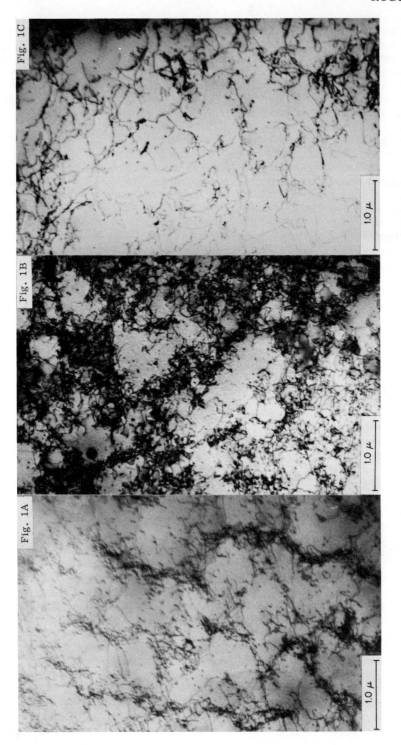

Fig. 1   Recovery of Cb-1W-1Zr (×25,000) (Ref. 17)
(A) Prestrained 10%
(B) 1 hr, 800°C
(C) 100 hr, 800°C

of increase in dislocation density with strain was less than that for material having a network distribution. The rate of change in dislocation density was found to be an exponential function of strain. The numerical value of the exponent varied dependent upon the initial grain size and substructure. These observations again demonstrate the importance of the entire thermal-mechanical history in governing the response to thermal-mechanical treatments.

Dispersed particles or second phases can have a major effect on the formation of dislocation substructures. In BCC metals, dispersed particles will readily force screw dislocations to cross slip and thereby act as multiplication centers. Many investigators have found that dispersed phases enhance the strain hardening behavior of metals. Since flow stress is affected by dislocation density, increased rates of multiplication should increase the work hardening rate. The particles also enhance strengthening by blocking dislocation motion, thus exerting a dual role like that of grain boundaries. Van Torne and Thomas (14) found rapid dislocation multiplication and tangling in grains of Cb where precipitates existed.

Kula (4) suggested that the shape of dispersed particles is an important factor in particle/dislocation interactions. Dislocations have a tendency to loop around spherical particles but may have to shear a platelet particle. The latter, therefore, may be more effective for enhancing strain hardening. Support for this argument was provided from strain hardening studies of Fe-C alloys by Kardonski, Kurdyumov, and Perkas in 1964 (21). Alloys with platelet carbides were observed to work harden at a much higher rate than those with spheroidized carbides. Kalns, Barr, and Semchyshen (22) found a similar effect for TZM-Mo alloys. Small acicular carbide particles increased the work hardening rate of this alloy.

The first experimental proof of a shape effect on particle/dislocation interactions was provided by Perkins and Lytton in 1965 (23). Transmission electron micrographs of a Cb-TZM alloy strained in tension at 2400°F are shown in Figure 2. In samples treated to produce spheroidized CbC particles prior to straining (Figure 2A), the dislocation lines were found to loop around the particles. No strong interactions were indicated. However, when the material was treated to produce a Widmanstätten array of CbC platelets (Figure 2B), a very strong interaction was observed. Massive pileups and tangles of dislocations were formed at the platelet particles after very little strain.

In a follow-on study, Perkins (24) observed that material with platelet carbides work-hardened more than material with spherical carbides. Other significant differences in behavior also were observed and will be discussed in other sections of the paper. The greater effectiveness of the platelet particle was concluded to be not only a result of the shape but also a direct result of the orientation and interfacial energy. The platelets were formed preferentially on the {100} planes (cube faces) and as such could effectively block cross slip on all slip systems. In addition, they had a very low interfacial energy which may have contributed to the interaction with dislocations.

Available evidence from several different materials indicates that the average cell size in the dislocation substructure also is an important factor in strengthening. Since tangled dislocation boundaries are barriers, it is reasonable to expect that strength would depend on cell size in much the same manner as it does on grain size. Recent observations on nickel and copper indicate this to be the case (5). The flow stress for these two metals was found to be inversely proportional to the square root of cell diameter. Extrapolation of the data for copper predicts the attainment of theoretical strength at a cell diameter of about 0.1μ.

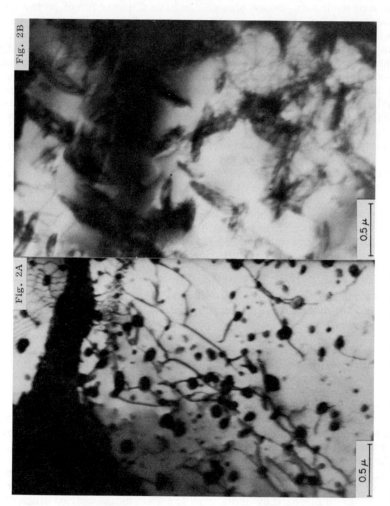

Fig. 2   Carbide particle – dislocation interactions in Cb–TZM alloy (×38, 000)
(Ref. 23)
(A) Spheroidized CbC
(B) Platelet CbC

The average cell size produced in cold worked refractory metals is about $0.5\ \mu$. As mentioned, a fine grain size promotes a fine cell size. Also, it has been reported that increasing strain rates and decreasing temperatures refine the cell structure of iron (25). The effect of strain rate was clearly demonstrated by the explosive deformation of nickel (26). Cell size decreased as the explosive force and strain rate increased. Cells $0.17\ \mu$ in diameter were produced compared with $0.5\ \mu$ diameter cells formed in slow straining of metals. Significantly, hardness of nickel increased in proportion to the decrease in cell size to the 1/2 power. While many investigators have studied the effect of temperature and degree of straining, few have looked at the influence of the type and rate of straining on substructure formation. These latter two factors are of great importance and are critical to the effective translation of any basic understanding to improved process technology.

Although these is general agreement that substructures are important, there still is considerable lack of agreement on how they form, how they can be controlled, and the precise role they play in strengthening metals. Based on available data, the following is a summary of knowledge on the formation of deformation substructures in refractory metals:

(1) Flow stress is directly proportional to the square root of dislocation density and inversely proportional to the square root of cell size. A high density of dislocations distributed in small uniform cells is desirable for maximum strain hardening.

(2) Large strains are required to produce high density cellular structures in refractory metals. Cell diameters of about $0.5\ \mu$ can be produced by cold working.

(3) Dislocation density can be increased and cell size decreased for any strain level by increased strain rate, decreased deformation temperatures, solid solution alloying, dispersed particles (particularly platelets), reduced initial grain size, and refined initial substructure.

Further research, particularly on the external as well as the internal factors that control substructure formation, is needed to clarify and resolve many aspects of behavior before this basic knowledge can be translated into processing modifications for the manufacture of improved alloys.

Substructure Stability

A major requirement for refractory metals is the retention of strength and ductility after exposure to elevated temperatures. The stability of deformation substructures is of vital importance where thermal-mechanical treatments have been used to strengthen the metals or to enhance low temperature ductility. Basically, the dislocation cell walls in strain hardened metals are not stable at elevated temperature. The stored energy of cold work is released through recovery and recrystallization processes. An understanding of what changes occur and how they may be controlled is essential to the development of useful thermal-mechanical treatments.

When a strain hardened metal is heated, vacancies and dislocations migrate to form low energy configurations. Initially, point defects are annealed out. This changes the electrical properties but has no effect on the dislocation substructure or mechanical behavior. With longer heating or higher temperatures, the tangled dislocations in the cell walls attract and interact with dislocations inside the cells. The dislocation density within the cells is reduced and the structure in the cell walls is sharpened. Figure 1B is typical of this early stage in the recovery cycle for a lightly strained (10%) Cb-alloy. This process will continue until a low density stable distribution of dislocations is formed (Figure 1C). As the lattice perfection continues to increase, stable dislocation networks are formed and distinct subgrains separated by low angle boundaries are developed. Subgrains formed during the recovery of strain hardened (10% strain) molybdenum at 900°C are shown in Figure 3. The initial substructure consisted of a cellular dislocation array $1-2\,\mu$ in diameter. After 1 hr at 900°C, sharp sub-boundaries had formed with the development of square subgrains $1\text{-}1/2-2\,\mu$ in diameter. The structure was sharpened on continued heating to 100 hr. Little growth in subgrain size was observed after the first hour at temperature (27). The primary grain configuration is not changed at this point.

The overall process of subgrain formation during recovery annealing is termed polygonization. In general, refractory metals polygonize readily. Subgrain structures formed in this manner from lightly strained metal are extremely stable at elevated temperatures. The residual strain energy in each subgrain is low and the structure has good resistance to recrystallization. In essence, this is a process in which the small diameter cells produced by deformation are replaced by slightly larger diameter subgrains having greatly enhanced stability at high temperature. The subgrain size is very small ($1-2\,\mu$) and the net effect on mechanical behavior should be similar to that produced by refining the grain size.

The recovery process in strain hardened metals is governed by a variety of external and internal factors. One of the difficulties encountered in studying the refractory metals is that temperatures for recovery are in a range where interstitial contamination and precipitation are prone to occur. The approximate recovery ranges for these metals based on a 1-hr anneal have been established as follows: $W-900/1100°C$, $Mo-500/800°C$, $Ta-700/900°C$, and $Cb-700/900°C$ (28). Dislocations in Cb often are found to be decorated with precipitates after a few hours at the recovery temperature (27). These effects tend to alter and obscure basic recovery processes.

The recovery and recrystallization behavior of all refractory metals is highly dependent on strain hardening and temperature. Basically, the temperature for the onset of recovery and recrystallization decreases with increasing cold work. This effect was demonstrated for a Cb-1Zr alloy by Stewart, Lieberman, and Rowe (29) as shown in Figure 4. Light reductions (strain hardening) give the highest recovery and recrystallization temperatures and the coarsest recrystallized grain size. Similar effects are observed for most refractory metals and alloys (7). The subgrain size of metals formed during recovery also tends to decrease with increased cold work and with increased recovery temperature (30).

Since the dislocation density decreases during recovery, it is reasonable to expect that the flow stress also will decrease. Although this is observed, no direct relationship between flow stress and dislocation density in recovered metals has been found (31). For molybdenum and columbium recovered at $800-900°C$, most of the drop in flow stress occurs within the first 1-hr at temperature (13, 27). Fractional flow stress recovery for Mo, Mo-1W, Cb, and Cb-1W prestrained 10% was in the range of 20 to 30% after 1 hr at $800-900°C$. This appears to be associated with a reduction of dislocation density within the cells and the motion of dislocations to join cell boundaries. The rate of flow stress recovery for these materials decreases on continued holding and in some cases tends to level off in the range of 30 to 40% recovery after 25 to 100 hr. A slight growth in cellular or subgrain size occurs within the first hour and little additional growth is observed on heating for 100 hr (27). Interstitial contamination and precipitation effects again tend to alter the basic process on long time exposure.

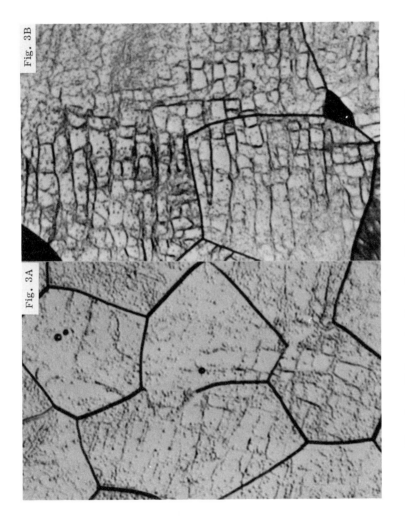

Fig. 3   Polygonization of unalloyed Mo (×3000) (Ref. 27)
(A) Prestrained 10%, recovered 1 hr, 900°C
(B) Prestrained 10%, recovered 100 hr, 900°C

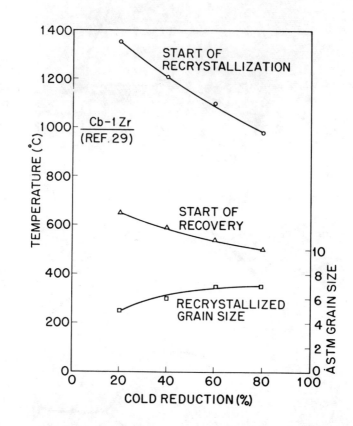

Fig. 4   Recovery and recrystallization of Cb-1Zr (Ref. 29)

At higher recovery temperatures or on the recovery of highly strained material, subgrain growth will continue until a new recrystallized grain structure is nucleated. Li (31) postulates that both sub-boundary and subgrain coalescence are involved in the nucleation of a recrystallized grain. Koo and Sell (32) observed subgrain coalescence for strain hardened tungsten at 1000 to 1200°C and suggest that this is a process common to many metals. Although the mechanism is still in doubt, it is clear that primary recrystallization must be prevented if refractory metals are to retain the enhanced properties imparted by thermal-mechanical treatments. Recrystallization reduces the dislocation density to minimum values and coarsens the effective grain structure by the elimination of subgrains.

Since precipitates and dispersed phases can be effective in blocking both dislocation movements and grain growth, it is reasonable to expect that they could be equally effective in blocking the nucleation of recrystallized grains by subgrain coalescence. A number of observations indicate that finely divided particles can inhibit the nucleation of new grains in many materials (33). Oxide precipitates in iron, SiO_2 and Al_2O_3 in Cu, Al_2O_3 in Al, and ThO_2 in Ni, all retard recrystallization. The basic mechanism appears to be that of inhibition of nucleation rather than of the growth of nucleated grains.

The best example of dispersed phase stabilization of strain hardened refractory metals is seen in the effects of carbides on the recrystallization behavior of molybdenum and tungsten. High purity molybdenum cold reduced about 90% will recrystallize in 1 hr at $900-1000°C$ (34). By the addition of a fine dispersion of titanium carbide in the TZM-Mo alloy, the recrystallization temperature can be increased to $1230-1345°C$ (34). At still higher carbon contents, a carbide dispersion will increase the 1-hr recrystallization temperature to $1455-1565°C$ (23, 24). Recent studies on W-Hf-C alloys reveal that a fine HfC dispersion in tungsten will raise the 1-hr recrystallization temperature above $2315°C$ (35). High purity tungsten can recrystallize as low as 1200°C.

The ability of particles to inhibit nucleation and recrystallization is dependent upon size, shape, and distribution. Certain types of particles actually accelerate rather than retard the process. Oxide inclusions larger than 100 μ in diameter were observed to speed up the recrystallization of deformed iron (33). English and Backofen (36) observed enhanced nucleation at inclusions in hot worked Si-Fe alloys. This behavior has been attributed to the local enhancement of dislocation density around inclusions. Recent studies by W. C. Coons at the Lockheed Palo Alto Research Laboratories (37) have provided experimental evidence that supports this theory. As shown in Figure 5A, a high concentration of dislocation etch pits is evident around massive particles of molybdenum carbide in unalloyed Mo strained 10% at room temperature and recovered 1-1/2 hr at 800°C. After 2 hr at 1240°C, definite polygonization is evident around the carbide particles but not in the rest of the grain interiors as shown in Fig. 5B. On continued heating, recrystallized grains form initially around these particles. The average diameter of the Mo_2C particles is about $3-4 \mu$.

The retardation of nucleation and recrystallization is strongly dependent on the shape, quantity, and distribution of particles as well as on the particle size. In molybdenum alloys for example, acicular or platelet carbides are much more effective in raising the recrystallization temperature of cold worked sheet than spherical or irregular particles (22, 24). This effect is illustrated for CbC dispersions in a Mo-1.5Cb-0.5Ti-0.3Zr-0.08C alloy (Cb-TZM) in Figure 6 (24). All materials have the same composition and were rolled to sheet by identical processes after the last recrystallization anneal. The materials identified as C, D, E, and F contain CbC plates 0.1 to 0.2 μ thick by 1 to 2 μ long precipitated

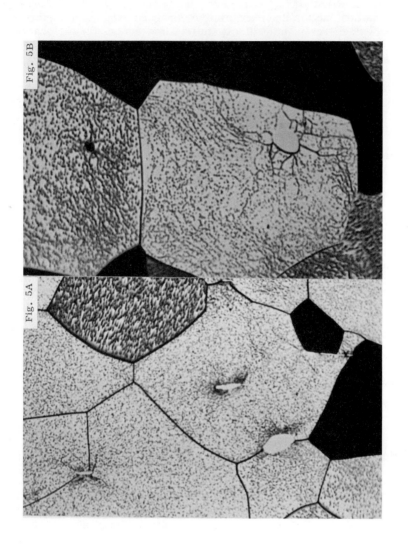

Fig. 5   Polygonization at carbide particles in Mo (×2000) (Courtesy W. C. Coons)
(A) Prestrained 10% and recovered 1 hr at 925°C
(B) Polygonized 2 hr at 1238°C

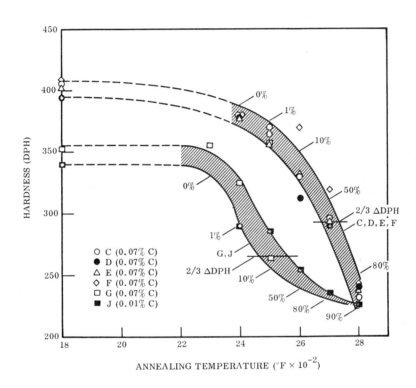

Fig. 6   Carbide stabilization of strain hardened Cb–TZM alloy (Ref. 24)

in a Widmanstätten array on {100} planes of the matrix. The material identified as G contains the carbide dispersed as random spheres, 0.1 to 0.2 μ in diameter. Not only did this material work harden less on cold rolling, but also the recrystallization temperature was 200°F below that of the sheet with platelet carbides. The material identified as J contains platelet carbides but has a low carbon content (0.01%). The lower recrystallization temperature in this case results from a major reduction in the amount of precipitate.

A further requirement for useful stabilization of substructures is that the precipitates used for this purpose be stable in the structure. A low interfacial energy, therefore, is desirable for maximum particle stability. If particles interact with dislocations, however, pileups and tangles may increase interfacial energy and reduce particle stability. This effect is illustrated in Figure 7 taken from the study by Perkins on Cb-TZM sheet processing (24). The platelet form of CbC has a low interfacial energy and high stability. No tendency for spheroidization was observed after 1000 hr at 1315°C. However, after cold rolling 89%, the platelets began to spheroidize in as little as 2 hr at 1315°C (Figure 7A). No evidence of subgrain formation was found in regions where platelet carbides remained. In regions where the carbides were spheroidized, however, stable dislocation networks and well defined subgrains formed after 14 hr at temperature (Figure 7B). The formation of high angle boundaries of recrystallized grains was observed in these regions after 44 hr at temperature (Figure 7C). The study was made during a creep test in which additional strain was being applied. The behavior observed is a type of strain induced recrystallization in which strain reduces the stability of a dispersed phase that in turn is acting to stabilize the deformation substructure.

Grain Structure and Texture

The mechanical properties and the deformation substructure of the refractory metals are strongly dependent on the size, shape, and orientation of the grains. The distribution of inclusions, dispersed phases, precipitates, and other defects also are important. The control of the macroscopic aspects of structure is as much a part of thermal-mechanical processing as the control of the dislocation substructure.

Most of the metal working operations used in processing and shaping metals tend to flatten and elongate the grains into fibered bundles or layered plate-like structures. Extrusion and wire drawing produce fibered structures while upset forging and rolling yield layered structures. A typical layered structure produced by 90% cold reduction of a molybdenum alloy in rolling is shown in Figure 8. The inclusions and dispersed phases will be strung out and layered in the same general manner. Preferred crystallographic orientations also will be developed by the deformation process. These changes will lead to a directionality of properties (anisotropy) that may be either helpful or harmful depending on the nature of the structure produced.

On annealing the cold worked metal above the recovery range, recrystallization and grain growth will occur. The refractory metals exhibit classic recrystallization behavior: the temperature and time for recrystallization and the recrystallized grain size decrease with increased levels of cold work (Figure 4). Unlike many metals, however, refractory metals do not tend to recrystallize to an equi-axed grain structure. Only the high purity (low interstitial) metals and alloys develop equi-axed grains. Virtually all of the commercial alloys will tend to recrystallize with flattened and elongated grains in a fibered or layered structure similar to that of the wrought material. An example for a cold worked recrystallized alloy is shown in Figure 8B. These materials also show a resistance

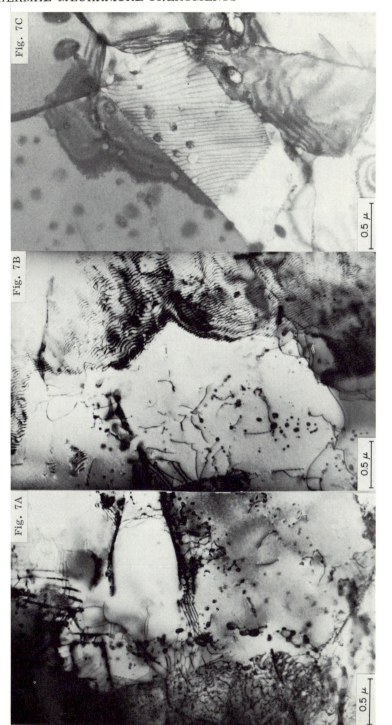

Fig. 7. Effect of carbide dispersions on the recovery and recrystallization of Cb–TZM alloy (×34,000) (Ref. 24)
(A) 2 hr, 1315°C, platelet CbC with carbide spheroidization and start of polygonization
(B) 14 hr, 1315°C, spheroidized CbC, polygonized
(C) 44 hr, 1315°C, spheroidized CbC, partial recrystallization.

Fig. 8   Grain structures in Cb-TZM alloy sheet (×100) (Ref. 23)
            (A) As-rolled, 89% cold reduction
            (B) Recrystallized 1 hr, 1705°C
            (C) Solution annealed 5 min, 2095°C

to grain growth and tend to reach a limiting grain size that does not vary greatly with temperature in the normal recrystallization range.

This behavior of commercial alloys is due to the existence of dispersed phases and precipitates that retard grain boundary migration. Most of the high strength alloys of W, Ta, Mo, and Cb contain fine dispersions of reactive metal (Ti, Zr, Hf) carbides or oxides. Grain growth control by finely dispersed phases is a well established mechanism in a wide variety of materials (33). The importance of this factor in refractory metals is that the dispersion also controls the shape of the recrystallized grain and does not permit complete refinement to small equi-axed grains by working and annealing. Thus, the texturing and anisotropy introduced by each step in working is carried through to all other steps.

Since the nature and distribution of the dispersions governs recrystallized grain size and shape in many of the refractory metals, redistribution of the precipitates can be used to effect grain control. If the alloy is heated to a sufficiently high temperature to take the blocking precipitates into solution, recrystallization to an equi-axed structure will occur. This is shown in Figure 8C for a molybdenum alloy solution annealed at 2095°C. Grain growth is very rapid due to the high solution temperatures required and coarse grained equi-axed structures are produced.

This treatment will completely erase any fibering and texturing effects from prior working operations. The coarse grain size, however, is not desirable for subsequent thermal-mechanical treatment. However, by using light reductions and holding directionality of the worked product to a minimum, followed by recovery treatments to produce a fine subgrain structure or low temperature recrystallization treatments, suitable structures can be produced. The production of fine equi-axed grain structures in refractory metals can be realized by the application of controlled thermal-mechanical processing techniques.

It is only in recent years that any efforts have been made to control the texture of refractory metal alloys. Like grain structure, texture is a consequence of the entire deformation processing history and can be altered many times during processing by thermal-mechanical treatments. A comprehensive review of textures in the refractory metals has been prepared by Cooper (38). He also demonstrated that the texture of tantalum can be changed from the normal $\{100\} < 011 >$ and $\{111\} <011>$ textures for cold worked refractory metals to the $\{110\} <001>$ (cube) texture by warm rolling at a temperature about half the absolute melting temperature. Perkins and Lytton (23) showed that a random structure with minimum texture (no preferred orientation) could be produced in heavily cold rolled molybdenum alloy sheet by controlled solution anneal and working cycles.

While many factors are known to influence the deformation and annealing textures, the interactions are not well defined. It is not possible at the present time to predict how specific textures can be produced and controlled (39). There is a major need with respect to deformation processing for a more complete understanding of both the factors that govern the origin of texture and the relationship of dislocation substructures and mechanical behavior to texture.

Effect on Properties

The most important characteristics of refractory metals are: (1) low temperature ductility and toughness — important to formability and resistance to brittle fracture; (2) high temperature strength and stability — important to dimensional control of load carrying structures; and (3) resistance to oxidation — important to the stability of surfaces and mechanical properties in all applications involving exposure to reactive atmospheres. The objectives of most alloy

development programs during the past decade have been to increase strength
without loss of formability, toughness, and low temperature ductility and to provide protection from oxidation by coatings without degrading the mechanical
properties over a broad temperature range.

Three methods for strengthening the refractory metals have evolved: solid
solution alloying, strain hardening, and dispersed phase strengthening. Most of
the Cb and Ta-base alloys are strengthened by a combination of solid solution
alloying and dispersed phases. The former provides good properties at low temperature while the latter confers resistance to deformation at high temperatures
in addition to enhancing low temperature strength. Molybdenum and W-alloys,
on the other hand, are strengthened by a combination of strain hardening and
dispersed phases. The former provides the necessary combination of low temperature strength and ductility while the latter acts to stabilize the strain hardened
state and to provide resistance to deformation at high temperatures.

By virtue of the basic strengthening mechanisms used, thermal-mechanical
treatments were first successfully applied to the development of improved Mo
and W-base alloys. Most of the examples of the effects of such treatments on the
properties of refractory metals are to be found in work on alloys of these metals.
Thermal-mechanical treatments are not necessarily restricted to Mo and W
alloys, however, and equally striking improvements are now being obtained with
Cb and Ta alloys. The recognition of the importance of dispersed phases in controlling resistance to deformation at high temperature has led to the consideration
of thermal mechanical treatments for improving the behavior of these materials.
The following sections of this paper summarize some of the more significant
effects of deformation processing on the low temperature ductility, strength, and
formability of these four metals.

Low Temperature Ductility

The ability to reduce the ductile-to-brittle transition temperature by cold
working of Mo, W, and their alloys is a well established fact. Siegle and
Dickinson (40) presented a comprehensive review of factors influencing the
ductile-to-brittle transition of these metals in 1962. It was known that the transition temperature could be progressively reduced by increased amounts of cold
or warm work. Highly worked structures were observed to have a transition to
brittle fracture in slow strain rate tests below -150°C. Cold work and stress
relief treatments were universally adopted to obtain low temperature ductility in
these materials.

The first detailed studies to correlate ductile-brittle behavior with structural details were conducted by Lement, et al. (9, 10, 11) from 1961 to 1963 and
by Opinsky and co-workers (40) in 1958. Both arrived at essentially the same
conclusions.

Based on the studies of Lement and co-workers (9, 10, 11), the following
behavior is exhibited. The ductile-to-brittle transition in tension decreased
progressively with cold work for both Mo and W. Although the critical crack
size corresponded to the first order subgrain size for both materials, work on
tungsten indicated that fiber width instead of subgrain size had the dominant effect
on fracture behavior. Increasing the fracture stress versus temperature relative
to the necking stress versus temperature is necessary to reduce the transition
temperature. Changes in the dislocation substructure by recovery anneals and
polygonization were found to increase room temperature strength without altering
the ductile-to-brittle behavior. The transition temperature increased, however,
on the nucleation of recrystallized grains. Reduction of the effective primary
grain size by mechanical deformation appeared to be the most important factor.

Siegle and Dickinson (40) point out, however, that a pronounced fiber texture is not necessarily the main factor since fiber texture is not always changed on recrystallization while the transition temperature invariably will increase. They also point out that a number of investigators have found a significant reduction in the ductile-to-brittle transition of W and Cr by very light strains (3 to 10%) at temperatures just above the ductile-to-brittle transition range. No fiber structures are produced by these treatments.

Tietz, Lytton, and Meyers (41) studied the effect of light prestrain and recovery anneals on the ductile-to-brittle transition of molybdenum in 1962. A 10% prestrain at 150°C lowered the DBTT of recrystallized Mo from +10°C to -22°C, a 32°C shift. Total elongation in the ductile range was decreased. After a 1-hr recovery anneal at 800°C, the DBTT temperature increased to -5°C and the ductility increased. Longer recovery treatments further improved ductility without increasing the DBTT. In continuing studies, Lytton and Tietz (42) found that 10% prestrain was not necessary to effect this change. Merely passing a Luders band through the material shifted the DBTT of commercial Mo sheet from +30°C to -10°C, a change of 40°C. The Luders band was initiated by exceeding the upper yield at 150°C and once formed could be passed through the specimen by straining at room temperature. Total strain in this case is of the order of 3-5%. This effect is shown in Figure 9. The DBTT of the Luders strained sheet is almost comparable to that of heavily cold worked and stress relieved sheet. The ductility, however, is much higher than that of strain hardened sheet and is comparable to that of recrystallized sheet. It should be noted that heavy cold working to reduce the DBTT also lowers useful ductility in the ductile range. Lowering the DBTT by means of light prestrain, however, retains the high ductility of recrystallized materials. This is a classic example of the use of a thermal-mechanical treatment to produce an improved material having a unique combination of properties.

Strength

Strain hardening is one of the simplest and most effective means of strengthening metals. Yield strength can be increased at least sixfold by this approach (5). Although widely used for many years to strengthen metals such as iron, molybdenum, and tungsten, strain hardening has developed to a point where only a fraction of its full potential has been realized in practice. With the fundamental understanding of the strain hardening process that has emerged in recent years, its continued development, and its application to engineering materials by controlled thermal-mechanical processing, significant improvements in materials properties can be anticipated.

The body centered cubic metals have a low work hardening rate at all temperatures. The work hardening rates for Cb, Ta, W, and Mo are roughly half those for face centered cubic metals (43,44). Strain hardening coefficients calculated from the slope of true stress strain curves at room temperature range from 0.2 to 0.25 for Fe, Cb, Ta, and Mo. ($\sigma = \kappa\epsilon^n$ where n = strain hardening coefficient.) The coefficient for FCC metals such as copper and brass is 0.50 (44). The work hardening rate of the refractory metals also tends to decrease markedly with increased amounts of strain. Cold rolled Cb and Mo (80-90% reduction) have strain hardening coefficients of 0.05. An increase in strain hardening rate with increased temperature is indicated. The coefficient for Mo at 200°C is 0.28 and for W at 400°C is 0.35. Very little true stress strain data are available from which reliable estimates of strain hardening behavior can be made.

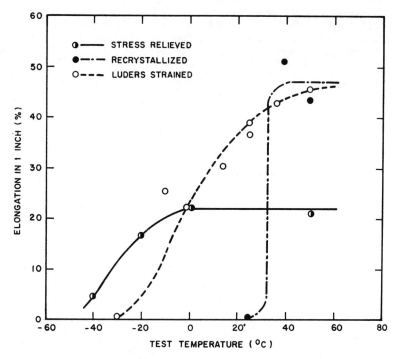

Fig. 9   Control of the ductile-brittle transition in Mo by thermal-mechanical
           treatments. (Ref. 42)

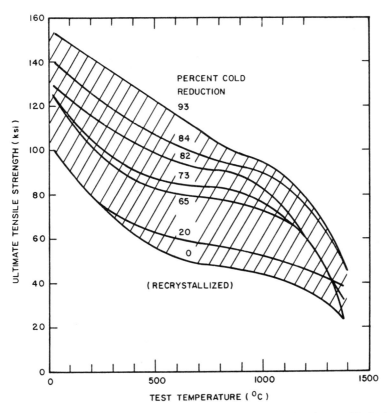

Fig. 10   Effect of strain hardening on the strength of Mo (Ref. 45)

The low strain hardening rate for these metals means that heavy reductions must be taken to achieve significant increases in tensile strength. The effect of reduction by cold rolling on the tensile strength of molybdenum over a broad temperature range is shown in Figure 10 (45). Molybdenum alloy processing today is dominated by reductions of 85% or more after the last recrystallization anneal to achieve a strength that is about double that of the recrystallized materials over a broad temperature range. A similar relationship exists for the other refractory metals. For example, a Cb-1Zr alloy has its yield strength more than doubled (from 23 to 53 ksi) by 20% cold work and tripled (from 23 to 64 ksi) by 60% cold work (29). The ultimate tensile strength was doubled (from 37 to 71 ksi) on 60% reduction by cold work.

Significantly, the greatest effect of strain hardening is on yield strength and not on ultimate strength. This is particularly true at low strain levels. The yield strength of Mo at room temperature was increased 56% by 7% strain and 84% by 13% strain (46). As shown in Figure 11, 5% prestrain of a Cb-TZM alloy followed by a 1-hr recovery treatment at 800°C increased the yield strength at 1315°C by 50% (from 26 to 40 ksi) without increasing the ultimate tensile strength. The major effect of thermal-mechanical treatments based on low strain and recovery to form stable polygonized structures will be to increase the yield strength over a broad temperature range. The effect on ultimate tensile strength may be negligibly small. This is in agreement with the structural analysis that predicts a relationship between flow stress and dislocation density, grain size, and cell size.

Since the major effect of thermal-mechanical treatments is to increase resistance to plastic deformation (flow stress) by introducing impediments to dislocation motion, its most significant contribution should be in the area of high temperature creep and rupture behavior. Temperatures, of course, are limited to those at which the substructures are stable. This does not, however, imply that only low temperature applications will be benefited. For example, Rubenstein (35) produced a sevenfold increase in the strength of tungsten at 1925°C (from 9 to 69 ksi ultimate strength) by stabilizing a fine dislocation cell structure with a HfC precipitate. The stress for rupture in 100 hr at 1925°C was 20 ksi and stress for a minimum creep rate of 10^{-6} second^{-1} was 23 ksi. This is a temperature where strain hardening could never be considered for strengthening but where a controlled thermal-mechanical treatment can be very successful.

Refractory metal technologists should regard strain hardening and stable substructure formation as a very important creep strengthening mechanism. The effect of polygonization on the creep of molybdenum is shown in Figure 12. In this study by Adams and Iannucci (47) it was found that 5% prestrain at room temperature followed by a high temperature polygonization anneal significantly reduced primary creep at 1000°C. In 8 hr, the unpolygonized sample had twice the creep extension of the polygonized sample. The secondary (minimum) creep rates, however, appeared to be similar for the two materials.

Columbium alloys are susceptible to extensive primary creep and the application of thermal-mechanical treatments should be of great benefit to these materials. Very little effort has been devoted to fundamental studies of creep behavior in Cb-alloys. Available creep data are poorly documented and often are confused by interstitial contamination and aging effects (48). In one study of the effect of deformation on the creep of Cb, Coldren and Freeman (49) found a significant reduction in total primary creep and secondary creep rates at 985°C by 20% prior warm working. As shown in Figure 13, the unworked material had 4% strain in 2 hr compared with less than 2% strain for the lightly rolled material at an applied stress of 3940 psi. The temperature of prior deformation

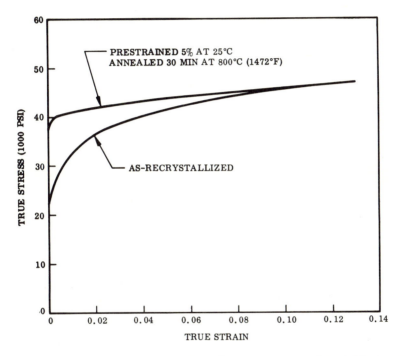

Fig. 11   Effect of thermal-mechanical treatments on the yield strength of
Cb-TZM alloy at 1315°C (Ref. 24)

ROGER A. PERKINS

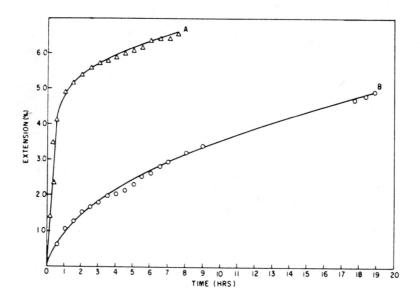

Fig. 12   Effect of polygonization on creep of Mo at 1000°C (Ref. 47)

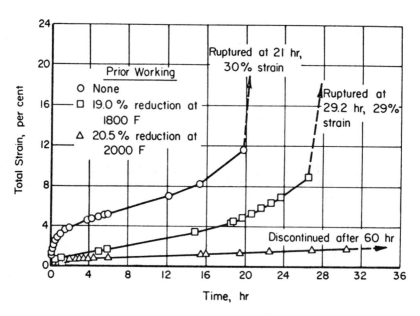

Fig. 13   Effect of strain hardening on the creep of Cb at 985°C (Ref. 49)

seemed important and better creep properties were exhibited by material rolled at higher temperatures.

The results of a more recent study by Bonesteel, et al. (17), show a significant effect of thermal-mechanical treatments on the creep of a Cb-1W-1Zr alloy (Figure 14). Although a prestrain and recovery treatment reduced the creep rate and more than doubled the rupture life at 1200°C, it was not until a dispersed phase (ZrO₂) was introduced by internal oxidation that a major improvement in creep properties was realized. Better creep properties were realized by internal oxidation prior to the introduction of a deformation substructure. On the basis of the time required to reach 0.05 strain at a stress of 13 ksi and a temperature of 2200°F, this material is a hundredfold improvement over the recrystallized base. In results of more recent work a three hundredfold improvement has been realized.

The results of several studies indicate that fine particle stabilization of deformation substructures is a prime requirement for major advances in creep and rupture properties. Recovery treatments per se cannot provide the structural stability that is needed in the temperature ranges of interest. As discussed earlier, particle size, shape, distribution, and interfacial considerations will dictate the degree of stabilization that can be achieved. The HfC particles which proved to be most effective in strengthening the W-Hf-C alloy at 1925°C are clearly indicated to be small platelets precipitated in a Widmanstätten array on preferred planes (35). Detailed studies of the effect of carbide morphology on the creep and rupture behavior of a Cb-TZM (Mo) alloy reveal the same effect (24). As shown in Figure 15, at stress levels above 30 ksi, a strain hardened matrix stabilized by CbC platelets in a Widmanstätten array had the lowest creep rates at all stress levels (points C, D, E, and F). The slope of the curve is characteristic of dispersion strengthened metals (m = 7). Similarly, strain hardened material (point G) stabilized by spheroidized CbC particles had a much higher creep rate and a slope characteristic of pure metals or solid solution alloys (m = 4). At low stress levels, the best creep properties are provided by a strain free (recrystallized) structure containing platelet carbides. This structure has good stability and resistance to deformation.

The relationships are very complex and certainly are not well understood at this time. Certain combinations of strain hardening and stabilization by recovery or precipitates can significantly reduce creep whereas other combinations can create conditions of marked instability. The onset of primary recrystallization during creep appears to have an accelerating effect on creep rates. Far more detailed studies of factors governing structural stability and creep behavior are needed to guide development efforts in the application of thermal-mechanical treatments to the refractory metals.

Formability

The use of thermal-mechanical treatments to control the texture of refractory metals has the potential to improve cold forming characteristics. One of the problems frequently encountered in the cold forming of tungsten and molybdenum sheet alloys is the tendency for these materials to delaminate or split lengthwise in the plane of the sheet. Lamination failures are prone to occur in shearing, punching, riveting, drilling, and bending operations. In a study of the processing of Mo-alloy sheet, Brentnall and Rostoker (50) found that the tendency to delaminate was related to the relative orientation of the {100} cleavage planes of the crystal and the plane of the sheet. In the cold worked state, a high intensity of cleavage planes are aligned in the plane of the sheet and the metal is weak in the short transverse direction. By a partial recrystallization treatment, the

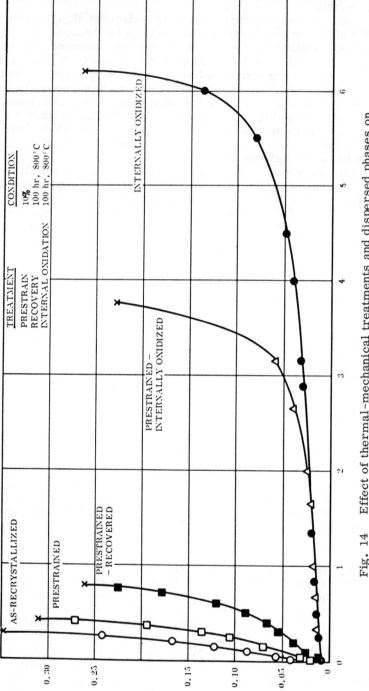

Fig. 14   Effect of thermal-mechanical treatments and dispersed phases on creep of Cb-1W-1Zr at 1200°C.   (Ref. 17)

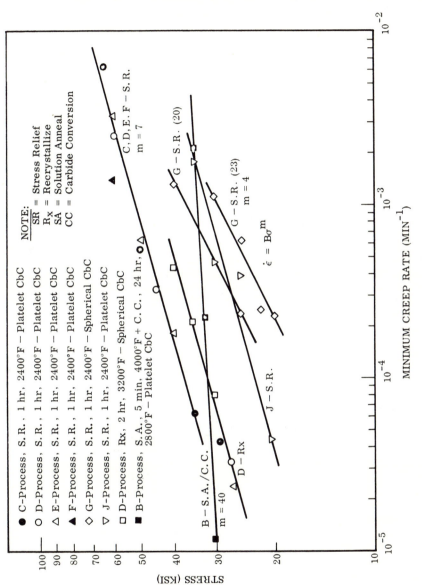

Fig. 15 Effect of strain hardening and carbide morphology on creep of Cb–TZM alloy at 1315°C (Ref. 24)

cleavage planes are reoriented and the laminating tendency is greatly reduced. Lytton and Perkins (23) found that a fully strain hardened Mo-alloy was resistant to delamination, even on reverse bending if processed to have a minimum texture (random orientations). The use of thermal-mechanical treatments to control texture considerably improved cold forming characteristics.

Brentnall and Rostoker (50) proposed the use of the plastic strain ratio, R, as a mechanical index of texturing and anisotropy. This index is the ratio of true strain in the width direction to that in the thickness direction (ϵ_w/ϵ_t). They found that heavily cold worked materials have a low R value and tend to split on forming. By reducing texture, the R value was increased and the short transverse properties were improved. Lytton and Tietz (42) reported R values of 0.6 for cold worked Mo sheet and 1.5 for fully recrystallized material. Recrystallized sheet that was Lüders strained to reduce the ductile-brittle transition maintained the high R value of 1.5.

This parameter defines the directionality of properties normal to the plane of sheet and also indicates resistance to thinning or puckering during deep drawing (44). Plastic strain ratios of 1.5 or above are generally sought for deep draw capabilities in steel. The refractory metals can be processed to have similar characteristics by controlled thermal mechanical treatments.

Role of Processing Variables

The basic processes used in the primary working, shaping, forming, and joining of refractory metals involve interactions between temperature, strain rate, and the type and amount of strain. Each step of processing is, in effect, a thermal-mechanical treatment that influences the dislocation substructure and hence the properties of the alloy. The interactions are very complex, with each step having an influence on or an interaction with the next. Therefore, the whole thermal-mechanical history is important.

The challenge that faces the refractory metal technologist today is to couple the concepts and understanding of thermal-mechanical treatments with the shop practices for working and shaping metals. Metals have to be produced in appropriate shapes to be useful and it is only by the regulation of all processing steps that thermal-mechanical treatments to control structure and properties can be applied. It is necessary, therefore, to establish the effect of processing variables on the structure as well as on the properties of metals and to develop a basic understanding that can be used to control processing. To date, most of the studies conducted on substructure formation and stabilization in metals have utilized simple tensile strain for mechanical deformation. The need today is to establish the effect of straining parameters characteristic of metal working processes on structure so that needed coupling of basic knowledge and applied technology can be achieved.

Initial steps in this direction were taken by the Air Force Materials Laboratory in 1963 with two programs to study the physical metallurgy of processing refractory metal sheet (23, 24, 50). Some of the more significant results in terms of structure and property effects have been reviewed in this paper. Basically these programs have demonstrated that production working schedules can be adjusted to provide a fine particle distribution, dislocation substructure, and texture that yields enhanced mechanical properties in a manner predicted from fundamental thermal-mechanical considerations. The early stages of processing are adjusted to affect major shape changes, control quality, refine and homogenize the structure, regulate the nature and distribution of dispersed phases, and to minimize texture. Since the dominant factor in thermal-mechanical treatments

Table 1. Dependence of Mechanical Properties on Microstructure
for Cb-TZM Alloy Sheet (Ref. 24)

| Property | Optimum Sheet Structures | |
	CbC Dispersion	Matrix
High Strength (Y.S., U.T.S.)		
75°F	Platelets	Strain Hardened
2400°F	Platelets	Strain Hardened
High Ductility (Formability)		
75°F	Spheres	Recrystallized
2400°F	Spheres or Platelets	Recrystallized
Low DBTT	Platelets	Strain Hardened
	or Spheres	Recrystallized
2400°F Creep and Rupture		
Low Stress (< 30 ksi)	Platelets	Recrystallized
High Stress (> 30 ksi)	Platelets	Strain Hardened

is the type and amount of strain after the last anneal and the subsequent thermal treatments to stabilize substructures, the finishing stages of processing hold the key to the successful applications of these principals.

Considering the complexity of interactions that occur and the different performance capabilities of various structures, it is evident that processing variables may have to be adjusted to produce specific structures that will have the required properties for specific applications. That is, no one treatment or structure is likely to provide the best combination of properties for all areas of use. This approach is illustrated by Table I where the optimum structure in terms of dispersed phases, strain hardening, and recrystallization is defined for the best properties in terms of strength, ductility and formability, transition temperatures, and creep-rupture behavior (24). The ideal material for long duration moderate stress applications would be a lightly strain hardened base stabilized by a platelet dispersion of CbC. This material would be useless, however, at high stress levels, even for short times. Heavily strain hardened material would perform best here, but this material would be inferior for long time service due to structural instability. Still another structure would be best for high formability.

Finally, it must be recognized that refractory metals will have to be coated for oxidation protection in reactive atmospheres at high temperatures. Much of the early alloy development did not carefully consider this fact and as a result many substrate/coating combinations are incompatible. The alloy must not unduly restrict the processing variables in coating or degrade coating performance and the coating process or coating/substrate interactions, in turn, must not degrade structure and properties. Since coatings in use today are brittle, thermal-mechanical treatments must be introduced before application of the coating. In addition, coating processes entail long time exposures at temperatures of 985 to 1425°C. The structure developed by thermal-mechanical treatments must be stable under these conditions. This interplay explains one reason why strain hardening has not been more widely used to strengthen refractory metals. However, as discussed in this paper, appropriate thermal-mechanical treatments show promise for greatly increased thermal stability to temperatures well above this range.

Acknowledgments

The author is grateful to Dr. T. E. Tietz and Dr. D. J. Rowcliffe for their helpful comments and suggestions during the preparation of this manuscript.

REFERENCES

1. J. E. Dorn, A. Goldberg, and T. E. Tietz, "The Effect of Thermal-Mechanical History on the Strain Hardening of Metals," Trans. AIME, 1948.

2. T. E. Tietz, R. A. Anderson, and J. E. Dorn, "Effect of Prestraining Temperature on the Recovery of Cold Worked Aluminum," Trans. AIME, 185, 1949.

3. Fundaments of Deformation Processing, Proceedings of 9th Sagamore Army Materials Research Conference, August 1962.

4. E. B. Kula, "Strengthening of Steel by Thermomechanical Treatments," Strengthening Mechanisms: Metals and Ceramics, Proceedings of the 12th Sagamore Army Materials Research Conference, August 1965.

5. G. Thomas, V. F. Zackay, and E. R. Parker, "Structure and High-Strength Metals," Strengthening Mechanisms: Metals and Ceramics, Proceedings of the 12th Sagamore Army Materials Research Conference, August 1965.

6. W. Rostoker, "Working of Refractory Metals," Fundamentals of Deformation Processing, Proceedings of the 9th Sagamore Army Materials Research Conference, August 1962.

7. A. G. Ingram and H. R. Ogden, "The Effect of Fabrication History and Microstructure on the Mechanical Properties of Refractory Metals and Alloys," DMIC Report 186, July 10, 1963.

8. J. A. Houck and H. R. Ogden, "Thermal-Mechanical Variables Affecting the Properties of Refractory Metals and Alloys," Metal Deformation Processing – Vol. I, DMIC Report 208, August 14, 1964.

9. B. S. Lement, D. A. Thomas, S. W. Weissman, W. S. Owen, and P. B. Hirsch, "Substructure and Mechanical Properties of Refractory Metals," WADD-TR-61-181, August 1961.

10. Ibid, WADD-TR-61-181- Part II, October 1962.

11. Ibid, WADD-TR-61-181-Part III, April 1963.

12. Symposium on the Role of Substructure in the Mechanical Behavior of Metals, ASD-TDR-63-324, April 1963.

13. A. S. Keh, Direct Observations of Imperfections in Crystals, Interscience, N. Y. (1962) 213.

14. L. I. Van Torne and G. Thomas, "Yielding and Plastic Flow in Niobium," Acta Met., 11 (1963) 881.

15. J. R. Stephens, "Dislocation Structure in Slightly Strained Tungsten and Tungsten-Rhenium and Tungsten-Tantalum Alloys," NASA-TN-D-4094, August 1967.

16. A. Berghezan and A. Fourdeux, "Deformation and Annealing Substructures of Niobium and Their Relation to the Mechanical Properties and Precipitation Phenomena, " Symposium on the Role of Substructure on the Mechanical Behavior of Metals, ASD-TDR-63-324, April 1963, p. 437.

17. R. M. Bonesteel, J. L. Lytton, D. J. Rowcliffe, and T. E. Tietz, "Recovery and Internal Oxidation of Columbium and Columbium Alloys, " AFML-TR-66-253, August 1966.

18. R. P. Jewett and E. D. Weisert, "Dislocation Morphology of Tantalum Deformed in Tension, " High Temperature Refractory Metals, AIME Metallurgical Society Conferences, Volume 34, 1964.

19. J. L. Walter, "Substructure and Recrystallization of Deformed Crystals of High Purity Silicon Iron, " Symposium on the Role of Substructure in the Mechanical Behavior of Metals, ASD-TDR-63-324, April 1963.

20. W. S. Owen, D. Hull, C. L. Formby, I. D. McIvor, and A. R. Rosenfield, "Yield Phenomena in Refractory Metals, " Substructure and Mechanical Properties of Refractory Metals, WADD-TR-61-181-Part III, April 1963.

21. V. M. Kardonski, G. V. Kurdyumov, and M. D. Perkas, "Effect of Size and Shape of Cementite Particles on the Structure and Properties of Deformed Steel, " Metal Science and Heat Treatment, January – February 1964, pp. 62 – 68.

22. E. Kalns, R. Q. Barr, and M. Semchyshen, "Effect of In-Process Solution Heat Treatments on the Properties of the TZM Alloy, " High Temperature Refractory Metals, AIME Metallurgical Society Conferences, Volume 34, 1964.

23. R. A. Perkins and J. L. Lytton, "Effect of Processing Variables on the Structure and Properties of Refractory Metals, " AFML-TR-65-234, July 1965.

24. R. A. Perkins, "Effect of Processing Variables on the Structure and Properties of Refractory Metals, " AFML-TR-65-234-Part II, May 1967.

25. J. T. Michalak and L. J. Cuddy, "Some Observations on the Development of Deformation Substructures in Zone Refined Iron, " Symposium on the Role of Substructure in the Mechanical Behavior of Metals, ASD-TDR-63-324, April 1963.

26. R. L. Nolder and G. Thomas, "Substructures in Plastically Deformed Ni, " Acta Met., 12 (1964) 227.

27. J. L. Lytton, K. H. Westmacott, and L. C. Potter, "Recovery Behavior of Pure Aluminum and Selected Body Centered Cubic Metals, " AFML-TDR-64-189, August 1964.

28. B. S. Lement, E. M. Passmore, S. Allen, J. E. Boyd, C. Anderson, and L. Vilks, "Ductile-Brittle Transition in Refractory Metals," Substructure and Mechanical Properties of Refractory Metals, WADD-TR-61-181, August 1961.

29. J. R. Stewart, W. Lieberman, and G. H. Rowe, "Recovery and Recrystallization of Cb-1 Zr Alloy," Columbium Metallurgy, AIME Metallurgical Society Conferences, Volume 10, 1960.

30. E. C. W. Perryman, "Recovery of Mechanical Properties," Creep and Recovery, American Society for Metals, 1956.

31. J. C. M. Li, "Recovery Processes in Metals," Recrystallization, Grain Growth, and Textures, American Society for Metals, 1965, p. 66.

32. R. C. Koo and H. G. Sell, "Subgrain Coalescence in Tungsten," Recrystallization, Grain Growth, and Textures, American Society for Metals, 1965, p. 97.

33. R. W. Cahn, "Recrystallization Mechanisms," Recrystallization, Grain Growth, and Textures, American Society for Metals, 1965, p. 110.

34. Molybdenum Metal, Climax Molybdenum Company, 1960.

35. L. S. Rubenstein, "Effects of Composition and Heat Treatment on the High Temperature Strength of Arc-Melted W-Hf-C Alloys," NASA-TN-D-4379, February 1968.

36. A. T. English and W. A. Backofen, Trans. AIME, 230, 1964, p. 396.

37. W. C. Coons, Private Communication.

38. T. D. Cooper, "The Effect of Rolling Temperature on Preferred Orientation in Tantalum," AFML-TR-64-375, February 1965.

39. W. F. Hosford, Jr. and W. A. Backofen, "Strength and Plasticity of Textured Metals," Proceedings of the 9th Sagamore Army Materials Research Conference, August 1967.

40. L. L. Seigle and C. D. Dickinson, "Effect of Mechanical and Structural Variables on the Ductile-Brittle Transition in Refractory Metals," Refractory Metals and Alloys, AIME Metallurgical Society Conferences, Volume 17, 1962.

41. T. E. Tietz, J. L. Lytton, and C. L. Meyers, "Recovery Behavior of Cold Worked Metals," ASD-TDR-62-984, November 1962.

42. J. L. Lytton and T. E. Tietz, "The Effect of Lüders Straining on the Tensile Flow Behavior of Arc-Cast Mo Sheet," Trans. AIME, 230 (1964) 241.

43. A. K. Mukherjee, "Work Hardening Mechanisms in Metals," Metal Deformation Processing, Volume II, DMIC Report 226, July 7, 1966.

44. R. A. Perkins, "Forming of Refractory Metal Sheet," The Science and Technology of Selected Refractory Metals, Pergamon Press, Oxford, England, 1964, p. 541.

45. E. D. Weisert, "The Present Status of Refractory Metals and Alloys," Metals Engineering Quarterly, February 1962.

46. M. A. Adams and H. Nesor, "Impuri:y Atom-Dislocation Interactions and Subsequent Effects on Mechanical Properties of Refractory Metals," ASD-TDR-62-11, March 1962.

47. M. A. Adams and Iannucci, WADD TR-59-441, Part II, December 1960.

48. Creep of Columbium Alloys, DMIC Memorandum 170, June 1963.

49. A. P. Coldren and J. W. Freeman, "Study of Substructure and Creep Resistance Using Nickel With Preliminary Data for Niobium," ASD-TR-61-440, November 1961.

50. W. P. Brentnall and W. Rostoker, "Application of Physical Metallurgy to Refractory Metal Processing," AFML-TR-64-345, Part II, November 1966.

INTERACTIONS IN COATED REFRACTORY
METAL SYSTEMS

Arthur G. Metcalfe[1]

and

Alvin R. Stetson[2]

SOLAR DIVISION OF INTERNATIONAL HARVESTER COMPANY

Abstract

Typical protective coating systems for refractory metals consist of a primary oxide barrier, a secondary barrier or reservoir, and a ternary barrier formed by surface alloying the substrate. Interaction between these various barrier layers and the substrate and atmosphere form the subject of this paper. Interactions discussed in detail include: expansion mismatch; interstitial sink effects; effect of coatings on mechanical properties; diffusion problems that control coating life; and metallic claddings. Methods to solve the complex interaction effects are discussed and illustrated by low temperature oxidation problems of refractory silicide coatings, crack arresting effects of porosity in coatings, coatings containing a molten phase, and low pressure effects. A review of future approaches is presented.

1. Associate Director of Research
2. Chief - Process Research Department

I

INTRODUCTION

Structural materials to operate in oxidizing environments above 2000° F are needed for more efficient power plants, for hypersonic flight, and for many high-temperature structures. Metal or ceramic matrix composites reinforced with filaments or whiskers are long-range possibilities[1], but coated refractory metals offer broader properties and a higher temperature capability than any composite currently available. Therefore, a critical examination is required of those properties of coated refractory metals that are significant in limiting the growth of these materials into reliable, efficient, structural materials, available to design engineers. In this paper, the properties of individual components of the coating system and substrate will be examined with particular reference to the interaction of these components, and the effect of such interaction on protection and mechanical properties. The paper will concentrate primarily on coated columbium and tantalum alloys because more extensive study has been made of these systems than of coatings for other metals.

A schematic drawing of a typical coating system is shown in Fig. 1. The coating is comprised of three major layers - the primary barrier (usually an oxide), the reservoir or secondary barrier from which the oxide is grown, and a diffusion zone or ternary barrier from which the reservoir is grown or which develops during the application of the reservoir. The performance of these three layers as oxygen barriers generally determines the time to failure of the coated system but not its mechanical efficiency. Interaction of the reservoir and/or diffusion zone with the substrate may strongly influence mechanical performance. Rapid coating-substrate interdiffusion, selective diffusion of interstitial elements from the substrate to the coating, craze cracks in the reservoir as a consequence of a mismatch in thermal expansion, and reservoir ductility can all be significant in determining mechanical performance. Rapid interdiffusion of reservoir and substrate, and craze cracking of the reservoir can also have significant influence on the oxidation performance.

The paper will present data on the various interactions of significance in the coating substrate system and will also briefly note some of the newer coating systems that have benefited from a better understanding of these interactions.

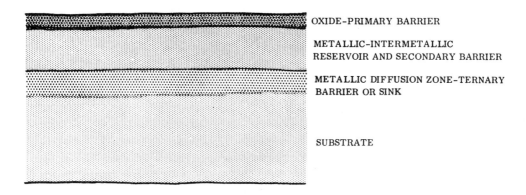

OXIDE-PRIMARY BARRIER

METALLIC-INTERMETALLIC
RESERVOIR AND SECONDARY BARRIER

METALLIC DIFFUSION ZONE-TERNARY
BARRIER OR SINK

SUBSTRATE

FIGURE 1. SCHEMATIC OF TYPICAL COATING ON A REFRACTORY METAL

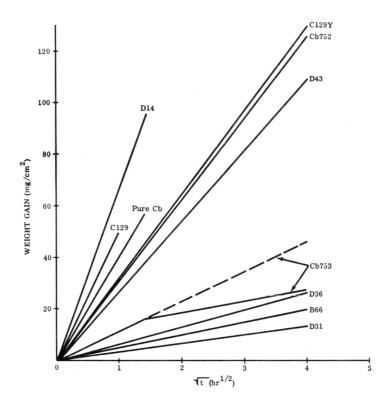

FIGURE 2. OXIDATION RATE PLOTS OF SEVERAL UNCOATED SUBSTRATE
ALLOYS; 1600°F

II

INTERACTION OF COATING COMPONENTS WITH ATMOSPHERE

The coating on a refractory metal is a complex multi-component system shown schematically in Fig. 1. Each of the components strongly interact to determine the effectiveness of the overall system in inhibiting oxidation of the substrate. In this section the properties of each component will be reviewed, i. e. , substrate, diffusion zone, reservoir, and primary barrier in relationship to their interaction with the environment. Emphasis will be primarily on silicide reservoir systems which appear most applicable to long-term protection below 2500°F, the area of greatest interest to the authors; however, aluminide and special coating systems such as hafnium-tantalum will also be briefly discussed. The data presented will be primarily on oxidation under static conditions at atmospheric pressure, but the influence of pressure, high flow, and thermal cycling will be noted.

2. 1 OXIDATION RESISTANCE OF SUBSTRATE ALLOYS AND MODIFIED ALLOYS

Coating reservoirs are formed primarily from the direct reaction of silicon or aluminum or their compounds with the modified or unmodified substrate. The coefficients of thermal expansion of most reservoirs are higher than the substrate. Therefore, craze cracks can be expected in the reservoir at room temperature, although they will tend to close at high temperatures. A slow oxidation rate at the base of these craze cracks, particularly at low temperatures, is believed to be required to allow sufficient time for the reservoir to repair the continuity of the primary oxide barrier.

The oxidation resistance at 1600°F of a number of commercially available columbium-base alloys is shown in Fig. 2.[2] The data show that in the amount present, zirconium, tungsten, and hafnium are deleterious, or have little effect, on the oxidation resistance of columbium; whereas, molybdenum, titanium, and vanadium markedly improve oxidation resistance. The substrates exhibiting the better oxidation resistance are also the substrates, that when silicided, provide the better oxidation resistance at least to temperatures of 2500°F. [3]

Extensive investigation of potentially ductile substrate alloys with improved oxidation resistance has been conducted by Stetson and Metcalfe. [2] These investigations, covering such additions as titanium, vanadium, molybdenum, chromium, beryllium, and aluminum to columbium, showed that significant improvement in 1600°F oxidation resistance is possible with addition of: Ti-Cr-Al, Ti-Cr, Ti-Mo, or Ti-V-Cr (Fig. 3); but also that there is little basis for optimism that ductile, oxidation-resistant, columbium alloys can be developed for use above 2000°F. Figure 4 shows the marked temperature sensitivity of one of the more oxidation-resistant ductile alloys tested.

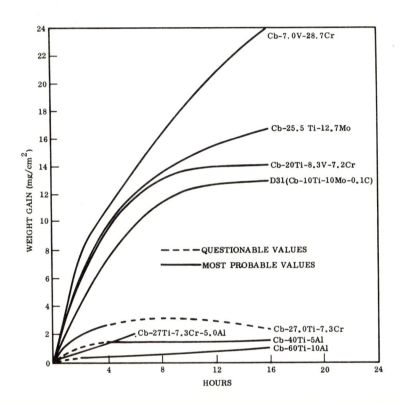

FIGURE 3. SUMMARY OF THE MOST OXIDATION RESISTANT DUCTILE COLUMBIUM ALLOY; 1600°F

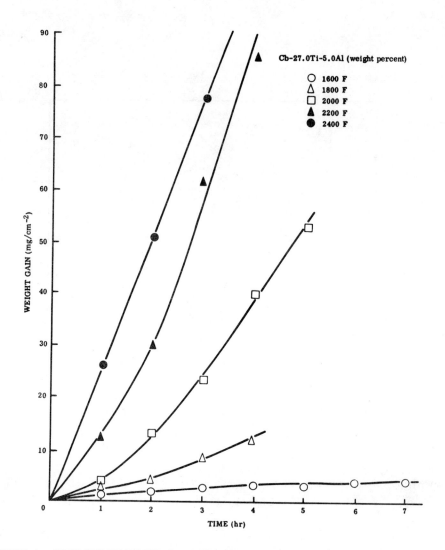

FIGURE 4.   OXIDATION RATE OF COLUMBIUM-TITANIUM-ALLUMINUM; One-
Hour Cycles at 1600 to 2400°F

Two current Air Force programs are investigating columbium alloys that combine strength and oxidation resistance. Geiselman, et al,[4] have found that the Cb-W-Hf alloys (15 to 20W and 30 to 40Hf) show only a modest weight gain (48 mg/cm^2) after 20 hours at 2000°F, but that oxygen penetration is at a rate of at least one mil/hour. Cornie[5] investigating a base of Cb-10W-10Ta-10Ti-1.8Hf with additions of titanium, molybdenum, vanadium, aluminum, chromium, yttrium, boron, rhenium, beryllium, and uranium, found that the best additions, titanium and aluminum, could limit weight gain to as little as 150 mg/cm^2 in 81 hours at 2200°F, but that oxygen penetration was at least 20 mils in 20 hours. These programs confirm the effectiveness of titanium and aluminum in improving the oxidation resistance of columbium and emphasize the difficulty of developing a ductile alloy that can act as a reservoir to form the primary barrier.

Columbium alloys that exhibit low weight gain in air are not particularly good barriers to oxygen contamination. However, as ternary barriers (Fig. 1), prealloyed sublayers may markedly decrease ingress of oxygen to the substrate. Two mechanisms contribute to this improvement. One is their intrinsic oxidation resistance at the intermediate temperatures, where the high-expansion silicides and aluminides begin to form open cracks, so that adequate time is available for the repair mechanisms in the primary and secondary barriers to become effective. The second mechanism arises from the fortunate circumstance that many of these compositions contain much titanium or hafnium; such compositions retain the oxygen in solution by the "interstitial sink" effect and do not permit transport to the substrate (par. 3.1.2). Under catastrophic damage to the primary and secondary barriers, life of the ternary barrier is measured in minutes to a maximum of an hour in the temperature range 2000 to 2500°F.

2.2 RESERVOIRS - SECONDARY BARRIERS

The reservoir is the phase that must oxidize at a slow enough rate to protect the substrate and possibly the ternary barrier from oxidation for a sufficiently long time to make use of a coated structure economical. However, this same phase must oxidize sufficiently rapidly for repair of defects such as craze cracks to minimize oxygen ingress to the substrate. Therefore, the most oxidation-resistant substances are not the best reservoirs in coating systems under all conditions. Silicides exhibiting extremely low oxidation rates, e.g., unalloyed $MoSi_2$ or WSi_2, tend to make coatings with short and inconsistent lives at low temperatures[6] where the oxidation rate appears to be too slow to heal craze cracks. Fig. 5 shows a comparison [7] of pure WSi_2 on the tantalum alloy, T222, with a disilicide containing 95W-5Ti (Solar TNV-13). The alloying additions control the oxidation rate to some compromise value and thereby control the properties of the oxide formed. Approximately 100 degrees F in refractoriness has been sacrificed by the compromise. Further improvement in low-temperature life is exhibited by the TNV-7 alloy which has 30 percent titanium plus vanadium, but at much greater sacrifice of refractoriness (Fig. 5).

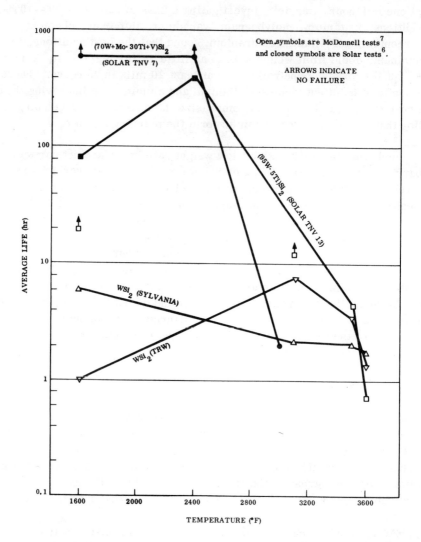

FIGURE 5.   EFFECT OF ALLOYING WSi$_2$ ON COATING LIFE ON T222 ALLOY

Compromises in reservoir selection may also be necessary between high- and low-temperature performance. Generally, a reservoir generating a pure oxide, such as SiO_2 or Al_2O_3, will provide the longest life at the highest temperature; however, low-temperature "pest" oxidation is usually most severe with compounds, e.g., $MoSi_2$ and WSi_2, [6, 8, 9, 10] which generate pure oxides at high temperature.

Silicides and aluminides form the principal reservoir material but, in very selected applications, other reservoirs, e.g., Hf-Ta show some utility.

2.2.1 Silicides

Two silicide types are of general importance in coatings. They are the MSi_2 and M_5Si_3 types. The MSi_2, having the highest silicon activity, is usually the initial silicide in the as-applied coating. As the coating oxidizes, the M_5Si_3 silicide theoretically forms at both the air-silicide interface and at the MSi_2-substrate interface. Identification of the M_5Si_3 silicide at the air-silicide interface is usually very difficult because the oxidation rate of this silicide is generally quite rapid.

MSi_2 Silicides

The most extensive recent investigation of the moderate temperature oxidation rate of MSi_2 silicides was performed by Stetson and Metcalfe. [2] In this study the arc-melted binary compositions — $CbSi_2$, $CrSi_2$, $TiSi_2$, VSi_2, $MoSi_2$, and WSi_2; various ternary silicides — $(V-Cb)Si_2$, $(Cr-Ti)Si_2$, $(Ti-Mo)Si_2$, $(Ti-W)Si$, and $(V-Cr)Si_2$; and quaternary silicides — $(Ti-Cr-Cb)Si_2$, $(Ti-Mo-Cb)Si_2$, $(Ti-W-Cb)Si_2$, and $(V-Cr-Cb)Si_2$ were oxidation tested at 1500 and 2400°F for approximately 96 hours or until failure, as judged by gross oxidation. The oxidation results are shown in Table 1.

With relatively few exceptions, the disilicides have excellent oxidation resistance at 1500°F. The exceptions are WSi_2 and $CbSi_2$, which fail catastrophically in less than 16 hours. Other more complex disilicides containing high tungsten and columbium (No. 7) or high tungsten only (No. 5) also oxidize rapidly. Most specimens retain a metallic luster or develop a very thin, almost interference, oxide film in the 96-hour test.

Oxidation rates at 2400°F are markedly different for the various disilicides. The $CbSi_2$ disilicide exhibits very poor oxidation resistance, while $MoSi_2$, WSi_2, $TiSi_2$, and VSi_2 are outstanding in oxidation resistance showing a weight gain of less than 1.0 mg/cm^2 (i.e., less than 0.2 mil penetration in 100 hours). Disilicides containing appreciable quantities of chromium usually lost weight and spalled on thermal cycling. Disilicide compositions containing both titanium and chromium

TABLE I

OXIDATION TEST RESULTS ON SEVERAL MSi_2 COMPOSITIONS

Number	Composition (at. %)(2)	Oxidation at 1500°F			Oxidation at 2400°F		
		16 hours (mg/cm²)	96 hours (mg/cm²)	Comments	16 hours (mg/cm²)	96 hours(1) (mg/cm²)	Comments
1	18.7Ti-14.6Mo-66.7Si (A)	-0.25	-0.18	Very good, metallic.	+6.78	+11.27	Very good, metallic.
2	13.3Ti-6.6Mo-80.1Si (A)	+0.14	+0.32	Very good, metallic.	+0.70	+1.37	Very good, black metallic.
2R	24Ti-8Mo-68Si (N)	---	+0.069	Very good, metallic.	---	5.87	Very good, metallic, no spalling.
3	8.7Ti-11.3Mo-10.6Cb-69.3Si (A)	+0.09	+0.12	Very good, metallic.	+0.42	+0.69	Very good, metallic.
4R	19.4Ti-8.7Mo-12.7Cb-59.2Si (A)	+4.06	Failed	Poor, failed 48 hours.	+1.25	+2.82	Good, thin black-yellow oxide.
4	16.5Ti-5.5Mo-10Cb-68Si (N)	---	+0.065	Very good, metallic.	---	+0.35	Excellent, metallic, no spalling
5	20.9Ti-15.1W-64.0Si (A)	Failed	---	Poor, failed 24 hours.	+1.02	+2.18	Very good, metallic.
6	11.2Ti-11.5W-77.4Si (A)	+0.01	+0.01	Very good, metallic.	+0.96	+3.19	Good, metallic.
6R	24Ti-8W-68Si (N)	---	-0.36	Very good, metallic.	---	+6.34	Very good, metallic, no spalling.
7	9.6Ti-10.7W-7.4Cb-72.4Si (A)	Failed	---	Poor, failed in 24 hours.	+0.27	+0.51	Very good, dark metallic.
8	16.1Ti-4.7W-14.2Cb-65.1Si (A)	+0.59	+0.96	Fair, yellow porous oxide.	+0.05	+0.34	Fair, yellow porous oxide.
9	16.7Ti-16.7Cb-66.6Si (A)	-0.47	-0.47	Very good, metallic.	+46.7	---	Poor, failed in 20 hours.
10	11.7Ti-9.0Cr-79.3Si (A)	+0.20	+0.30	Very good, metallic.	+8.90	+8.14	Poor, local melting.
10R	24Ti-8Cr-68Si (N)	---	+0.20	Very good, metallic.	---	+39.1	Poor, heavy oxide, spalling.
11	9.6Ti-23.3Cr-67.2Si (A)	0.00	0.00	Very good, metallic.	+30.19	Failed	Poor, failed 20 hours.
12	14.2Ti-11.1Cr-10.4Cb-64.3Si (A)	+0.02	+0.03	Very good, metallic.	+28.87	+36.16	Fair, green to grey oxide.
13	2.4Ti-5.5Cr-10.7Cb-81.4Si (A)	+0.00	+0.00	Very good, metallic.	+0.61	+1.51	Good, black oxide, some glazing.
13R	16.5Ti-5.5Cr-10Cb-68Si (N)	---	0.12	Very good, metallic.	---	+20.7	Severe spall, poor.
14	7.2Ti-14.9Cr-10.9Cb-67.0Si (A)	+0.01	+0.02	Very good, metallic.	-4.15	+9.38	Fair, green to grey oxide.
15	14.4Ti-17.7Cr-67.9Si (A)	+0.20	+0.30	Very good, metallic.	-0.37	-0.11	Good, black oxide.
16	8.3V-24.7Cr-67.0Si (A)	+0.02	-0.03	Very good, metallic.	+0.32	-016	Fair, green to black oxide.
17	24.8V-7.7Cr-67.5Si (A)	+0.05	+0.06	Very good, metallic.	+0.19	+0.23	Good, brown oxide.
18	10.1V-10.9Cr-9.8Cb-69.1Si (A)	+1.28	+2.33	Good, black oxide.	+1.30	+10.11	Fair, green to brown oxide.
19	5.5V-17.0Cr-9.7Cb-67.8Si (A)	+1.19	+2.47	Very good, metallic.	+4.15	+19.66	Fair, green to brown oxide.
20	15.5V-9.5Cr-10.0Cb-65.0Si (A)	+0.71	+0.90	Good, black oxide.	+0.48	+1.20	Good, black-brown oxide.
21	14.2V-16.6Cb-69.1Si (A)	+1.81	+4.57	Fair, yellow oxide.	+1.60	+0.77	Good, metallic.
25	31.7Mo-68.3Si (A)	+0.03	+0.03	Fair, porous black oxide.	+0.33	+0.86	Very good, metallic.
26	30.5W-69.5Si (A)	Failed	---	Failed 24 hours - yellow powder	Very low(4)	Very low(4)	Very good, metallic.
27	30.4V-69.6Si (A)	+0.81	+1.21	Excellent, metallic.	+0.34	0.77	Good, black film.
28	33Cb-67Si (A)	Failed	---	Failed 24 hours - grey powder	Failed	---	Poor, failed in 12 hours at 2000 F.
29	33Ti-67Si (N)	+0.23	+0.44	Very good, thin adherent oxide.	+0.3	+0.8(3)	Very good, thin adherent oxide.
30	33Cr-67Si (N)	+1.5	+1.5	Very good, metallic.	+1.5	+5.5(3)	Fair to good, but oxide spalled continuously.

1. First 12 hours at 2200°F, plus 96 hours at 2400°F. (Weight change is from beginning of 2400°F hour test.)
2. A - analyzed, N - Nominal.
3. Tested 2400°F only (no 2200°F testing).
4. Irregular specimen shape did not permit accurate area calculation.

show rapid weight gain and rarely last 96 hours, even when silicon is in excess of the disilicide ratio. Additions of columbium to the Cr-Ti-Si compositions greatly retard the oxidation rate. In contrast to the Cr-Ti-Si compositions, disilicides in the V-Cr-Si system have excellent oxidation resistance.

Of the disilicides containing columbium, those containing both titanium and molybdenum, followed on a selective basis by those containing titanium plus tungsten and vanadium plus chromium are the most oxidation resistant. Tolerance for columbium by a disilicide is extremely important in coatings formed by diffusion into columbium alloys.

M_5Si_3 Silicides

The oxidation rate of M_5Si_3 silicides, including binary silicides Ti_5Si_3, Cr_5Si_3, Cb_5Si_3, V_5Si_3, and Mo_5Si_3; the more complex silicides $(Ti-Cr)_5Si_3$, $(Ti-Cr-Cb)_5Si_3$, $(TiMo)_5Si_3$, $(Ti-Mo-Cb)_5Si_3$, $(V-Cr)_5Si_3$, and $(V-Cr-Cb)_3Si_5$ are presented in Table II after exposure to 1500 or 2400°F for 101 hours. [2]

The binary silicide, Cb_5Si_3 has essentially no oxidation resistance at either test temperature. The binary silicide, Mo_5Si_3, shows relatively severe exfoliation at 1500°F and catastrophic oxidation at 2400°F. At 1500°F, Cr_5Si_3 is the most oxidation resistant, followed closely by V_5Si_3 and Ti_5Si_3 in that order.

The ternary silicides, $(Ti-Cr)_5Si_3$ and $(V-Cr)_5Si_3$ exhibit good to excellent oxidation resistance at 1500 and 2400°F. The $(Ti-Cr)_5Si_3$ composition has smaller positive weight gains at both temperatures.

The three quaternary M_5Si_3 silicides exhibit some of the more interesting oxidation behavior. The $(Ti-Cr-Cb)_5Si_3$ composition (No. 37) has equal or superior oxidation resistance to the disilicide containing these elements. Other quaternary silicides containing Ti-Mo-Cb or V-Cr-Cb are rapidly oxidized at both 1500 and 2400°F.

A significant point in the oxidation of the M_5Si_3 silicides is the unusual performance of the $(Ti-Cr)_5Si_3$ and $(Ti-Cr-Cb)_5Si_3$ compositions at both 1500 and 2400°F. These compositions have oxidation rates similar or lower than similar compositions with the disilicide metal-silicon ratio. None of the other M_5Si_3 silicides show performance comparable to their MSi_2 counterparts. Tolerance for columbium is particularly lacking in silicides that do not contain titanium and chromium. The significance of having comparable oxidation performances for both the M_5Si_3 and MSi_2 compositions is quite marked. Coating life (Sec. III) is not only a function of the oxidation rate of the initial reservoir, but of the rate of conversion of the reservoir to a less oxidation resistant composition. Using the data of Bartlett, [11] two mils of $CbSi_2$ would be converted to four mils of

TABLE II

OXIDATION OF SEVERAL M_5Si_3 SILICIDES

Number	Nominal Composition (at. %)	Oxidation at 1500°F			Oxidation at 2400°F		
		21 hours (mg/cm²)	101 hours (mg/cm²)	Comments	21 hours (mg/cm²)	101 hours (mg/cm²)	Comments
31	62.5Ti-37.5Si	+ 3.98	+ 3.80	Tan, good	+ 30.4	+271.8	Tight oxide, but rapid oxidation.
32	62.5Cr-37.5Si	+ 0.51	, 0.78	Very thin green-oxide, excellent	3.39	-16.0	Slow exfoliation, corners sharp.
33	62.5Cb-37.5Si	Failed 4 hours	---	Castastrophic oxidation	Failed 4 hours	---	Catastrophic oxidation
34	62.5V-37.5Si	+ 0.82	- 1.96	Glazed, gray to yellow, excellent.	- 8.45	- 9.82	Blue-black, less glazing than at 1500°F.
35	62.5Mo-37.5Si	- 6.24	-10.27	Poor, exfoliation	Failed 4 hours	---	Catastrophic oxide
36	31.3Ti-31.3Cr-37.4Si	+ 1.24	+ 1.99	Blue-green, excellent sharp corners.	+ 1.80	+ 1.66	Gray-brown, excellent
37	20.8Ti-20.8Cr-20.8Cb-37.6Si	+ 0.83	+ 1.65	Light brown, sharp corners, excellent.	+ 1.31	+ 5.90	Not quite as good as No. 36, but excellent.
38	31.3Ti-31.3Mo-37.4Si	-10.49	---	Specimen disintegrated from internal stress not oxidation.	+ 2.37	- 3.05	Slight glaze, sharp edges, good.
39	20.8Ti-20.8Mo-20.8Cb-37.6Si	- 5.75	+75.1 (61 hours)	Very rapid oxidation	+ 36.91 Failed		Rapid oxidation, poor
40	31.3V-31.3Cr-37.4Si	+ 2.01	+ 3.52	Blue-brown, thin oxide good.	- 1.40	- 5.20	Glazed, blue-gray good, sharp edges.
41	20.8V-20.8Cr-20.8Cb-37.6Si	+ 3.87	+20.48	Loose oxide.	+162.35 Failed		Rapid oxidation, poor.

Cb_5Si_3 at 2400°F in approximately 51 hours. Assuming a similar performance for disilicide coatings containing Ti-Mo-Cb, V-Cr-Cb, and Ti-Cr-Cb, only the Ti-Cr-Cb composition, forming an M_5Si_3 silicide with excellent oxidation resistance, could be expected to have reliable oxidation performance beyond the time necessary for conversion of all of the MSi_2 to M_5Si_3. The past success with the (Ti-Cr)-Si coating on columbium alloy may, in part, be due to the excellent oxidation resistance of the subsilicide layer and alloy layer formed beneath the disilicide.

2.2.2 Aluminides

Early efforts [12-15] showed that $CbAl_3$ was a promising protective coating material for columbium. At temperatures above 2000°F, this compound is moderately oxidation resistant, forming a scale that is predominantly Al_2O_3. At lower temperatures or during slow thermal cycling, the compound undergoes catastrophic attack. Alloying the aluminide with titanium, chromium, or silicon can mitigate low-temperature failure and extend high-temperature life. [13]

A recently completed investigation by Rausch[2] had as its goal the investigation of MAl_3 and MAl aluminides for use as reservoirs for coatings on columbium. The MAl aluminides were included primarily because they had exhibited a degree of toughness not associated with silicides or MAl_3 aluminides. Compositions tested were in the form of pieces from arc-melted buttons. Rausch found that: VAl_3 showed very large weight gains at 2200 to 2500°F; that $CbAl_3$ failed by "pest" at 1600°F in 90 hours, but was moderate in oxidation resistance at 2200 to 2500°F; and that $TiAl_3$ was excellent in oxidation resistance at 1600°F for at least 90 hours (weight gain 0.93 mg/cm^2) and fair at 2500°F (weight gain 16 mg/cm^2 in 40 hours). Additions of chromium or molybdenum to $CbAl_3$ improved the 1600°F performance, but did not significantly improve the 2500°F performance. All trialuminides investigated by Rausch were markedly less oxidation resistant than the more oxidation-resistant disilicides at either the 1500 to 1600°F or 2400 to 2500°F temperature ranges.

The gamma phase, TiAl, which borders on being ductile, was regarded as a candidate sublayer beneath the trialuminide, or as a primary reservoir if the oxidation resistance could be improved. Rausch[2] studied additions of columbium, tantalum, molybdenum, and chromium to this compound. Overall substitutions of columbium and molybdenum for part of the titanium provided the best overall performance. Representative data on these compounds tested at 1600 and 2400°F are given in Table III. No evidence of pest and only extremely thin oxide layers were formed in the 16-hour test.

TABLE III

OXIDATION DATA FOR GAMMA ALLOYS

Nominal Composition		Weight Gain (mg/cm^2)					
		1600°F			2400°F		
(At. %)	(wt %)	1 hour	4 hours	16 hours	1 hour	4 hours	16 hours
$(Ti_{0.8}, Cb_{0.1}, Mo_{0.1})Al$	Ti-11Cb-11.4Mo-32Al	0.34	0.50	3.7	0.80	1.2	1.9
$(Ti_{0.7}, Cb_{0.15}, Mo_{0.15})Al$	Ti-15.7Cb-16.2Mo-30.4Al	1.2	1.1	0.50	0.54	1.2	1.6
$(Ti_{0.7}, Cb_{0.2}, Mo_{0.1})Al$	Ti-21Cb-10.8Mo-30.4Al	0.28	1.0	0.11	0.71	1.0	1.9
$(Ti_{0.7}, Cb_{0.1}, Mo_{0.2})Al$	Ti-10.4Cb-20.6Mo-30.3Al	0.89	0.72	7.6	1.4	1.5	1.4
$(Ti_{0.8}, Cb_{0.15}, Mo_{0.05})Al$	Ti-16.1Cb-5.6Mo-31.2Al	0.07	0.40	2.4	3.3	9.3	14.5

The actual use of aluminides as coating systems has been limited primarily to applications in which life, and particularly cycle life, is quite limited. Several factors limit the effectiveness of such coatings:

- The extreme sensitivity of the oxide generated on the reservoir to thermal strain (par. 2.3)

- The rapid diffusion rate of aluminum into columbium and to a lesser extent into tantalum[3]

- The mismatch in thermal expansion

To overcome this problem of poor cyclic performance of the aluminide coatings, the group at Sylvania[16] undertook to support the aluminum oxide on a molten layer. Two coating systems, Sn-Al-X (X for a refractory metal, e. g. , molybdenum or tantalum, added primarily to prevent flow off of the coating during firing) and Ag-Al-Si were developed. The expected improvement in performance was attained with these two coatings; cyclic lives were comparable to or superior to silicide coatings and were truly ductile at room temperature. In concept the coating was composed of a MAl_3 phase on the substrate, a liquid phase containing tin or silver saturated with aluminum (or silicon) and the outer surface oxide floating on the alloy. The obvious difficulty with these types of coatings was their poor resistance to aerodynamic shear. Stetson and Moore[3] showed that rapid burn through resulted from exposure of the Sn-Al type coating to a high flow rate plasma at only 2500°F. Although the coatings were of interest academically, since they provided proof that aluminum oxide could be equivalent or even superior to silica as a primary oxygen barrier, the liquid phase coatings have essentially no use in the aerospace environment.

2.2.3 Specialty Reservoirs

One of the more interesting reservoir materials to be developed in the last several years is a cladding of the ductile hafnium alloy containing 20 to 30 percent tantalum. The oxidation rate of the unmodified alloy is approximately one mil/minute at 3500 to 3700°F. [17] No internal oxidation occurs in the substrate before the complete consumption of the cladding. The coating also is pressure insensitive because of the extreme thermodynamic stability and low volatility of HfO_2; thus, for systems that can tolerate the weight, the cladding provides a very reliable system for short time usage.

Hill, et al, [18] are investigating additions to the Hf-Ta system to improve the composition for use in the 1600 to 2500°F temperature range to the equivalency of silicides. Complex alloys, e. g. , Hf-Ta-Ir-Al-Si and Hf-Ta-Cr-B have exhibited 100 hours of protection in this range. These alloys may be useful both as cladding and as structural materials.

2.3 PRIMARY BARRIERS

Oxides constitute the only successful primary barrier for oxygen and other aggressive atmospheric components. Silica and silica glasses are the most effective primary barriers for use on the refractory metals at temperatures below 2500°F for extended periods of time. Beryllia, alumina, and hafnia are far inferior. With hafnia, diffusion of oxygen through the oxide is its limitation as with all oxides with the fluorite structure, and with all Group IV oxides which tend to be oxygen deficient at elevated temperatures. This is not the problem with BeO and Al_2O_3, which have dense stoichiometric lattices. These crystalline oxides have elastic moduli between 50 to 60 x 10^6 psi. Minor strains due to differential expansion between oxide, reservoir, or substrate; oxide growth stresses; or differential temperature produce cracks in the oxide that lead to rapid reservoir burn up.

Silica on the other hand has a low elastic moduli ($\sim$10 x 10^6 psi) which minimizes the effect of minor strain. Silica can be vitreous and readily vitrified at temperatures below 2500°F by minor additions, particularly vanadium or boron. These same additives, and others also, lower the softening temperature of the glass to approximately 1200°F, almost completely eliminating strain in the primary oxide as a causative factor in coating failure.

A low softening temperature primary barrier has other advantages. There is frequently a marked expansion mismatch between silicide or aluminide reservoirs and columbium, tantalum, molybdenum, or tungsten substrates. Craze cracks develop in the reservoir on cycling. During every thermal cycle oxide builds up in these craze cracks. Unless the oxide can be extruded on heatup, shear failure will eventually occur between the reservoir and the substrate. Aluminum oxide, which doesn't soften until above 3000°F, cannot be extruded from craze cracks, greatly limiting the cycle performance of aluminide reservoirs. Silicide coatings on columbium have withstood multiple thousands of cycles from room temperature to 2200 to 2500°F without reservoir spall. [6, 19]

There has been increasing interest recently in combining either silicide or other reservoirs and controlled composition silica glasses as primary barriers. The composition of the glass formed on a silicide barrier is a function of the reservoir composition, temperature, and pressure; the composition thus varies with these three parameters. Application of a controlled composition primary barrier lessens the dependence on the reservoir and can markedly improve the oxidation resistance of the reservoir. Wimber and Stetson[6], impregnated a tungsten silicide coating on T222 tantalum alloy with a barium borosilicate glass, increasing the maximum time to failure at 1600°F from 40 to 160 hours without decreasing the life at 2400°F.

2.4 THE EFFECT OF ENVIRONMENT ON COATING-ATMOSPHERE INTER-
     ACTION

The majority of data cited in this section was in reference to individual
phases within a coating and how the phases resist oxidation at intermediate temperatures at atmospheric pressure. It should be noted, however, that at temperatures above approximately 2500°F, silicides or silicide coatings are very sensitive to oxygen pressure. Active oxidation (loss of the protective SiO_2 layer and
evolution of SiO gas) can occur reducing the lives of coatings from hours to minutes. This problem has been extensively investigated and the active-passive
boundaries determined for WSi_2, $MoSi_2$, $CbSi_2$, $TaSi_2$, and complex modified
disilicides. [11,20,21] Regan, et al, [20] showed that additives such as vanadium,
titanium, and various borides markedly lengthened the life of columbium disilicide under low-pressure conditions, and that test results obtained on monolithic
silicides were effectively reproduced when similar compositions were tested as
coatings on columbium alloys.

Catastrophic failure at low temperature is another problem with silicides
(and other intermetallic compounds such as aluminides and beryllides). The
phenomenon known as "pest" is characterized by the formation of SiO_2 and a
metallic oxide, e.g., MoO_3 or WO_3, and complete disintegration of the silicide
with only minor weight gain. The problem, first reported by Fitzer[8] has been
more recently studied by Bartlett[10] and Berkowitz-Mattuck. [9] Berkowitz-
Mattuck has provided evidence that the catastrophic "pest" failures are stress
induced. Evidence with coatings would tend to confirm this mechanism, since
such additives as boron or vanadium to $MoSi_2$ or even a borosilicate overlay on
WSi_2 markedly reduces the problem. Boron, by forming a glass, would tend
to decrease the stress generated even at temperatures as low as 1000°F (the
pesting temperature for $MoSi_2$). Pesting in coatings is a relatively minor problem and has effectively been overcome or delayed in coatings for all refractory
metals except possibly WSi_2 on tungsten or tantalum. [6]

Furnace testing was one of the early techniques to evaluate both candidate
reservoirs and coating systems. In this test the specimens are injected into a
furnace stabilized at the test temperature, held at temperature for a set period
of time, and then air cooled. Both heating and cooling are quite rapid. Few
potential aerospace applications have this type of heating cycle and consequently
the results are not necessarily representative. Since no reentry-type vehicle
will be subjected to an environment in which it is heated or cooled extremely
rapidly, as in furnace testing, Moore and Stetson[3] evaluated the performance
of coatings on columbium alloys during slow cycling. The cycle used is shown in
Fig. 6 and had a T_{max} of 2500°F and a total duration of one hour. Only five minutes of the cycle was at T_{max}. A stress was applied to the specimen in the slow

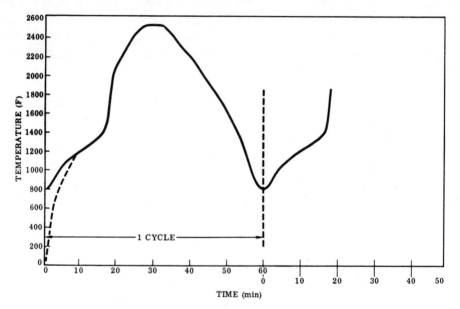

FIGURE 6.   TYPICAL TIME-TEMPERATURE PROFILE; Obtained with
the Environmental Simulator

cyclic test to effect approximately 2.5 percent creep in 22 cycles. Results for four coated alloys showed:

- Longer lives were obtained with the more oxidation-resistant colum-bium alloy, B66, and the poorest oxidation-resistant substrate (C129) gave the shortest life.

- Modifying the silicide gave no improvement over the simple silicided columbium alloys (Cb752, B66, D43, and C129).

- Modifying the substrate by the (Cr-Ti)-Si coating gave the best improve-ment in slow cycle performance (nearly 50 percent lasted longer than the test duration of 22 cycles against average lives of 8.9 and 8.7 hours for the simple silicide and modified silicide).

The severity of this test is emphasized when it is considered that only a total of 220 minutes in 22 cycles is at or above 2400°F; whereas, the life of the (Cr-Ti)-Si coating is 35 to 100 hours in cyclic furnace oxidation tests. However, even this coating gave inconsistent results showing that oxidation of the substrate at craze cracks must be carefully controlled to achieve reproducible results.

Although this test is severe and reproduces reentry conditions well, it may not be applicable to long life service conditions such as in jet engine components because there is insufficient exposure time at T_{max}.

III

COATING-SUBSTRATE INTERACTIONS

Coating-substrate interactions include both the thermodynamic and mechan-ical incompatibilities. True thermodynamic compatibility is probably not attain-able or desirable. For example, even in the simple molybdenum-silicon system, the usual rates of siliciding lead to an almost pure molybdenum substrate and molybdenum disilicide; whereas, the appropriate tieline in the binary system gives a molybdenum solid solution containing approximately one percent of sili-con in equilibrium with Mo_3Si. On the other hand, kinetic compatibility, mea-sured by adequate life of the thermodynamically incompatible $Mo-MoSi_2$ system, must be achieved. Mechanical compatibility is also essential and is attained by adequate strength of components and interfaces to resist failure, and adequate matching of properties to minimize stresses.

3.1 KINETIC COMPATIBILITY

Kinetic compatibility implies control of diffusional processes so that ade-quate service life is attained. Because of the complexity of modern protective

systems, full analysis of all the diffusional processes is not possible, and simplified approaches have been used. One approach has been to determine the rate
of consumption of the reservoir phase by the multiple processes of interaction
with atmosphere (evaporation, oxidation, and other reactions), interaction with
substrate, and internal reactions with products. A rather simplified approach
has been taken where one species diffuses much faster than the others. For
example, interstitial element diffusion has been studied on the simplified basis
that negligible substitutional diffusion occurs. However, even simplified theoretical treatments have not been attempted in most cases, and results have been
expressed by behavior in standard types of oxidation tests.

Three aspects of kinetic compatibility will be discussed:

- Coating life

- Interstitial sink effect

- Metallic claddings

3. 1. 1 Coating Life

Coating life is governed by the consumption of the reservoir by the twin
processes of diffusion into the substrate and transfer into the protective oxide.
The transfer process may be accelerated if active oxidation is occurring with
loss of silicon as SiO. Fig. 7 from Perkins[22] shows the processes in consumption of coatings, and appropriate rate equations were derived to describe the
life. However, these equations do not reflect the maximum coating life (burn-
out life) for several reasons:

- Failure is controlled by thinnest portion of coating.

- Failure may initiate at cracks, pinholes, or other defects before diffusional processes are completed.

Barriers to reduce the rate of silicon loss from the disilicide to the substrate would increase overall coating life. Data on the growth of disilicides[11]
shows that rates for most of the refractory metals differ by less than 50 percent
so that barriers do not appear to offer much promise.

Prealloyed sublayers such as the Cb(Ti-Cr) alloy interdiffuse with the
columbium substrate with adverse effects:

- Weakening by the interstitial sink effect or loss of unalloyed substrate

- Loss of oxidation-resistant sublayer

- Increase of undesirable columbium in the disilicide layer

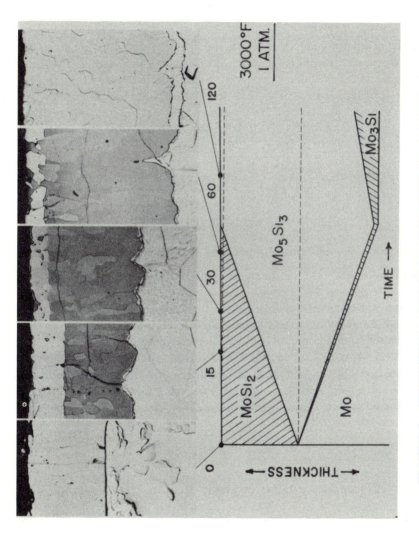

FIGURE 7. DIFFUSIONAL PROCESS GOVERNING COATING LIFE

Diffusion barriers of tungsten, molybdenum, and tantalum after 100 hours at 2400°F were evaluated[2] by comparing the weight of the barrier required/unit area with the weight of the affected columbium alloy substrate in the same time. Only tungsten and molybdenum were reasonably effective but the demonstrated advantage was not sufficient to warrant further work.

3. 1. 2 Interstitial Sink Effect

This term was introduced by Metcalfe[23] to describe the movement of interstitial elements to a component of a coating (or a braze alloy or diffusion bonding intermediate) in which the partial molal free energies of the interstitials are initially much lower than in the substrate alloy. Early observations on the effect were reported in 1965.[24] It was shown that loss of strength on coating an alloy such as D43 (Cb-10W-1Zr-0. 1C) resulted from segregation of carbon to the titanium-rich phase of the (Ti-Cr)-Si coating.

The loss of strength could exceed 50 percent. The problem was analyzed theoretically by considering three equilibria:

- Between zirconium-rich carbides and columbium alloy matrix
- Between columbium solid solution and titanium solid solution
- Between titanium solid solution and titanium-rich carbides

Only the second equilibrium was shown to be important because: (1) removal of small amounts of carbon from the columbium solid solution to the titanium makes the zirconium carbide particles unstable and they begin to go into solution; and (2) when TiC appears in the titanium solid solution, the partial molal free energy is given by that of the carbon saturated titanium solid solution. For the ideal solid solution of interstitial X in metal A at concentration C and saturation concentration S in equilibrium with the compound AX,

$$\Delta F_{X \text{ in } A} = \Delta F^{\circ}_{AX} + RT \, \ell n \frac{C}{S}$$

A typical plot of these functions for oxygen is presented in Fig. 8.

Further study of this problem has been made by Metcalfe, Klein, and Brentnall. [25-27] The increase in creep rate of the D43 alloy when exposed to a titanium sink has been related to the loss of carbides. Klein[27] has shown that where the creep process is controlled by the climb of edge dislocations over particles, the creep rate $\overset{\circ}{\varepsilon}$ is given by

$$\overset{\circ}{\varepsilon} \propto \frac{\text{interparticle spacing}}{\text{volume fraction of particles}} \qquad \propto \frac{1}{\text{carbon content}}$$

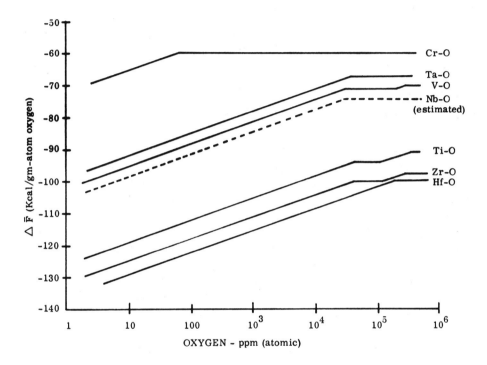

FIGURE 8. PARTIAL MOLAL FREE ENERGY OF FORMATION OF SOLUTION OF
OXYGEN IN VARIOUS TRANSITION METALS

until particles begin to dissolve completely and increase the particle spacing. Between the initial content of 1000 ppm and 100 ppm carbon (Fig. 9), the creep rate at 2200°F (min^{-1}) was given by

$$\overset{\circ}{\epsilon} = 1.7 \times 10^{-2} \ \frac{1}{\text{carbon content}}$$

Other findings from these studies are that good agreement is obtained between calculated and observed interstitial sink effects for Ti-Nb and Ti-V alloys in contact with the D43 alloy[26, 27], that prior thermal mechanical processing effects are obliterated by the powerful effect of titanium sinks[26], that coarse carbides in an alloy such as D43 reduce the rate of carbon removal but make the initial strength lower[26], and that hafnium carbide strengthened alloys (e. g., SU16) behave similarly. These findings are illustrated in Fig. 10. The time-temperature parameter schematically may be of the Dorn type when the activation energy will approach that of zirconium diffusion in columbium (i. e., solution of zirconium carbide).

The interaction between dispersion strengthened columbium alloys and interstitial sinks has been discussed at length because thermomechanical processing and aging are being applied to the refractory metals to achieve the strengths and stabilities required for future applications. However, the thermodynamic origin of the interstitial sink effect causes it to dominate all of the minor energy terms employed to achieve alloy stability. Two trends can be identified in coating development to combat these interstitial sink effects:

- Reduction of titanium content of coating compositions. Excellent examples are provided by the GTE R512A and R512E coatings with compositions of 25Cr-5Ti-75Si and 20Cr-20Fe-60Si, and by the Solar TNV coatings. [6]

- Coating by less reactive metals prior to application of (Cr-Ti) layer. The Solar V-(Cr-Ti)-Si coating is of this type. [2]

The interstitial sink effect is used in coating application technology in other ways. For example, pack chromizing from unalloyed chromium is not satisfactory, partly because of the limited solubility of columbium for chromium, but mainly because columbium is a sink for the oxygen in the atmosphere or in the chromium. However, an 80Cr-20Ti composition retains oxygen in the pack so that the trend has been to reduce, but not eliminate, the titanium content of such coatings.

An unusual example of the interstitial sink effect is provided by the oxidation resistant Hf-Ta alloys[17, 18] that are finding application as claddings for tantalum-base alloys. Fig. 8 shows that hafnium will act as a powerful sink for

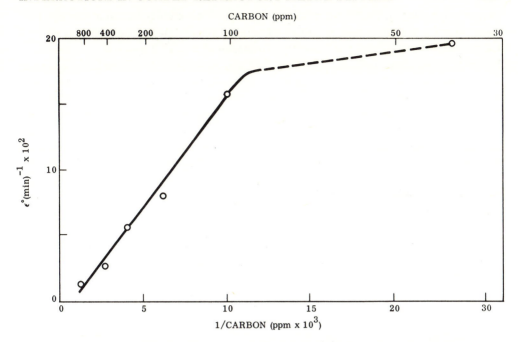

FIGURE 9.   CREEP STRAIN RATE AS A FUNCTION OF THE CARBON
CONCENTRATION IN D43 ALLOY[27]

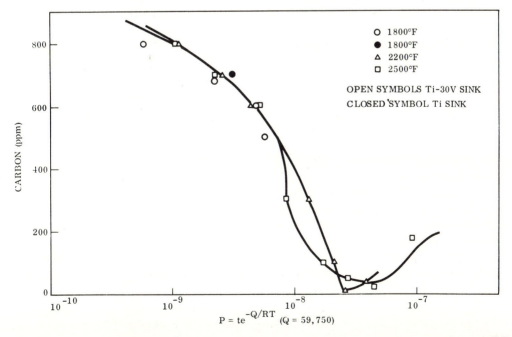

FIGURE 10.   PARAMETRIC PLOT FOR CARBON REMOVAL FROM D43 ALLOY[26]

oxygen when clad on tantalum alloys. Hafnium itself cannot be used because the oxide is not protective as a result of expansion mismatch and the monoclinic-tetragonal transformation at 3100°F. Additions of 20 to 30 percent tantalum is believed to stabilize orthohombic HfO_2[17] and slow the rate of attack by causing alpha-beta segregation in the matrix during oxidation giving a lamellar structure normal to the surface. Fig. 11 from Reference 17 shows this powerful effect. The length-to-thickness ratio of the lamellae is typically greater than 100:1, so that the faster diffusing oxygen travels along the length of the lamellae; whereas, tantalum and hafnium diffuse transversely to satisfy the thermodynamic requirements such as expressed in Fig. 8. Addition of a beta stabilizer such as molybdenum increases the beta in the reaction zone and slows oxygen penetration further. [17, 18] No oxygen penetration to the tantalum alloy substrate is found until the cladding is completely oxidized. The concept is valuable for temperatures up to 3800°F, but the coating thickness and weight are much higher than for most coating systems, and the oxidation resistance is not good at intermediate temperatures. Application to radiation cooled thrust chambers has given excellent results for typical short-time service conditions.

3. 1. 3 Metallic Claddings

The Hf-Ta claddings of tantalum alloys, described under interstitial sink effects because of the unusual nature of the protection mechanism, is a successful example of an all-ductile system. Such systems are inherently reliable. An extensive study of platinum alloy claddings on FS85 and TZM[28] has shown that the very low predicted oxidation rates are obtained at first, but give way to breakdown and failure after a short period. Oxidation rates indicate that 0. 002 inch of platinum should give 100 hours life at 3000°F, but typical results for clad molybdenum were 15 hours at 2550°F for 0. 003 inch and 50 hours at 2550°F for 0. 005 inch. Compound formation, transport of the base metal through the platinum to the surface, blistering, and other problems beset the development. No diffusion barriers could be found.

3. 2 MECHANICAL COMPATIBILITY

Mechanical compatibility requires that coating-substrate stresses be minimized by expansion match and that the strengths of individual components and interfaces be adequate to resist the stresses. Oxide coatings for tungsten were developed on this basis. [29] It was found that thoria had a good expansion match for tungsten. It was strengthened by a three-dimensional tungsten grid and allowed to deform plastically above 3800°F to reduce stresses. This system flew successfully on ASSET and was recovered intact from the Atlantic Ocean. Similarly, hafnia was used successfully by partial stabilization of the cubic hafnia by yttria. Contraction over a wide temperature range near 3100°F offsets the expansion difference.

Such detailed investigation of expansivities to design coatings with properties matched to the substrate has not been characteristic of most coating development. And yet studies of coatings in a late stage of development often show that
this was a critical factor in the success. Reference has already been made to
the stabilization of orthorhombic hafnia by tantalum[17] as a critical reason for
the success of the Hf-Ta system. Strengthening of the oxide by the lamella of
tantalum-rich alloy is another factor. Marnoch[17] reports that oxidized Hf-Ta
can be deformed without spalling of the oxide.

Three distinct approaches have been adopted to meet mechanical compatibility in coatings on columbium and tantalum:

- Coatings with a melt present at temperature so that stress buildup is
  limited. The Sylcor Sn-Al coating on tantalum was an early example
  of this type and consists of solid tantalum aluminide and a tin-rich
  melt. In 1966, Lux, et al[30] introduced impregnation of silicided
  columbium with Sn-Al-Si alloys to solve the expansion mismatch
  problem.

- Accept incompatibility but incorporate a repair mechanism to heal
  defects. The NRL coating based on Nb-Zn[31] provided such repair
  by zinc transport but was limited to approximately 2100°F. Another
  approach has been through moderately oxidation-resistant prealloyed
  substrates such as (Ti-Cr)-Si coatings.

- Coatings with matched expansions based on measurements of the expansion of the components of coatings.

The molten phase approach may be successful for specific applications,
but such coatings may fail at low pressures or under conditions of high aerodynamic shear. The prealloyed substrate approach represented a major step for-
ward when first introduced approximately eight years ago. The TRW coating,
(Ti-Cr)-Si, consists of a Cb-Ti-Cr alloy layer with complex silicides as the outer
protective layer.

The approach used successfully at Solar[29] to develop oxide coated tungsten
and tantalum nosecaps for reentry was also used at Solar to develop coatings for
columbium.[2] The most desirable coating constituents are selected based on
data for expansivity, interdiffusion, oxidation, emittance and any other property
desired. As coatings become more complex, such an approach will be used to
a greater extent in coating technology because coating techniques do not have to
be developed until the constituents are selected. As an example, Regan, Baginski,
and Krier[21] subsequently used this technique to select coating compositions with
improved low-pressure performance at 2800°F.

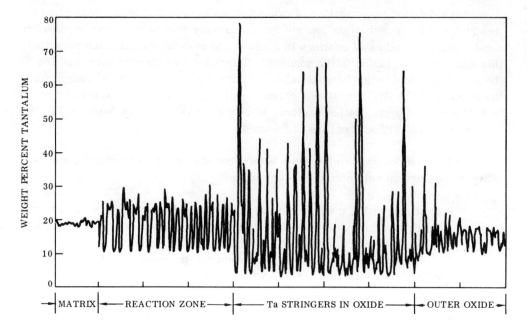

FIGURE 11. ELECTRON MICROPROBE STUDIES; Tantalum Traverse

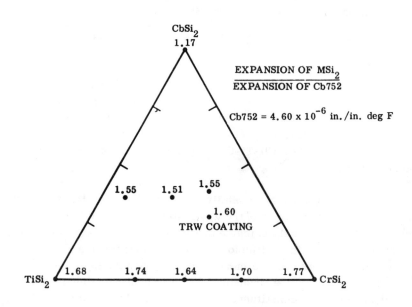

FIGURE 12. RATIO OF DISILICIDE THERMAL EXPANSION TO EXPANSION OF
TYPICAL COLUMBIUM-BASE ALLOYS

The most extensive study to date was made on the expansivity of disilicides, trisilicides, trialuminides, monoaluminides, prealloyed substrates, and columbium alloys at Solar in 1964-65. [2] The only other significant study was by Bartlett, [11] principally by X-ray diffraction, but these data do not show good agreement with other published data. The reasons are felt to be that powders were used in a vacuum below 2×10^{-2} mm Hg (so that it is certain that oxidation and even volatilization would occur) and that the planes chosen to calculate the A and C parameters of the uniaxial cells were not very sensitive. Fig. 12 shows the expansivities from the Solar study for the pseudo ternary system, $CbSi_2$-$TiSi_2$-$CrSi_2$, compared with the columbium alloy, Cb752. The point marked "TRW coating 1.60" is for the silicides removed from TRW-coated Cb752 specimens, and is plotted at a typical disilicide composition given by electron microprobe analysis. It is evident that agreement is reasonable between the arc-melted disilicides and the disilicides from the coating.

The significance of this expansion difference can be seen in typical (Ti-Cr)-Si coatings that show craze cracking at room temperature, although the underlying prealloyed Cb-Ti-Cr layer provides sufficient protection until the cracks are resealed by a silica-rich glass after heating to high temperatures. The expansion of molybdenum disilicide is 4.9×10^{-6}/degree F up to 1800°F or nearly seven percent higher than that of the Cb752 alloy. Fig. 13 shows the Cb752 alloy clad with molybdenum by diffusion bonding with a thin layer (0.0005 inch) of titanium and almost completely silicided. The crack-free specimen is unusual in silicide-coated metals. Examination of the free surface for cracks revealed:

Coating	Difference in Expansion of MSi_2 with Cb752 (%)	Cracks/Inch
$(Cb-Ti-Cr)Si_2$	60	545
$MoSi_2$	7	75

These figures are in agreement with predictions based on similar mechanical properties of the two silicides. The temperature at which the silicide cracks on cooling is determined by:

- The nonstress temperature such as temperature of siliciding or the temperature of service prior to cooling.

- The temperature of loss of plasticity, and hence the temperature where elastic stresses cease to be attenuated by deformation. On the other hand, elastic stresses can begin to build up at higher temperatures than this, depending on the cooling rate.

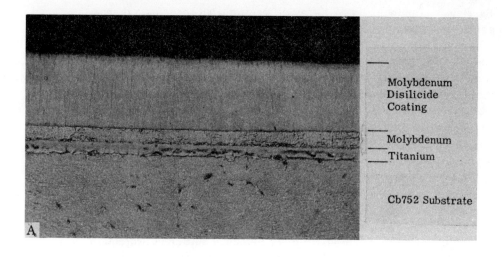

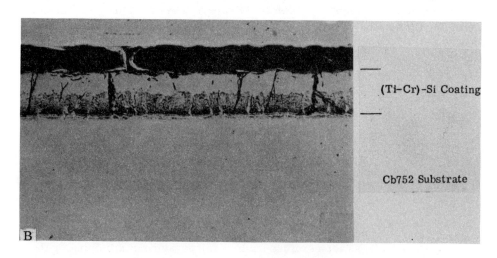

FIGURE 13.   COMPARISON OF CRAZE CRACKING IN MoSi$_2$ AND (Ti-Cr)-Si
              COATED Cb 752 ALLOY

- The temperature of cracking. This equals the temperature where the residual stress that increases on cooling equals the tensile strength of the silicide. If the temperature of loss of plasticity is 2200°F, the modulus and strength of the silicide are 30,000,000 and 30,000 psi, respectively; and the expansivity is 50 percent higher than that of the columbium substrate, the temperature of cracking is equal to 1765°F.

Because the cracks in the silicide (or aluminide) must open to an appreciable extent before serious oxidation can occur at the root, Stetson and Metcalfe[2] used 1500°F as a standard temperature to evaluate the oxidation resistance of sublayers and other compounds that might be exposed as a result of cracking of the primary protective barrier.

Table IV presents a comparison of the expansion of typical protective compounds for columbium. The reasons for restricted use of aluminides are clear from this Table. The columbium silicides have poor oxidation resistance although fair expansion match. The good oxidation resistance of $(Cb-Ti-Cr)Si_2$ in the TRW coating is obtained at the expense of expansion match. Increased use of $MoSi_2$ and WSi_2 with both expansion match and oxidation resistance is a logical development from this data and is exemplified by coatings such as the Solar TNV series.[6]

Coating service is such that even a perfect expansion match will not avoid cracking under thermal shock conditions. A recent development[6] has been the introduction of matched expansion coatings with a small amount of porosity to act as crack arrestors (Fig. 14). These coatings contain at least 50 percent of tungsten and/or molybdenum and are sintered to the columbium or tantalum surface. Siliciding reduces the porosity but leaves enough pores to arrest cracks as shown in the figure.

IV

EFFECT OF COATINGS ON MECHANICAL PROPERTIES

Ductile solid solutions are generally limited in the amount of reservoir element that they can contain. * Recognition of this limitation has led to coatings containing intermetallic compounds. Very successful coatings for superalloys

---

* The only exception appears to be Hf-Ta alloys where 80 percent of hafnium is present to provide a reservoir. However, this alloy is never used as a thin coating.

TABLE IV
COMPARISON OF EXPANSIVITIES

Compound	Expansivity/degree F $(x10^{-6})$	Expansion Relative to Cb752
$CbSi_2$	5.4	1.17
Cb_5Si_3	4.9	1.07
$TiSi_2$	7.7	1.68
$CrSi_2$	8.15	1.77
$(Cb, Ti, Cr)Si_2$	7.3	1.60
$CbAl_3$	7.1	1.54
$(Cb_{0.2}Ti_{0.8})Al$	7.5	1.63
$MoSi_2$	4.9	1.07
WSi_2	4.7	1.02

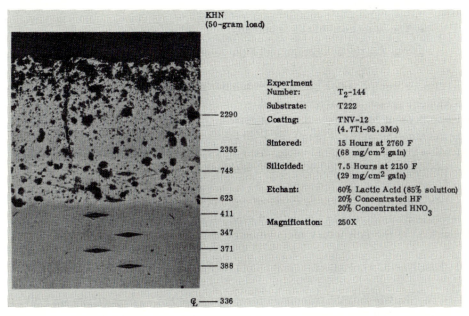

FIGURE 14.   CRACK ARRESTING PROPERTY OF POROUS SILICIDE COATING;

TNV-12

contain as much as 50 atomic percent aluminum at the surface and contain compounds such as Cr_3Al_2 and NiAl. [32] Reasonable life above 2400°F for thin coatings makes use of silicides or aluminides mandatory for the refractory metals. The mechanical properties of the refractory metal are influenced strongly by such coatings. This section will review the major effects.

Coatings based on intermetallic compounds usually contain fine hairline cracks as a result of differential contraction on cooldown after coating. Based on this model, the mechanical properties of refractory metals fall into three groups governed by:

- The plasticity of the coating

- The plasticity of the substrate

- The oxidation resistance of the system

Under compression and at high temperatures the cracks close up and the mechanical properties are governed by the plasticity of the coating. At low temperatures or at high rates of tensile strain, the ability of the substrate to flow may become the critical factor in determining the mechanical properties. The oxidation resistance of the system becomes the controlling factor when the cracks become wider than the hairline size referred to above.

In addition to the three types of control mentioned above, reference has already been made to the interstitial sink effect where a strength–conferring dispersion in the refractory metal may be made unstable by a coating.

4. 1 MECHANICAL PROPERTIES CONTROLLED BY PLASTICITY OF COATING

The ductile–brittle transition temperature (DBTT) of coatings was discussed in Paragraph 3. 2 in connection with the formation of hairline cracks on cooling. Based on an assumed loss of ductility at 2200°F (DBTT), it was shown that for a typical silicide coating, cracking would occur at approximately 1750°F. Unfortunately, few data are available on the deformation characteristics of coatings, yet such information is essential to understand certain mechanical properties.

Comparison of the behavior of 20Cr–20Fe–60Si coating on 0. 020–inch Cb752 alloy under tensile and compressive loads has been made by Metcalfe and Rose. [33] Fig. 15 shows that the buckling load reaches a maximum at 1650°F. This behavior is characteristic of a ductile–brittle transition (cf. mild steel at -150°F or tantalum beryllide at 2400°F) and arises because the stress concentrations at defects rise faster (because of low ductility) than the intrinsic strength. The dashed curve shows the expected strength increase in the absence of the ductile–brittle transition. It must be borne in mind that these coatings had formed fine hairline cracks in cooldown after coating so that the compression behavior also reflects

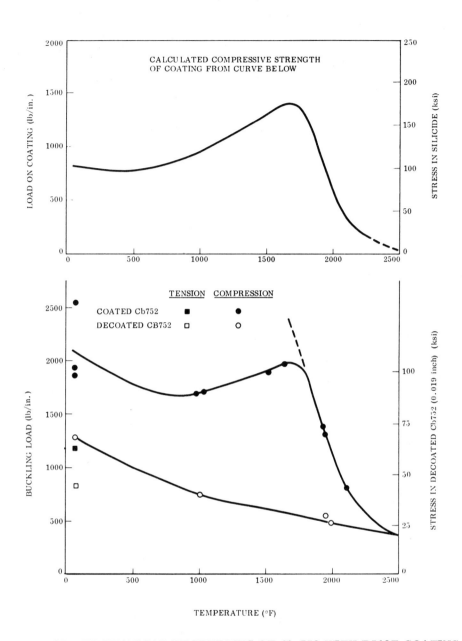

FIGURE 15. MECHANICAL PROPERTIES OF Cb 752 WITH R512E COATING

the closing of these cracks. In tension, on the other hand, the pre-existing cracks eliminate any strength contribution from the disilicide, so that the compressive strength of the decoated Cb752 alloy and the tensile strengths of either the coated or decoated Cb752 alloy are approximately equal. The compressive strength curve of decoated material is approximately that of the original Cb752 sheet. The strengths appear to come together at 2500°F where the silicide will have very little remaining strength.

The upper plot of Fig. 15 gives the load borne by the coating calculated from the difference of buckling stresses of coated and decoated Cb752. The stress in the silicide is approximate, based on an average thickness of 0. 008 inch of coating.

Plasticity of coatings is shown also by creep tests on coated refractory metals in air. Elongations up to 10 percent without oxidation attack have been measured and can only occur if the coatings flow plastically with the refractory metal. There appears to be a maximum creep rate to preserve coating continuity. [34] The mechanism of flow is not known, but the temperature is high enough for flow to be by diffusion (above half of the absolute melting point) in which case the mechanisms of creep flow and high-temperature coating repair may be similar.

4. 2 MECHANICAL PROPERTIES CONTROLLED BY PLASTICITY OF SUB-STRATE

Ability of the substrate to deform plastically at the root of the hairline cracks must control low-temperature mechanical properties. Tensile and fatigue strengths of refractory metals change in markedly different ways as a result of coating.

Room-temperature tensile tests show that yielding occurs at 70 to 90 percent of the yield strength of the substrate alloy for coatings such as (Ti-Cr)-Si or V-(Cr-Ti)-Si [2, 19] on 0. 012-inch columbium alloys such as Cb752, D43, B66, or D36. This reduction is explained largely by the reduction of thickness of the substrate in the coating process. On the other hand, the ductility is markedly reduced, probably because plastic deformation of the substrate is localized at the roots of the hairline cracks. The ultimate tensile strength is generally reduced more than the yield strength in accordance with the lower elongation.

The notches caused by the hairline cracks are readily attenuated by plastic deformation in the case of tensile tests, but much less readily under alternating stress conditions. The more ductile alloys with lower strength, such as D36, can deform more readily, so that the reduction in fatigue strength on coating is 40 percent at 10^6 cycles. Fig. 16 shows this trend. It can be seen that (Ti-Cr)-Si and V-(Cr-Ti)-Si coatings cause the same reduction in fatigue strength, supporting

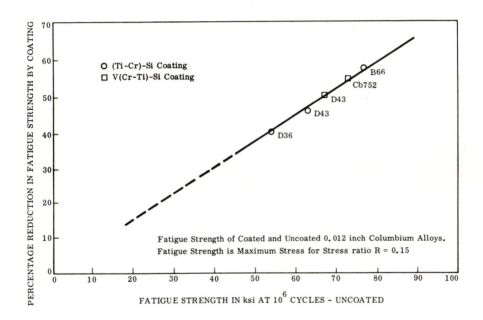

FIGURE 16. RELATION BETWEEN FATIGUE STRENGTH OF UNCOATED CO-
LUMBIUM ALLOYS AND REDUCTION OF FATIGUE LIFE BY COATING, ROOM
TEMPERATURE.

the view that the silicides are cracked and act as stress raisers. It is noteworthy that composition of the silicide has no influence on the fatigue life. Four coatings on 0. 006-inch B66 (Chromizing, G. T. & E. , Tapco, and Pfaudler) gave identical fatigue strength in an earlier study[3] to support this view.

V

FUTURE TRENDS

In the Introduction to this paper, the typical layers of a more advanced coating system were presented in Fig. 1. The interactions, interplay, and properties of the various layers of this relatively complex model coating were noted. Much is known about the layers and their interaction; however, the ideal coatings, based on existing knowledge, cannot be defined for all applications. Ductile coatings would be optimum, but none are available with the required diffusional stability and long life to make them of interest except for very short-time, high-temperature applications, e. g. , Hf-25 Ta for leading edges or rocket nozzles. For the application of greatest interest to the authors, i. e. , blading for gas turbine engines or long life re-usable heat shields for reentry vehicles, there appears to be a route, however complex, that must be taken to secure reliable performance for 1000 plus hours protection combined with moderate impact tolerance and retention of the majority of mechanical properties of the substrate. The three-layer structure of Fig. 1 appears to be essential, and the following properties will most likely be the ones for each of the layers.

Primary Barrier (outer oxide).

This layer will not be formed initially by consumption of the reservoir, but will be applied as a separate coating in a thickness of 0. 001 to 0. 002 inch. Very high softening point glasses, closely matching the expansion of the substrates, are the logical primary barriers. Until damaged, the vitreous coating will greatly minimize the consumption of the reservoir. Upon damage to the primary barrier, the reservoir will retain hundreds of hours of life if not also damaged.

Reservoir (secondary oxygen barrier).

Tungsten and molybdenum disilicides offer the best potential as reservoirs. Modification by amounts of titanium, vanadium, boron, or iron will be desirable to decrease the elastic moduli, to increase their ability to deform plastically, and to eliminate pest oxidation. For coatings on columbium and tantalum these silicides have the best expansion

match, the best diffusional stability, and are the most oxidation resistant of all candidate reservoirs. Only silicides of tungsten and molybdenum with modifiers have demonstrated 1000 plus hours of protection
in the temperature range from 1600 to 2400°F. [6] Reservoir thickness
will be in the range of 0. 004 to 0. 008 inch.

Ternary Barrier (ductile, moderately oxidation resistant under layer)

With the development of primary barriers with close expansion match
to the substrate, the primary role of the ternary barrier will be to provide survival capability in the event of catastrophic damage to the primary barrier and reservoir such as may occur upon solid object impact.
The ternary barrier is perhaps the most difficult to create by the material scientist because the barrier must have:

- Ductility

- Oxidation resistance at 2400°F for 10 to 20 hours in a thickness of
  0. 002 to 0. 004 inch

- Diffusional stability in contact with the substrate and reservoir for
  1000 plus hours at 2400°F

- Oxygen or nitrogen sink capability to prevent substrate embrittlement during 10 to 20 hours of exposure without the reservoir and
  primary barrier cover

Moderately oxidation resistant columbium alloys appear to be the near-
est to being adequate ternary barriers without being severely detrimental
to the stability of carbide dispersoids.

For the near term on an experimental basis, coating usage will be dictated
by simplicity of application techniques and moderate life. Slurry silicides and aluminides will remain the most popular. These coating systems do not have the life or
reliability potential for economical use in either military or commercial applications. To fully exploit the properties of the refractory metals, particularly columbium and tantalum, major efforts in the future will be required to develop primary,
secondary, and ternary barriers; to measure their properties and to balance a coating-
coating-substrate system.

ACKNOWLEDGEMENT

The authors wish to express their indebtedness to the many contributors at Solar and elsewhere who have advanced the state of knowledge in the field of coatings and phase interactions. Particular thanks are extended to the Air Force Material Laboratory and to Mr. N. M. Geyer for their support on several programs and, more recently, to the NASA Lewis Research Center and Messrs R. E. Oldrieve and S Grisaffe.

REFERENCES

1. Ohnysty, B. and Stetson, A. R. , Evaluation of Composite Materials for Gas Turbine Engines. Part I, June 1966 - Part II, December 1967. Contract AF33(615)-2574, AFML-TR-66-156.

2. Stetson, A. R. , and Metcalfe, A. G. , Development of Coatings for Columbium Base Alloys - Part I - Basic Property Measurements and Coating System Development. Part I September 1967. Contract AF33(615)-1598, AFML-TR-67-139.

3. Moore, V. S. and Stetson, A. R. , Evaluation of Coated Refractory Metal Foils. Part I September 1963 - Part II December 1964.

4. Geiselman, D. , Roche, T. K, and Graham, D. L. , Development of Oxidation Resistant, High Strength, Columbium-Base Alloys. AFML Contract AF33(615)-3856. (Progress reports only have been issued.)

5. Cornie, J. A. , Development of a Ductile Oxidation Resistant Columbium Alloy. AFML Contract AF33(615)-67-C-1689. (Progress reports only have been issued.)

6. Wimber, R. T. , and Stetson, A. R. , Development of Coatings for Tantalum Alloy Nozzle Vanes. July 1967. Final Report, NASA CR-54529, Contract NAS3-7276.

7. Jackson, R. E. , Tantalum Systems Evaluation, February 1967. Interim Progress Report No. 3. AF33(615)-3935.

8. Fitzer, E. , Plansee Proc. 2nd Seminar, Reutle/Tyrol, 1955, Pergamon Press, London 1956.

9. Berkowitz-Mattuck, J. , Blackburn, P. E. , and Felten, E. J. , Trans. AIME (Met. Soc.) (1965) 233, 1093.

10. Bartlett, R. W. , et al, Investigation of Mechanisms for Oxidation Protection and Failure of Intermetallic Coatings for Refractory Metals, Part I - Molybdenum Disilicide. June 1963. ASD-TDR-63-753.

11. Bartlett, R. W. , Investigation of Mechanisms for Oxidation Protection and Failure of Intermetallic Coatings for Refractory Metal. September 1965. ASD-TDR-63-753, Part III.

12. Carlson, R. G. , Oxidation Resistance of Aluminum Dip Coated Columbium Alloys. Columbium Metallurgy, Interscience Publishers, New York, 1960.

13. Rausch, J. J. , Low Temperature Disintegration of Intermetallic Compounds. August 31, 1961. ARF-2981-4.

14. Sama, L. and Ludkin, M. F. , Preparation and Properties of CbAl$_3$, General Telephone and Electronic Laboratories, Inc. , Bayside, New York.

15. Wukusich, C. S. , Oxidation Behavior of Intermetallic Compounds in the Cb-Ti-Al System. July 31, 1963. GEMP-218.

16. Sama, L. , High Temperature Oxidation-Resistant Coatings for Tantalum-Base Alloys, February 1963. ASD Technical Documentary Report 63-160, General Telephone and Electronics.

17. Marnoch, K. , High Temperature Oxidation Resistant Hafnium Tantalum Systems. July 1965. AFML-TR-65-240.

18. Hill, V. L. , Development of Oxidation-Resistant Hafnium Alloys. Interim reports on Contracts N00019-67-C-0403 and NOw66-0212-d.

19. Moore, V. S. , Stetson, A. R. , Development of Coating for Columbium Base Alloys - Part III, Evaluation of the V-(Cr-Ti)-Si Coating for the Turbine Environment, January 1968. AFML-TR-67-139, Part III.

20. Perkins, R. A. , and Packer, C. M. , Coatings for Refractory Metals in Aerospace Environments, September 1965. AFML-TR-65-351.

21. Regan, R. E. , Baginski, W. A. and Krier, C. A. , "Oxidation Studies of Complex Silicides for Protective Coatings", Ceramic Bulletin, Vol. 46, pp 502-508.

22. Perkins, R. A. , Coatings for Refractory Metals; Environmental and Reliability Problems. 1964 Golden Gate Metals Conference, San Francisco, Vol. II, Pages 579-590.

23. Metcalfe, A. G. , Unpublished Work, 1964.

24. Metcalfe, A. G. , and Dunning, J. S. , Interstitial Sink Effects in the Refractory Metals, October 1965. AIMME Symposium on Refractory Metals, French Lick, Indiana, (in publication).

25. Brentnall, W. D. , Klein, M. J. , and Metcalfe, A. G. , Structural and Mechanical Effects of Interstitial Sinks. October 1966. First Progress Report on Contract NAS 7-469.

26. Brentnall, W. D. , and Metcalfe, A. G. , Interstitial Sink Effects in Columbium Alloys. Final Report on Contract AF33(615)-5233, (in preparation).

27. Klein, M. J. , Structural and Mechanical Effects of Interstitial Sinks. October 1967. Fourth Quarterly Report on Contract NAS 7-469.

28. Girard, E. H. , Clarke, J. F. , and Breit, H. , Study of Ductile Coatings for the Oxidation Protection of Columbium and Molybdenum Alloys. May 1966. Final Report on Contract NOw 65-0340-F.

29. Stetson, A. R. , and Ohnysty, B. , Oxidation-Resistant Coatings for Use on Tungsten Above 4000°F. December 1963. AIMME Conference on Refractory Metals and Alloys III: Applied Aspects, Los Angeles, Part II, p. 765-778.

30. Lux, B. , Stecher, P. , and Perlhefter, N. , "Protection Contre L'Oxydation du Niobium au Moyen d'une Couche de NbSi$_2$ Impregnee de SnAl " July-August 1966 Annales des Mines, p. 1-20.

31. Brown, B. F. , et al, Protection of Refractory Metals for High Temperature Service. January 1961. NRL Report 5581.

32. Moore, V. S. , Mazzei, P. J. , Stetson, A. R. , Evaluation of Coatings for Cobalt and Nickel-Base Superalloys. September 1967   Task I report on Contract NAS 3-9401.

33. Metcalfe, A. G. , and Rose, F. K. , Production Tool for Diffusion Bonding. Final Report on Contract AF33(615)-2304 (in preparation).

34. Metcalfe, A. G. , Welty, J. W. , and Stetson, A. R. , The Application of Refractory Metal Foils to Reentry Structures. February 1964. ASM Golden Gate Metals Conference, San Francisco, California pp. 667-735.

INTERACTIONS OF REFRACTORY METALS WITH ACTIVE GASES
IN VACUA AND INERT GAS ENVIRONMENTS

H. Inouye

Abstract

The state of our knowledge of the interactions of refractory metals with active gases in vacuum and inert gas environments was reviewed. At temperatures from 600 to 2000°C and at pressures from 10^{-3} to 10^{-10} torr, contamination, degassing, decarburization, and sublimation reactions occur between the refractory metals and absorbed gases.

The extent and kinetics of these reactions are governed by the deviation of the interstitial concentration of the refractory alloy from the equilibrium concentration dictated by the temperature and pressure of the active gas. The reaction rates in inert gases are equivalent to vacua of approximately 10^{-5} to 10^{-7} torr. These interactions can be reduced by minimizing the surface-to-volume ratio of the metal and minimizing the degree of equilibrium disparity. Added control is gained by the use of barrier foil envelopes and the use of low-pressure CH_4 to neutralize the oxidizing gases.

H. Inouye is a Metallurgist with the Metals and Ceramics Division, Oak Ridge National Laboratory, Oak Ridge, Tennessee, 37830, Operated by Union Carbide Corporation for the U.S. Atomic Energy Commission.

Introduction

A decade of intensive research and development has shown that refractory alloys based on niobium, molybdenum, tantalum, and tungsten meet many of the requirements of high-temperature service. A remaining technological problem of considerable importance is their interaction with active gases even at low pressures. Use of the high vacuum equipment available has reduced the extent of the gas-metal reactions to levels approaching the limits of detection. In spite of this progress, the reactions still occur, for the interstitial content of refractory alloys changes when held at high temperatures.

Because of the limited amounts of active gases available in a vacuum furnace, the reaction rates are correspondingly low and are usually of little consequence for short exposure times; but these low reaction rates extended over a long time could cause interstitial changes that might significantly affect their properties. The purpose of this study was to determine the nature and the state of knowledge of the interaction of refractory metals with active gases in vacuum and inert gas environments so that methods might be adopted to minimize this universal problem.

A review of the published data shows that the interstitial concentration in refractory alloys changes as a result of contamination, degassing, and decarburization. Because many refractory alloys react with oxygen to form volatile oxides, sublimation is also encountered at the higher temperatures. At temperatures from 600 to 2000°C and at active gas pressures from 10^{-3} to 10^{-10} torr these reactions occur between refractory alloys and gases absorbed on their surfaces — that is, the reactions are characterized by the absence of a surface reaction product.

This paper illustrates by example the principles governing the extent and the kinetics of these processes in terms of the temperature, pressure, reaction time, alloying effects, and specimen geometry. In the first section, we deal with the nature of the environment and consider the contamination and degassing processes in terms of gas-metal equilibria. The second section is a consideration of the reaction kinetics that result from the departure of the interstitial concentration of the refractory metal from the equilibrium value. Decarburization is a special case of the contamination process and is discussed separately in the third section because of its importance to the use of carbide-strengthened alloys. Sublimation, the topic of the fourth section, is considered in this review because it becomes increasingly important in the compatibility of refractory alloys with oxides at the higher temperatures. Because of the lack of sufficient experimental data, the use of inert gases as a substitute for vacuum environments is considered only briefly in the fifth section. In the final section, we suggest methods for controlling the above reactions and describe techniques that reduce the extent of contamination.

Gas-Metal Equilibria

A knowledge of the composition and concentration of the active gases is necessary to understand the high-temperature reactions between refractory metals and their environments. The major impurities in inert gases and in the residual gases in vacuum systems are H_2O, O_2, N_2, CO, CO_2, H_2, and CH_4. These gases originate from system leaks, outgassing of the hot furnace surfaces, the pumping system, and the metal being heated. The pressure of each of these gases is greatly influenced by the history of the hot surfaces and the test conditions. Therefore no specific composition can be assigned to the environment for a given set of conditions.

During the normal heating cycle of the metal in a vacuum, the total pressure of these gases increases to a maximum value, then decreases to a relatively constant low value characteristic of the pumping system. If no attempt is made to regulate the pressure, the interstitial content of the refractory metal will change in the direction of the equilibrium concentration dictated by the temperature and the partial pressures of the various gaseous components.

The extent of the interstitial instability due to nonequilibrium conditions can be determined precisely from the pressure-temperature-composition relationship if the gas pressure is known at a given temperature. As an example, Fig. 1 shows the equilibrium conditions for solutions of nitrogen in niobium (1). A hypothetical alloy (containing 1 at. % N) is in equilibrium with nitrogen at 5×10^{-4} torr and 1760°C, but at a nitrogen pressure of approximately 1×10^{-5} torr the alloy would degas until the nitrogen content reached 0.16 at. %. At the same temperature but at a nitrogen pressure of 10^{-2} torr, the nitrogen content of niobium would increase eventually to 4 at. %. The nitrogen instabilities arising from temperature fluctuations at a given pressure can be deduced similarly. These data also show that nitrogen will tend to migrate from high to low temperatures in a temperature gradient.

The precipitation or the solution of interstitial phases is a form of interstitial instability that has an important effect on the mechanical properties of refractory alloys. Gas-metal equilibria such as that shown in Fig. 2 precisely define the conditions of stability of the Ta_2N precipitated phase (2). Note, however, that in this case the nitrogen pressures over the tantalum-nitrogen alloys in the two-phase regions are constant. The difficulty in controlling the quantity of Ta_2N phase at a given temperature can be illustrated by determining the equilibrium nitrogen pressures of the circled points at A and B at 1500°C. These points on the diagram are shown by the arrows in the upper left corner of Fig. 2. A hypothetical alloy containing Ta_2N would not dissolve nor increase in quantity so long as the nitrogen pressure was equal to the decomposition pressure of the nitride (estimated to be approximately 2.5×10^{-4} torr). If the nitrogen pressure were 2.0×10^{-4} torr as in point A, the nitrogen content would decrease to 7 at. % N, and the Ta_2N would go into solution; at a pressure of 3×10^{-4} torr, the nitrogen content would increase to 25 at. % N, and the alloy eventually would be converted to Ta_2N.

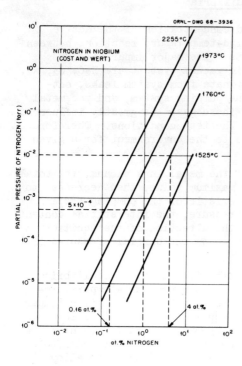

Fig. 1. Isothermal equilibrium for niobium–nitrogen terminal solid solutions.

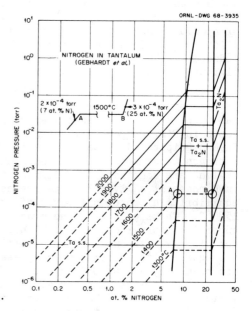

Fig. 2. Pressure composition relationships at equilibrium for the tantalum-nitrogen system between 1300 to 2000°C.

The affinity of refractory metals for gases and their solubility in them are altered markedly by alloying elements, as Table 1 (3) indicates. These data show that oxygen solubility in niobium is lowered from 0.5 wt % at 1000°C to about 0.2 wt % by magnesium and is in the range 0.01 to 0.02 wt % in the presence of calcium. On the other hand, the affinity of niobium for oxygen is increased markedly by these elements. These data further show that niobium containing 100 ppm O would be contaminated by most oxides with oxygen. The alloying components — zirconium, titanium, and hafnium — present in many of the current alloys are also expected to lower the oxygen solubility since their free energies of formation are also high.

Whereas the solubility limit of oxygen in niobium is lowered by zirconium (4), the nitrogen solubility limit is increased from 0.43 wt % in niobium (5) to approximately 0.75 wt % in Nb—0.86 wt % Zr at 1400°C (6). The solution characteristics of nitrogen in unalloyed niobium and in Nb—1% Zr are significantly different, as Fig. 3 (7,8) shows. These data show that 1 wt % Zr lowers the equilibrium nitrogen pressure over the alloy at given nitrogen concentrations.

It is estimated that zirconium increases the concentration of nitrogen in solution by a factor of about 4 (> 267/75) at a given temperature and pressure. The oxides and nitrides of molybdenum and tungsten have a lower free energy of formation than oxides and nitrides of niobium and tantalum. Therefore, the interactions of molybdenum and tungsten with oxygen and nitrogen are expected to be less extensive than those of the latter metals. Molybdenum and tungsten alloyed with active alloying elements such as titanium, zirconium, and/or hafnium, however, are expected to be as reactive toward the above gases as are the alloys of niobium and tantalum containing these elements.

Reaction Kinetics

As discussed in the previous section, the conditions of equilibrium between a gas and the interstitial concentration in a metal are defined precisely. Since this state is rarely attained under a normal heating cycle, it follows that the interstitial concentration will tend to change as required by thermodynamic equilibrium. The rate of this change and the parameters that control the extent are covered in this section.

The factors governing the reaction kinetics between low-pressure gases and metal surfaces that result in contamination can be deduced from a model based on the kinetic theory of gases. For example, the mass incident rate, Q_i, of a gas at a surface in a vacuum is given by $5.833 \times 10^{-2} \, P \sqrt{M/T}$. Let it be assumed that a gas molecule in a vacuum system incident on the metal surface is either absorbed or rebounds. The rate of the surface reaction is then given by

$$Q_a = \alpha Q_i = \alpha [5.833 \times 10^{-2} \, P (MT^{-1})^{\frac{1}{2}}] \, , \qquad (1)$$

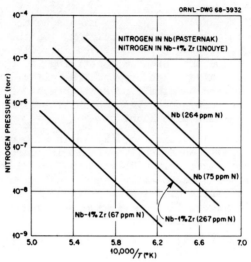

Fig. 3. Comparison of the solution equilibrium of nitrogen in niobium with nitrogen in Nb–1% Zr.

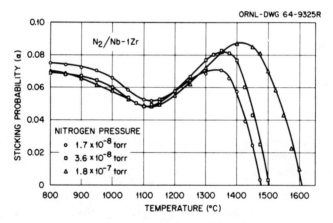

Fig. 4. Variation of the sticking probability versus
temperature for the reaction of nitrogen with
Nb–1% Zr for pressures between 10^{-8} to 10^{-7} torr.
Reaction time, 10 min; initial nitrogen concentration, 51 ppm.

Table 1. Changes of the Thermodynamic Potentials Associated
with the Formation of Nb-O Solutions at 1000°C[a]

Phase in Equilibrium with Nb-O Solution	Oxygen in Solution, wt %	$-\Delta F°$ (1273°K) (kcal/mole O_2)
CaO	0.01–0.02	242
MgO	~ 0.20	224
NbO	~ 0.50	168

[a]G. A. Meerson and T. Segorcheanu, Affinity of Niobium for
Oxygen, USAEC Translation AEC-tr-5910 (Aug. 22, 1963).

Table 2. Contamination of Tantalum Stress-Rupture Specimens
in Argon at 2600°C[a]

Time at Temperature (hr)	Specimen Thickness 0.010 in.		Specimen Thickness 0.020 in.	
	ΔO	ΔN	ΔO	ΔN
1.05	498	84		
3.45	399	78		
12.67	266	276		
1.10			131	87
7.58			178	154
13.60			187	259

[a]High Temperature Materials Program Report, GEMP-25A,
No. 25 (July 1963).

where

Q_a = absorption rate, g cm^{-2} sec^{-1},
Q_i = incident rate, g cm^{-2} sec^{-1},
P = gas pressure, torr,
M = molecular weight of the gas, g, and
T = absolute temperature, °K.

The parameter, α, is dimensionless and is the probability that the impinging gas will become absorbed; hence this parameter is called the sticking probability. Since the pressure in the vacuum system is P_1 when measured at an ambient temperature of T_1 (°K), its pressure at T_0 is greater by the temperature ratio $(T/T_0)^{\frac{1}{2}}$; that is, $P = P_0(T/T_0)^{\frac{1}{2}}$. Substituting this pressure correction due to thermal transpiration in Eq. (1) gives

$$Q_a = \alpha[5.833 \times 10^{-2} \, P_0 (MT_0^{-1})^{\frac{1}{2}}] \; . \tag{2}$$

Let it be further assumed that the absorbed gas diffuses into the interior of the metal of surface area A (cm^2) and weight W (g). The change in concentration, ΔC_m (g of gas/g of metal) of the interstitial in the alloy for a reaction time, Δt (sec) is then

$$\Delta C_m = Q_a \, \Delta t \, A/W \; . \tag{3}$$

Since A/W is equal to A/ρV and A/V is equal to 1/X, Eq. (3) can also be expressed by

$$\Delta C_m = Q_a \, \Delta t \, A/\rho V \; , \tag{4}$$

or

$$\Delta C_m = Q_a \, \Delta t/\rho X \; , \tag{5}$$

where

ρ = density of the metal, g/cm^3,
V = volume of metal exposed to the gas, cm^3, and
X = thickness of metal exposed to the gas, cm.

Equations (1) and (2) state that the contamination rate of a metal by a gas is proportional to the sticking probability and the test pressure. Further, the extent of the contamination [Eqs. (3–5)] is shown to be proportional to the reaction time and surface-to-volume ratio and inversely proportional to the thickness. The derived expressions do not include the effect of temperature, which is known to be an important variable, because the incident rate of the gas is independent of the temperature as Eq. (2) shows. Thus, although the sorption expressions do not explicity show the effect of temperature, it is evident that this variable must influence the value of α, which ranges between 0 and 1.

Equations (4) and (5) illustrate the relationship between the extent of contamination and the geometry of the metal, an important factor frequently overlooked by those concerned with contamination control. Experimental verification of the geometry effects predicted by these two expressions is contained in the data tabulated in Tables 2 and 3. Table 2 shows that the oxygen contamination of tantalum (9) was greater for the 0.010- than the 0.020-in.-thick specimens. Note that the results for nitrogen contamination after about 13 hr do not agree with the predictions. This we believe to be due to the attainment of equilibrium between the nitrogen in the argon and in the metal, in which

case the geometry effects stated in Eqs. (4) and (5) are invalid. The
data in Table 3 show more clearly the effects of the specimen thickness
of Nb–1% Zr on the extent of contamination (10). In this case the
geometry effects of the specimen on contamination are compared for the
same conditions of environment and reaction time. As in the previous
example, the extent of contamination is greater in the thinner specimens
than in the thicker specimens at temperatures from approximately 850 to
1200°C. The last column in Table 3 shows that $\Delta C_1/\Delta C_2 = X_2/X_1$, a relationship that can be obtained from Eq. (5) for specimens of different
thicknesses.

In light of the various derived expressions above, the kinetics of
the contamination process are seen to depend on the value of α, which is
the dimensionless parameter defined in Eqs. (1) and (2) as the sticking
probability and which is equal to the ratio of Q_a/Q_i. Typically the
values of α are low at high temperatures and do not vary uniformly with
temperature. As an example, the initial reaction rate between Nb–1% Zr
(initially containing 51 ppm N) and nitrogen expressed in terms of α
between 800 and 1600°C is shown in Fig. 4 (11). At 800°C the values of
α are about 0.07 and decrease slowly to minima of about 0.05 at 1125°C.
Above 1125°C, α increases to a maximum value of approximately 0.09,
depending on the pressure, then decreases rapidly to zero at the highest
temperatures. For the test conditions shown, the maximum value of α was
less than 0.10 — that is, less than 10% of the incident nitrogen was
sorbed by Nb–1% Zr. Another example is shown in Fig. 5. In this case a
maximum of about 4% of the incident oxygen reacts with tungsten (12) and,
like the nitrogen reaction with Nb–1% Zr, α shows both minima and maxima.

The intersections of the separate curves with the abscissa at
$\alpha = 0$ in Fig. 4 correspond to the equilibrium temperatures and pressures
for Nb–1% Zr containing 51 ppm N (see Fig. 6). Since the equilibrium
conditions change with the nitrogen content of the alloy, α would be
zero at other temperatures and pressures as the nitrogen content varied.
Variations in the initial nitrogen content also influence the values of
α for this alloy, as shown below:

Initial Nitrogen Content of Nb–1% Zr, ppm	α at 1200°C
10	0.089
11	0.071
51	0.057
67	0.052
90	0.038

Thus, Fig. 4 and these data show that α is dependent on the interstitial
content of the alloy, the temperature, and the pressure of the gas.
Therefore, the curves shown in Fig. 4 would shift to the left and down-
ward if the nitrogen content were higher and to the right and upward if
the nitrogen content were lower.

With the continued sorption of nitrogen over an extended period of
time, α continuously decreases, as would be expected from the data
tabulated above. The manner by which α varies with the concentration is
illustrated in Fig. 7. These data represent the kinetics of the reaction

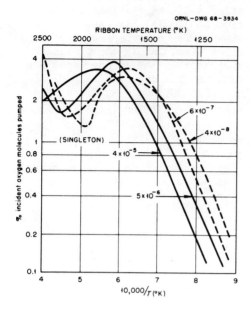

Fig. 5. Variation of the sticking probability versus temperature for the reaction of oxygen with tungsten. Initial P_{O_2} (torr) before heating tungsten is indicated for each curve.

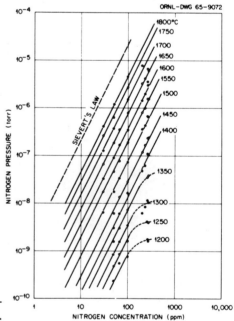

Fig. 6. Pressure-composition relationship at equilibrium for solutions of nitrogen in Nb−1% Zr between 1200 to 1800°C.

Table 3. The Effect of Specimen Thickness on
the Contamination of Nb–1% Zr[a]

Test Temperature (°C)	Specimen Thickness, in.		Contamination[b] ppm (wt)		Ratio	
	X_1	X_2	ΔC_1	ΔC_2	X_2/X_1	$\Delta C_1/\Delta C_2$
856	0.0201	0.0391	724	378	1.94	1.91
988	0.0200	0.0308	966	646	1.54	1.50
1081	0.0200	0.0384	1396	696	1.92	2.01
1135	0.0198	0.0393	1506	766	1.98	1.97
1201	0.0197	0.0382	2464	1354	1.94	1.72

[a]H. Inouye, Refractory Metals and Alloys III: Applied Aspects, Part 2, Vol. 30, p. 871, Gordon and Breach Science Publishers, New York, 1963.

[b]Sum of oxygen, carbon, and nitrogen contamination at an average pressure of 5.6×10^{-7} torr over 984 hr.

Table 4. Effect of Alloying Elements on the Oxygen
Contamination of Mo, Nb, and Ta at 1200°C[a]

Alloy Designation	Alloying Element(s) (wt %)	Alloy Base	Oxygen Contamination,[b] ppm	
			Alloy	Base
TZM	0.5 Ti + 0.08 Zr	Mo	340	—9
FS-80	1 Zr	Nb	320	100
T-111	2 Hf + 8 W	Ta	180	29

[a]H. Inouye, Refractory Metals and Alloys III: Applied Aspects, Part 2, Vol. 30, p. 871, Gordon and Breach Science Publishers, New York, 1963.

[b]1000 hr at 2.5×10^{-7} torr.

when the absorbed nitrogen remains in solution. The later stages of
sorption are represented by straight lines whose equation is

$$\alpha = \alpha_o (1 - C_m/C_e) \, , \tag{6}$$

where

α_o = sticking probability for a hypothetical nitrogen-free alloy,
C_m = mean nitrogen concentration in solution, ppm, and
C_e = equilibrium nitrogen concentration, ppm.

The ratio C_m/C_e corresponds to the fractional saturation of the alloy
with nitrogen, and the quantity $(1 - C_m/C_e)$ corresponds to the deviation
from equilibrium. Therefore the reaction rate of a gas with a metal
whose surface is free of a reaction product is seen to be proportional
to the "equilibrium disparity" — that is, the deviation of the instantaneous interstitial concentration from the interstitial concentration the
alloy would contain at equilibrium.

When the sorption process is continued so that the interstitial
concentration in the alloy exceeds the solubility limit, the kinetics
are altered as shown in Fig. 8. These data show that the value of α for
the reaction between niobium and nitrogen decreases continuously with C_m
as in Fig. 7; at the solubility limit, however, the smooth curve
exhibits a kink as α becomes constant (7) because the concentration of
nitrogen in solution, C_m, remains constant due to the precipitation of
Nb_2N. When this condition is attained, the quantity $(1 - C_m/C_e)$ in
Eq. (6) becomes constant and therefore accounts for the constancy of α
beyond the solubility limit.

Since it has been determined that the value of α is governed by the
"equilibrium disparity" of the interstitial in solution, the experimental
data which follow can now be explained. For example, Eq. (1) predicts
that the reaction rate of a gas with a metal will increase linearly with
the test pressure. This relationship is experimentally verified by the
data of Fig. 9 for the reaction of nitrogen with Nb–1% Zr in runs of
10 min (11). Similar results are obtained for the reaction of oxygen
with Nb–1% Zr, as Fig. 10 (13) shows. The reason for the linear relationship between the reaction rate and the test pressure in Fig. 9 is
that the change in C_m for a reaction time of 10 min under these conditions of temperature and pressure varies from 2.8×10^{-4} to 0.77 ppm for
the geometry of the sample tested and therefore has no significant
effect on the quantity $(1 - C_m/C_e)$ in Eq. (6). The linear relationship
between the reaction rate and pressure, seen in Fig. 10, is obeyed for
a different reason. In this case we believe that, due to the low oxygen
solubility in Nb–1% Zr, zirconium oxide is being precipitated within the
alloy at a rate equal to the rate of oxygen uptake. Since the matrix is
saturated with oxygen at a constant concentration, C_m, and since C_e is
fixed by the temperature and pressure, the factor $(1 - C_m/C_e)$, which
governs the reaction rate, is also constant. This result is analogous
to the example shown in Fig. 8.

The occurrence of the maxima in the values of α at the intermediate
temperatures of approximately 1300°C shown in Fig. 4 can also be
explained in terms of Eq. (6). Since the quantity $(1 - C_m/C_e)$ was
determined to be constant for reaction times of 10 min, α_o therefore

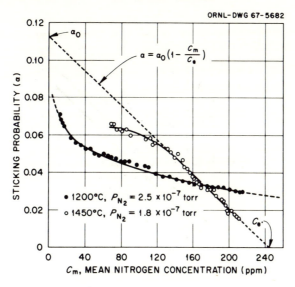

Fig. 7. Relationship between sticking probability and the mean nitrogen concentration of Nb-1% Zr at a nitrogen pressure of 2.5×10^{-7} torr and 1200°C and at a nitrogen pressure of 1.8×10^{-7} torr at 1450°C.

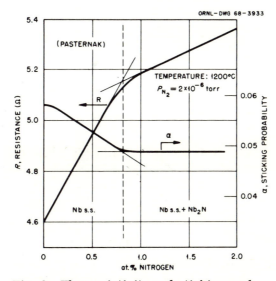

Fig. 8. The resistivity and sticking probability as a function of the nitrogen concentration for the reaction of niobium with nitrogen at 1200°C.

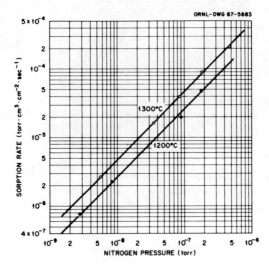

Fig. 9. Relationship between sorption rate of nitrogen by Nb-1% Zr and the nitrogen pressure at 1200 and 1300°C.

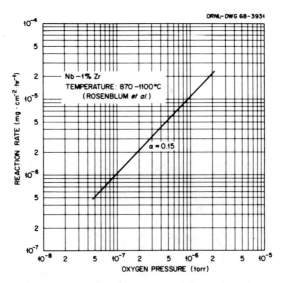

Fig. 10. Oxidation rate of Nb-1% Zr as a function of the oxygen pressure between 870 to 1100°C.

must increase with temperature. Above the maxima, α is influenced principally by C_e, whose value at a given pressure decreases with increasing temperature as required by the equilibrium relationships in Fig. 6. Thus, when C_m is not significantly less than C_e, small increases in temperature rapidly cause the ratio C_m/C_e to approach unity, and α therefore correspondingly decreases.

The reaction rates of a gas with a metal over an extended period of time can also be deduced from Eq. (6). At a given temperature and pressure, C_m will increase continuously with the reaction time until its value ultimately equals C_e. When the sorption process results in the solution of the interstitial, Q_a will decrease since α decreases; and, as may be seen from Eq. (3), ΔC_m would not be expected to increase linearly with the reaction time as experimentally verified by the data shown in Fig. 11. In contrast, when the pressure and temperature conditions favor the formation of an interstitial compound, C_m will remain constant at the solubility limit. Figure 12 shows the expected linear relationship between the weight gain of Nb—1% Zr in oxygen as a function of time (14) due to the low oxygen solubility in the alloy.

The expected effects of alloying elements on the reactivity of refractory metals with gases can be deduced from the above examples. If the alloying element lowers the solubility limit, it is more probable that the gas-metal interaction for a given temperature and pressure condition will result in the precipitation of a second phase. Figures 8 and 12 show that when these conditions prevail the reaction rates remain constant; and over a long reaction period, it would be expected that the alloy would be contaminated to a greater extent than the pure metal. Moreover, alloying elements that lower the solubility limit increase the thermodynamic potential of the interstitial in the alloy, as illustrated in Table 1, and therefore would likely increase the reaction rates. On this basis, zirconium, titanium, and hafnium, which are the principal constituents in many of our current alloys, would be expected to increase the reactivity of the base alloy toward oxygen. Table 4 compares the oxygen contamination of unalloyed molybdenum, niobium, and tantalum with the corresponding alloys containing the reactive elements titanium, zirconium, and hafnium. In each case, these data confirm the alloying effects postulated above.

It is obvious that there are test conditions that could also cause degassing of the interstitial. According to Eq. (6) degassing would occur when $C_m > C_e$, and negative values of α would be obtained. An example of the reversibility of the nitrogen reaction with niobium (7) is given by Fig. 13. In this experiment nitrogen was allowed to react with niobium at 2×10^{-6} torr at 1367°C almost to saturation. The reaction rate expressed in terms of α decreased as in segment (a). When the nitrogen pressure was increased to 4×10^{-6} torr, additional nitrogen reacted with niobium as in (b); then, when the nitrogen pressure was decreased to the initial pressure of 2×10^{-6} torr, the sample degassed as in segment (c).

Decarburization

All refractory alloys contain carbon as an impurity, and several contain up to 0.1 wt % as an intentional alloying component to enhance

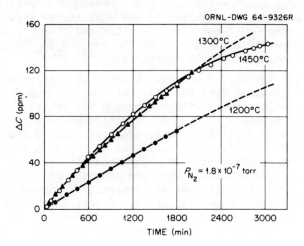

Fig. 11. The increase in nitrogen content of
Nb–1% Zr at a nitrogen pressure of 1.1×10^{-7}
torr and temperature of 1200, 1300, and 1450°C.

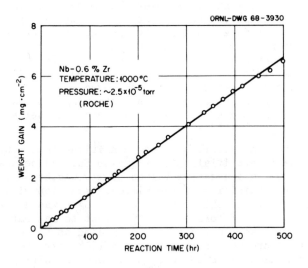

Fig. 12. Weight gain of Nb–0.6% Zr as a function of time at a pressure of 2.5×10^{-5} torr and
a temperature of 1000°C.

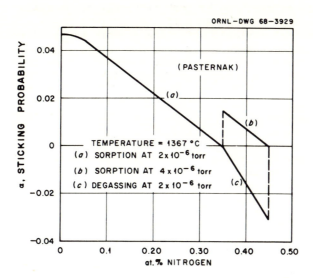

Fig. 13. Sticking probability as a function of nitrogen content for the reaction of nitrogen with niobium at 1367°C.

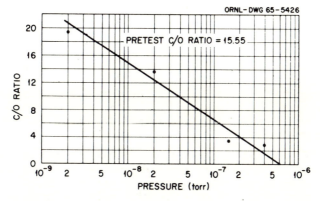

Fig. 14. The variation of the carbon–oxygen ratio of D-43 as a function of the test pressure at 1500°C.

their strength through the formation of carbides. Although the carbon concentration can be controlled within small variations in the fabricated state, this element can be most difficult to maintain at the desired level at high temperatures if the environment is not under control. A principal variable affecting carbon stability according to Table 5 is the test pressure (15). For the examples given, decarburization occurred under all test conditions and was most severe at the higher test pressures of approximately 10^{-7} torr. Note from the tabulated data that the large carbon losses at 10^{-7} torr are accompanied by a corresponding increase in the oxygen and nitrogen contents. Posttest chemical analyses showed that the carbon and the oxygen contents of the alloys expressed as a carbon-to-oxygen ratio was a function of pressure (Fig. 14) and temperature (Fig. 15). Note in Fig. 14 that D-43 with an initial carbon-to-oxygen ratio of 15.5 will decarburize at 1500°C when the pressure in the vacuum system exceeds 1×10^{-8} torr; Fig. 15 shows that the carbon-oxygen interaction in Nb–1% Zr, which increases with temperature, becomes a significant factor above 1200°C (15). These correlations suggest that decarburization is due to the formation of carbon monoxide that is removed from the furnace by the vacuum pumps.

A carbon-oxygen interaction has been established to be a mechanism for the decarburization of tungsten by Becker et al. (16) who studied the factors governing the reaction rate. Their results are summarized in Figs. 16 and 17. Figure 16 shows that the P_{CO} and P_{CO_2} resulting from the decarburization reactions decrease with the time of heat treatment even though the oxygen pressure increases. The decarburization rate under conditions where sufficient oxygen is available is governed by the diffusion rate of carbon in the alloy. Inasmuch as the diffusivity of carbon in tungsten increases with temperature (17), the rate of CO generation should also increase with temperature, as shown in Fig. 17. In this example, the generation rate of CO approaches a saturation value at the higher temperature, since the rate is now governed by the pressure of oxygen — that is, by the rate of arrival of oxygen to the surface.

Sublimation

The phenomenon of sublimation results from the interaction of refractory alloys with oxidizing gases to form volatile oxides and is therefore a competitive reaction with the decarburization and the contamination processes. The process is complex because the amount and species of volatile oxides that are formed depend on the temperature and the oxidation pressure.

Figures 18, 19, and 20 identify the volatile oxide species formed by the reactions of oxygen with tungsten, molybdenum, and tantalum, respectively, over a wide range of temperatures (18,19,20). In each example, the generation rate of a specific volatile oxide is seen to be temperature dependent and to pass characteristically through a maximum. The decrease in the reaction rates at temperatures above those at which the maximum occur is attributed to a decrease in the number of adsorbed oxygen atoms at the surface. The temperature at which the maximum rate of sublimation occurs in molybdenum and tungsten increases with the oxygen pressure in the manner shown in Fig. 21. For a given pressure,

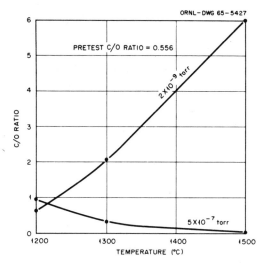

Fig. 15. The variation of the carbon-oxygen ratio of Nb-1% Zr as a function of the test temperature at 10^{-9} and 10^{-7} torr.

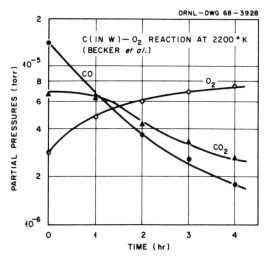

Fig. 16. Pressures of CO_2 and CO as a function of time due to the decarburization of tungsten with oxygen.

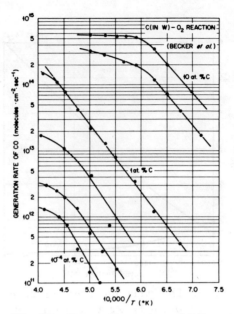

Fig. 17. Generation rates of CO as
a function of 1/T due to the decarburization of tungsten with carbon.

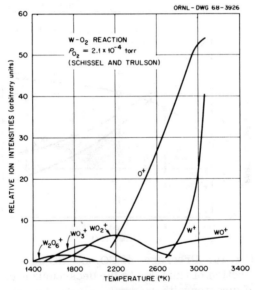

Fig. 18. Temperature dependence of ion
intensities: tungsten-oxygen reaction.

note that molybdenum attains its maximum sublimation rate at a lower temperature than tungsten.

The sublimation rates of tungsten and molybdenum are constant at a given temperature and pressure — that is, the extent of the reaction increases linearly with time. In contrast, the sublimation rate of tantalum varies markedly with time (20). Typical examples of the sublimation behavior are given in Fig. 22. In Fig. 22(a) the generation rate of TaO(g) increases continuously with the reaction time to a saturation value. This behavior results from the competition between the solution of oxygen in tantalum and the formation of a volatile oxide. When the oxygen pressure is increased as in Fig. 22(b), additional oxygen is taken into solution; and when saturation is reached, the sublimation rate again becomes constant. When the temperature is changed as in Fig. 22(c) and (d), a momentary change in the TaO intensities results from changes in the oxygen concentration on the surface and in solution. The sublimation phenomenon is of minor consequence in terms of material losses in the normal course of heat treatment under ultrahigh vacuum conditions. But in a practical application in which the metal is exposed to an oxygen source such as an oxide fuel or other ceramic oxides, the saturation of the metal with oxygen and the sublimation process can, over an extended reaction time, be seen to be a potentially serious problem at very high temperatures.

Interaction of Refractory Metals in Inert Gases

Inert gas environments are being considered as the working media in advanced high-temperature power plants that use refractory metals. Inert gas environments are also a logical substitute for vacuum environments. Unfortunately, the lack of sufficient data precludes a detailed characterization of the interactions similar to that presented in the previous sections. From some available data, it appears that the parameters that govern the interactions of refractory metals with the impurities in vacua also apply in the case of the inert gases (see Table 2). Some differences do exist, however, and are discussed qualitatively below.

A major disadvantage of the use of an inert gas is the great difficulty of measuring precisely the concentrations of the active impurity gases. For example, if the decarburization of D-43 is to be prevented during annealing at 1500°C in 1 atm pressure of inert gas, the permissible pressure of active gases according to Fig. 14 is 1×10^{-8} torr. This concentration in the inert gas is approximately 10^{-11} atm or 10^{-5} ppm. It is readily apparent that impurity concentration in this range cannot be detected easily.

Excluding the technological difficulty of impurity gas analysis, there is an advantage in the use of inert gases since according to theory their presence will lower the reaction rate between the refractory alloys and the active gas impurities because the impurity gases collide with and must diffuse through the inert gases to reach the alloy surface — that is, the incident rate for a given pressure of an active gas at the alloy surface is lower in an inert gas than in a vacuum.

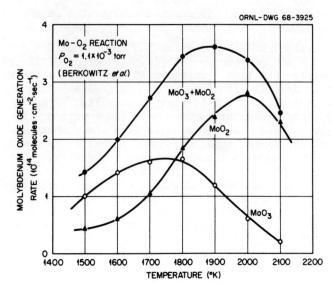

Fig. 19. Generation rates of volatile molybdenum oxides as a function of temperature during sublimation of molybdenum at 10^{-3} torr.

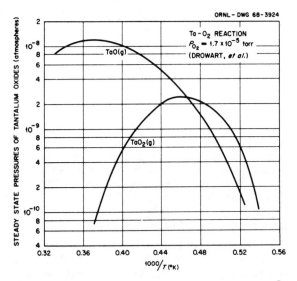

Fig. 20. Sublimation of tantalum at 1.7×10^{-5} torr: pressures of volatile tantalum oxides as a function of temperature.

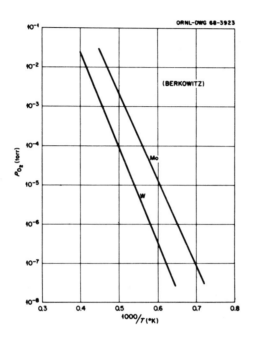

Fig. 21. Temperature-pressure relationship for maximum sublimation
rates of tungsten and molybdenum.

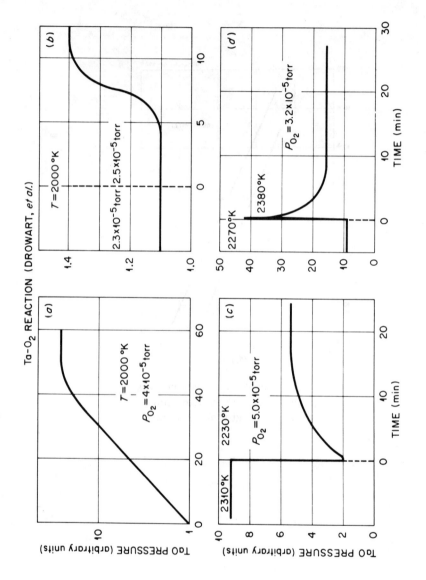

Fig. 22. Variation of the pressure of TaO(g) with time: (a) on first exposure of tanta-lum to oxygen; b) on increasing oxygen pressure after attainment of a steady state; (c) on lowering the temperature after attainment of a steady state; and (d) on increasing the temperature after attainment of a steady state.

Table 5. Effect of Pressure and Temperature on the
Stability of Interstitials in Refractory Alloys[a]

Alloy Designation	Temperature (°C)	Pressure (torr)	Change in Interstitial Concentration, ppm[b]			
			Carbon	Oxygen	Nitrogen	Hydrogen
D-43	1500	2×10^{-9}	−50	−16	−56	−3
	1500	2×10^{-8}	−50	−13	−66	−3
	1500	1.5×10^{-7}	−400	+101	+351	−3
	1300	1×10^{-9}	−20	−12	−63	−4
	1300	5×10^{-7}	−380	+347	+244	−4
FS-80	1500	3×10^{-9}	−70	−175	−53	−6
	1500	6×10^{-7}	−97	0	+92	−6
	1300	5×10^{-9}	−10	−137	−32	−7
	1300	5×10^{-7}	−48	−40	+6	−4
T-111	1500	2×10^{-9}	−10	−56	−8	−3
	1500	5×10^{-7}	−30	+40	+171	−3

[a]D. T. Bourgette, Trans. Vacuum Met. Conf. 8th, New York, 1965, 57–73 (1966).

[b]500- to 750-hr tests.

Table 6. Comparison of the Contamination Rates of
Niobium and Tantalum in Helium, Argon, and Vacua

Test Temperature (°C)	Test Material	Exposure (hr)	Environment	Reaction Rate (mg cm^{-2} sec^{-1})
982	Nb	2538	Static argon	8×10^{-5}[a]
1010	Nb	56	Static argon	1×10^{-2}[a]
1150	Nb and Ta	500	300 psi He, high velocity	4×10^{-3}[b]
1000	Nb	5	$P_{O_2} = 1 \times 10^{-5}$ torr	1×10^{-2}[c]
1000	Nb and Ta	1000	Vacuum, $P = 2 \times 10^{-7}$ torr	5×10^{-5}[d]

[a]H. E. McCoy and D. A. Douglas, Columbium Metallurgy, Vol. 10, p. 85, Interscience Publishers, New York, 1961.

[b]L. A. Charlot et al., Corrosion of Superalloys and Refractory Metals in High Temperature Flowing Helium, BNWL-SA-1137 (March 1967).

[c]H. Inouye, Columbium Metallurgy, Vol. 10, p. 649, Interscience Publishers, New York, 1961.

[d]H. Inouye, Refractory Metals and Alloys III: Applied Aspects, Part 2, Vol. 20, p. 871, Gordon and Breach Science Publishers, New York, 1963.

Table 6 compares the reaction rates of niobium and tantalum in helium, in argon, and in vacua. Due to the difficulty of analysis, the impurity contents of the inert gases for the examples given are unknown at the test temperatures. By comparing the reaction rates in the various environments, we can see that these inert gas environments correspond to vacua in the range between 10^{-5} to 10^{-7} torr. Since 1×10^{-3} torr is about 1 ppm, the effective impurity gas pressures in the argon tests are calculated to be between 10^{-2} to 10^{-4} ppm. Because inert gases are not usually this pure, we concluded that the incident rates of the active gases are lowered by the inert gas.

Controlling Gas-Metal Reactions

A gaseous environment that is in equilibrium with the interstitials in the refractory alloy at the heat-treating conditions is probably an unattainable goal. Since many of the parameters that control the reaction kinetics are now known, however, some practical methods can be employed to suppress many of the undesirable reactions.

First is the geometry variable and its influence on the extent of the reaction. Experimental data conclusively confirmed that the extent of contamination was proportional to the surface-to-volume ratio and inversely proportional to the thickness. Thus, if it is known that non-equilibrium conditions prevail, the surface-to-volume ratio of the refractory alloy should be held to a minimum and its thickness to a maximum.

Secondly, the heat-treating conditions should be chosen, if possible, to minimize the degree of "equilibrium disparity." To do this it would be necessary to have gas-metal equilibrium data and the interstitial analysis of the alloy. Data for most alloys do not exist; however, it is pointed out that maxima in the reaction rates occur at intermediate temperatures. Therefore, these temperatures are not desirable.

Further reductions in the interactions of active gases with refractory alloys are possible through the use of protective foil wraps on refractory metals. Metals that are the most reactive toward the active gases would appear to be the suitable choice. As an example, Table 4 suggests that the preferred foil materials should contain the more reactive elements, such as titanium, zirconium, and hafnium, so that they will react preferentially with the noxious gases.

Figure 23 compares the effects of molybdenum, tantalum, and Nb-1% Zr envelopes on the extent of contamination of Nb-1% Zr at 2.7×10^{-7} torr for 1000 hr (10). These data show that foil wraps markedly reduce the extent of contamination; note, however, that their efficiency with respect to minimizing the extent of contamination was in the order molybdenum, tantalum, then Nb-1% Zr, which is the reverse of the order of their reactivity according to Table 6. Further evaluation of this technique showed that the effectiveness of protective envelopes was independent of the foil composition but dependent on the extent of the leaks in the seams of the envelope. Therefore, the function of a foil wrap is not to getter the active gases but to serve as a physical barrier to the incident noxious gases. This concept of protective foils

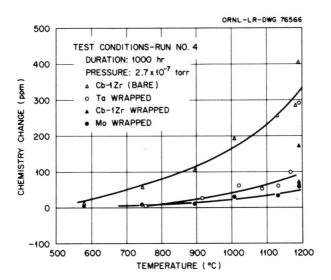

Fig. 23. Influence of foil wraps on reducing the contamination of Nb−1% Zr at 2.7×10^{-7} torr.

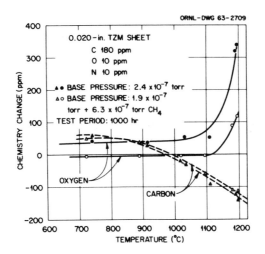

Fig. 24. Effect of CH_4 on the oxygen contaimination of TZM.

necessarily requires that the sticking probability must be low as was shown to be the case in a prior section.

The dominant presence of carbon monoxide in the heat-treating environment, the prevalence of oxygen contamination, and the observed phenomena of decarburization above 1200°C and 1×10^{-8} torr (see Figs. 14 and 15) indicate that the oxidizing gases are the most aggressive active gases with respect to the refractory alloys. Perkins and Lytton (24) were successful in preventing decarburization or restoring the carbon content of molybdenum alloys during annealing by introducing low-pressure CH_4 into the vacuum environment. On the other hand, CH_4 has been used to reduce the extent of oxygen contamination of TZM (10) as shown in Fig. 24. These data show that CH_4 at 6×10^{-7} torr was capable of lowering the extent of the oxygen contamination at a base pressure of 2×10^{-7} torr by a factor of approximately 3 at 1200°C but was not effective in preventing decarburization of the alloy.

Summary and Conclusions

A review of the published data shows that the interactions of refractory alloys based on niobium, tantalum, molybdenum, and tungsten with the active gases in vacua and inert gas environments can be grouped into processes that result in contamination, degassing, decarburization, and sublimation of the refractory alloy as volatile oxides. From 600 to 2000°C these reactions occur in the pressure range between 10^{-3} and 10^{-10} torr without the formation of a surface reaction product. The extent and kinetics of these reactions are governed precisely by the degree of "equilibrium disparity"; that is the deviation of the interstitial concentration of the alloy from the concentration the alloy would have when equilibrium is obtained at a given temperature and pressure.

For the contamination process, the reaction rate is given by $Q_a = \alpha Q_i$, where α is the sticking probability and Q_i is the incident rate of the gas at the alloy surface. In general, less than 10% of the incident gas reacts with refractory alloys at high temperatures — that is, α has a maximum value of approximately 0.10. Since the reaction proceeds toward equilibrium, α initially is a maximum for a given temperature and pressure but decreases as equilibrium is approached in accordance with the empirical relationship $\alpha = \alpha_0 \left(1 - C_m/C_e\right)$, where α_0 is the sticking probability for an interstitial free alloy and C_m and C_e are the mean and equilibrium interstitial content, respectively. The effects of pressure, temperature, reaction time, and alloying components on the reaction kinetics are explained in terms of this relationship.

A decarburization process results from an oxidation reaction that produces carbon monoxide as the reaction product and is encountered above 1200°C and approximately 10^{-8} torr. The decarburization rate is proportional to the diffusion rate of carbon in the alloy at the low temperatures and in the presence of sufficient amounts of oxygen but at high temperatures is dependent on the oxygen pressure.

Sublimation of refractory alloys as volatile oxides results from their interaction with oxidizing gases at the higher temperatures and is a competitive reaction with the decarburization and contamination processes. The reaction rates increase with pressure but characteristically go through maxima for a given pressure as the temperature is increased. For each metal-oxygen reaction, several volatile oxide species are formed.

The contamination rates of refractory alloys by the active gases in helium and argon are equivalent to the rates observed in vacua in the range 10^{-5} to 10^{-7} torr. It appears from the experimental data available that inert gases significantly lower the incident rate of the active gas at the alloy surface, as predicted by theory.

The principles governing the reaction kinetics suggest that the extent of the several reactions can be reduced by minimizing the surface-to-volume ratio and selecting heat-treating conditions to minimize the degree of "equilibrium disparity." Other techniques that have been successful are the use of protective foil envelopes and the use of low-pressure CH_4 to neutralize the oxidizing gases.

REFERENCES

1. J. R. Cost and C. A. Wert, Acta Met. 2, 231–242 (April 1963).

2. E. Gebhardt et al., Z. Metallk. 52, 464–473 (1961).

3. G. A. Meerson and T. Segorcheanu, Affinity of Niobium for Oxygen, USAEC Translation AEC-tr-5910 (Aug. 22, 1963).

4. D. O. Hobson, High Temperature Materials, Vol. II, Interscience Publishers, New York, 1963, p. 325.

5. W. M. Albrecht and W. D. Godde, Jr., Reaction of Nitrogen with Niobium, BMI-1390 (July 1959).

6. E. de Lamotte et al., Equilibrium Solutions of Nitrogen in Columbium Base Alloys, Air Force Materials Laboratory, Report No. ML-TDR-64-134, Technical Documentary Report (June 1964).

7. R. A. Pasternak et al., J. Electrochem. Soc. 113(7), 731–735 (1966).

8. H. Inouye, Equilibrium Solid Solutions of Nitrogen in Nb–1% Zr Between 1200 to 1800°C, ORNL-TM-1355 (February 1966).

9. High Temperature Materials Program Report, GEMP-25A, No. 25 (July 1963).

10. H. Inouye, Refractory Metals and Alloys III: Applied Aspects, Part 2, Vol. 30, p. 871, Gordon and Breach Science Publishers, New York, 1963.

11. H. Inouye, High-Temperature Sorption of Nitrogen by Nb–1% Zr in Ultrahigh Vacuum, submitted for publication in Transactions of the American Society for Metals.

12. J. H. Singleton, J. Chem. Phys. 45(8), 2819–2826 (1966).

13. L. Rosenblum et al., Oxidation of Columbium, Tantalum, and Their Alloys, NASA-SP-131, pp. 170–187 (Aug. 23–24, 1966).

14. T. K. Roche, Refractory Metals and Alloys III: Applied Aspects, Part 2, Vol. 30, p. 901, Gordon and Breach Science Publishers, New York, 1963.

15. D. T. Bourgette, Trans. Vacuum Met. Conf., 8th, New York, 1965, 57–73 (1966).

16. J. A. Becker et al., J. Appl. Phys. 32(3), 411–423 (March 1961).

17. I. I. Kovenskii, Diffusion in Body-Centered Cubic Metals, American Society for Metals, Metals Park, Ohio, 1964, p. 283.

18. P. O. Schissel and O. C. Trulson, J. Chem. Phys. 43(2), 737–743 (July 1965).

19. J. B. Berkowitz-Mattuck et al., J. Chem. Phys. 39(10), 2722–2730 (November 1963).

20. J. Drowart et al., Mass Spectrometric Study of the Oxidation of Tungsten and Tantalum, Brussels University (Belgium), AD-622809 (August 1965).

21. H. E. McCoy and D. A. Douglas, Columbium Metallurgy, Vol. 10, p. 85, Interscience Publishers, New York, 1961.

22. L. A. Charlot et al., Corrosion of Superalloys and Refractory Metals in High Temperature Flowing Helium, BNWL-SA-1137 (March 1967).

23. H. Inouye, Columbium Metallurgy, Vol. 10, p. 649, Interscience Publishers, New York, 1961.

24. R. A. Perkins and J. L. Lytton, Effect of Processing Variables on the Structure and Properties of Refractory Metals, Part I, AFML-TR-65-234 (July 1965).

HYDROGEN EFFECTS IN REFRACTORY METALS

W. T. Chandler and R. J. Walter

Abstract

The effects of hydrogen on the refractory metals molybdenum, tungsten, columbium, and tantalum are reviewed. Solubility, permeability, and diffusion of hydrogen in the refractory metals, refractory metal-hydrogen phase diagrams, and the effects of absorbed hydrogen and hydrogen environments on mechanical properties are covered. Molybdenum and tungsten have very low solubilities for hydrogen and are essentially unaffected by hydrogen. Columbium and tantalum can absorb large quantities of hydrogen and form hydrides, and are greatly embrittled by relatively small amounts of hydrogen at low temperatures. However, the solubility of hydrogen in columbium and tantalum decreases to low values above approximately 1600 F (870 C), the hydrides are stable only at relatively low temperatures, and relatively large quantities of hydrogen are required to cause embrittlement at elevated temperatures. Embrittlement of columbium and tantalum by hydrogen at room temperature and below is usually associated with hydride formation. No investigations have been made of the mechanism of embrittlement by hydrogen at higher temperatures. Columbium and tantalum specimens will fragment when cooled from elevated temperatures in hydrogen or, under certain conditions, when exposed to hydrogen at room temperature.

W. T. Chandler is a Principal Scientist and R. J. Walter is a Member of the Technical Staff at Rocketdyne, a Division of North American Rockwell Corporation, Canoga Park, California. This paper was presented at the Symposium on Metallurgy and Technology of Refractory Metal Alloys, Washington, D.C., 25-26 April 1968, Sponsored by the Metallurgical Society of AIME and the National Aeronautics and Space Administration.

197

INTRODUCTION

The group of elements called refractory metals, here defined as metals having a melting point above 3500 F (1927 C), include metals which can absorb large quantities of hydrogen and whose properties are greatly affected by hydrogen and other metals which can absorb only very little hydrogen and whose properties are, under most circumstances, not significantly affected by hydrogen.

Metals in general have often been classified into endothermic and exothermic occluders of hydrogen (1, 2) according to whether the heat of absorption is negative or positive. Although this classification has no particular fundamental significance, it serves to classify metals into two groups having different general characteristics with regard to hydrogen absorption. With endothermic occluders, the amount of hydrogen which can be absorbed is small and increases with increasing temperature, and hydrides or other second phases are not formed. Molybdenum and tungsten are refractory metals which have low solubilities for hydrogen and are probably endothermic occluders. Nonrefractory metals which have low solubilities for hydrogen are: iron, nickel, chromium, copper, platinum, and gold. Even though the amount of hydrogen which can be absorbed by endothermic occluders is small, some are susceptible to hydrogen embrittlement; a prime example is iron.

In contrast, the exothermic occluding metals can absorb large quantities of hydrogen, the amount of hydrogen which can be absorbed increases with decreasing temperature, and hydrides are formed. Columbium, hafnium, and tantalum are refractory metal exothermic occluders. Other examples of exothermic occluders are: palladium, thorium, titanium, uranium, vanadium, and zirconium. It is evident that the exothermic occluders include the metals whose properties are the most seriously affected by hydrogen. Undoubtly because of this, hydrogen effects have been investigated to a much greater extent for exothermic occluders, with the important exception of iron.

For the refractory metals not listed above, i.e., iridium, osmium, rhenium, rhodium, ruthenium, and technetium, there is very little information on the absorption of hydrogen or its effects on properties, and data from different investigators are conflicting.

Hydrogen which is absorbed into a metal may affect the mechanical properties as a result of: (1) the presence of the hydrogen in solid solution, (2) the formation of a hydride (or other second phase), and (3) a chemical reaction with alloying elements in the metal. The last of these is, in a sense, a secondary effect of hydrogen but can be serious. An example is the decarburization of steels with the formation of methane which collects in voids at high pressure and can result in cracking. Also, the removal of carbon will, in itself, affect properties, possibly seriously. Such chemical reactions will occur most readily at elevated temperatures and high hydrogen pressures, but the resultant removal of alloying elements (including other interstitials) or the formation of hydrides with them could then affect properties at any temperature.

The hydrides formed in the exothermic occluders are more brittle and less dense than the pure metal. Consequently, the metal lattice is strained and plastic flow is inhibited. The brittle hydrides are favored sites for crack initiation and propagation. Of the metals that form stable hydrides, titanium and zirconium have been the most thoroughly investigated and it has been found (3, 4) that very small amounts of hydride will reduce the ductility of these metals. The embrittling effect of the hydride phase increases with an increase of strain rate provided, of course, that the hydride is present at the beginning of testing.

The formation of a stress-induced hydride has been observed in titanium alloys (5, 6) and in zirconium (7) and may occur in the refractory metals. This process requires hydrogen diffusion during the application of stress, e.g., during testing, and thus the embrittlement due to stress-induced hydride will be more severe with slow strain rates.

Hydrogen, as well as other interstitials, in solution will affect the properties of metals. The effects are greatest at low temperatures and serious manifestations are the raising of ductile-brittle transition temperatures and the effect on yielding behavior of the body-centered-cubic (bcc) metals. A well-known example of embrittlement by hydrogen in solution is that for steels.

The hydrogen embrittlement of steels, which has been most thoroughly investigated, is characterized by delayed brittle failure in static loading tests and by more severe embrittlement with slow strain rate. The exact embrittling effect of hydrogen in steel is still not adequately defined, but it is accepted that diffusion of hydrogen to certain regions, e.g., regions of high stress, is required. Cracking results when a critical combination of stress and hydrogen concentration is attained in these regions. The delayed failure and more severe embrittlement at slow strain rates results, of course, from the requirement for hydrogen diffusion. Similar dependency of hydrogen embrittlement on temperature and strain rate as found for steels has been observed for vanadium (8). As will be discussed, a ductile-brittle-ductile behavior has been observed in columbium and tantalum. This behavior has been attributed to a diffusion-controlled mechanism for embrittlement of Group V A metals by hydrogen in solution, but another explanation has recently been proposed.

This paper will be concerned mainly with the effects upon properties of hydrogen in the refractory metals. However, it should always be remembered that under certain conditions, hydrogen simply as an environment can affect properties. Such environmental effects are much more serious at high hydrogen pressure, but even a 1-atmosphere pressure hydrogen environment has been shown to have a decided effect upon crack propagation, at least in high-strength steel (9). Again, such effects have been investigated much more extensively for steels than for other metals.

A few general comments concerning difficulties experienced in investigating hydrogen effects in metals are felt to be in order. The mobility of dissolved hydrogen in metals is very high. Thus, in some cases, even at relatively low temperatures, hydrogen may enter or leave a metal so rapidly that, unless great care is exercised, hydrogen effects may be studied under non-reproducible, transient conditions and they may not correlate with measured hydrogen concentrations. On the other hand, at lower temperatures, the introduction of hydrogen into even the metals which occlude large quantities of hydrogen is not so simple as is sometime envisioned. The rate of hydrogen absorption is a function of prior mechanical and thermal treatments, purity of the hydrogen, and, in particular, the condition of the surface. Most natural oxides are effective barriers to hydrogen.

As noted above, property changes in a metal attributed to the influence of the hydrogen per se may actually be the result of secondary effects, e.g., the interaction with and possible removal of other solutes in the metal or environmental effects such as alteration of sintering rates. Recent investigations have shown that hydrogen in iron may be immobilized by being trapped at various structural irregularities and it is possible that similar hydrogen trapping occurs in other metals. Such trapping can influence solubility, permeability, and diffusion results, depending upon the type of experiment used. The method of charging a metal with hydrogen must be selected with care. With relative ease, metals can be subjected, by design or otherwise, to hydrogen at very high fugacities, e.g., by acid attack or electrolytic charging. Such high fugacities lead to structural changes, such as the introduction of traps or even cracks, and other non-equilibrium conditions which make it impossible to extrapolate results to other conditions of hydrogen absorption. A number of the above points are considered in greater detail in a recent review of hydrogen in transition metals by Oriani (10).

In the following sections, the effects of hydrogen upon the four most commonly used refractory metals, molybdenum, tungsten, columbium, and tantalum, will be discussed, in that order. The little information available on the effects of hydrogen upon the other refractory metals is covered in standard reference books. The effects of hydrogen upon metals has been reviewed by Smith (1) and Cotterill (2), and the nature of metal hydrides has been reviewed by Gibb (11).

MOLYBDENUM

SOLUBILITY, PERMEABILITY, AND DIFFUSION OF HYDROGEN IN MOLYBDENUM

The solubility of hydrogen in molybdenum at a 1-atmosphere pressure was investigated by Sieverts and Bruning (12) and by Hill (13). The results of these investigators are reasonably consistent with each other and are shown in Fig. 1. The results of Martin (14), indicating a maximum in the solubility at approximately 1500 F (816 C), do not agree with the other results and would appear to be in question. The solubilities of hydrogen in molybdenum as calculated by Jones et al (15), from permeability data, and by Moore and Unterwald (16), from their data on thermal dissociation of hydrogen on molybdenum, are also shown in Fig. 1. No formation of hydrides or second phases has been reported in the molybdenum–hydrogen system.

The data are very limited, but the solubility of hydrogen in molybdenum–base alloys is also expected to be small. D. W. Jones et al (17) have reported that there is no solubility of hydrogen in binary alloys of titanium with molybdenum having an electron/atom ratio greater than 5.6 (assuming values of 4 for titanium and 6 for molybdenum). Jones et al (17, 18) suggest

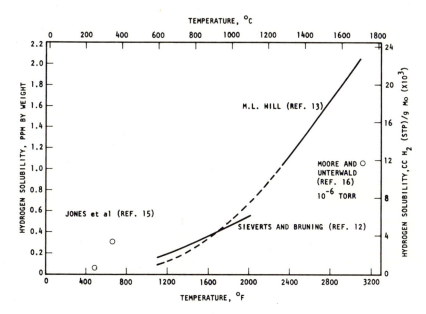

Fig. 1. Solubility of hydrogen in molybdenum.

that this is a general effect and, at least for columbium-base and molybdenum
base alloys, the solubility for hydrogen is low when the electron/atom ratio
is between 5.6 and 6.0. Jones (19) has shown that the addition of rhenium
to molybdenum (thus raising the electron/atom ratio above 6.0) increases
the hydrogen affinity as measured by the heat of solution of hydrogen in the
alloys. Interestingly, the heat of solution begins to increase at an
electron/atom ratio of about 6.05 (approximately 10 weight percent rhenium),
rises to a maximum at 6.25 (approximately 40 weight percent rhenium), and
again falls to zero at 6.3 (approximately 50 weight percent rhenium). Jones
associates the minimum in the hydrogen solubility for electron/atom ratios
between 5.6 and 6.0, with a minimum in the density of energy states curve
in the same range.

It should be noted that Booth (20) found that for chromium, which should
behave similarly to molybdenum, hydrogen solubility remained fairly constant at about 2 ppm at 2282 F (1250 C) and 1 atmosphere pressure when
approximately 10 and 30 weight percent rhenium was added.

The permeability of hydrogen through molybdenum is quite low as is shown
in Fig. 2, which contains data obtained by a number of investigators (15
and 21-25). The results of the different investigators agree quite well
and the equation giving the best fit for all data is:

$$P\left(\frac{cc(STP)mm}{sec\ atm^{1/2}\ cm^2}\right) = 0.166\ exp\left[-\frac{19,300\ cal/mole}{RT(°K)}\right]$$

Limited data are available on the diffusion of hydrogen in molybdenum.
Table 1 contains diffusivity data obtained by various investigators (13,
15, 16, 26, 27) from different types of investigations.

THE EFFECT OF HYDROGEN ON THE PROPERTIES OF MOLYBDENUM

Hydrogen itself apparently has little effect upon the mechanical properties of molybdenum and molybdenum-base alloys. This is attested to by
the fact that many operations, such as consolidation or heat treatment,
are conducted in hydrogen or hydrogen-containing environments. However,
hydrogen environments may indirectly affect properties. Studies were
conducted at General Electric (28-32) on the stress rupture of various
refractory alloys including molybdenum and molybdenum-50 percent rhenium
alloy in argon and in hydrogen at 4000 F (2200 C). It was found that the
stress-rupture properties were the same in hydrogen as in argon for the
Mo-50 Re alloy and for arc-cast molybdenum. However, unalloyed molybdenum
produced by powder metallurgy had a higher rupture strength in hydrogen
than in argon. This effect was also found for other powder metallurgy-
sintered refractory metals and was related to the original sintering
temperature. These studies indicated that the stress-rupture strength of
sintered materials at temperatures exceeding the sintering temperature
will be higher in hydrogen than in argon. This was attributed (31) to
additional sintering promoted by the hydrogen environment. Also, significant decarburization of molybdenum or molybdenum alloys can occur in hydrogen
(28, 33) or even in argon containing 2 percent hydrogen at temperatures above
3200 F (1760 C), as shown by Perkins and Lytton (33).

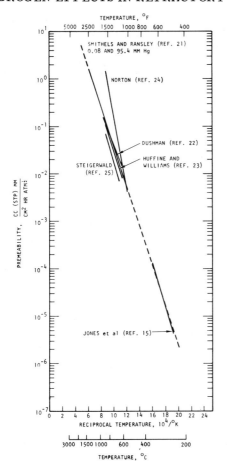

Fig. 2. Permeability of hydrogen through molybdenum.

Table 1. Diffusion of Hydrogen in Molybdenum

D, cm^2/sec	Temperature, F	D_0, cm^2/sec	Q, cal/mole	Temperature Range, F	Reference	Method of Obtaining
8.7×10^{-8}	660	7.6×10^{-5}	8,400	480 to 660	15	Permeability
1.23×10^{-9}	480				15	Permeability
		5.9×10^{-2}	14,700	1067 to 1800	13	Evolution of hydrogen from molybdenum in vacuum
2.8×10^{-7}	3115				16	Thermal dissociation of hydrogen on molybdenum
10^{-9}	210				26	Trapping of hydrogen by molybdenum
		0.158	22,200	2912 to 4172	27	Evolution of hydrogen from molybdenum in vacuum; gas pressure determined by magnetic mass spectrometer

The profound effect of interstitial elements upon the low-temperature
properties, e.g., the ductile-brittle transition, of bcc metals is well
known. In particular, the Group VI A refractory metals, molybdenum, tungsten, and chromium, because of their very low lattice solubilities for
interstitial impurities, are affected more by small amounts of interstitial
impurities at low temperature than the Group V A refractory metals, columbium
and tantalum (34). This great effect of interstitials has been demonstrated
with molybdenum for oxygen, nitrogen, and carbon but not for hydrogen. Hydrogen may be embrittling to molybdenum but it is difficult to introduce,
is soluble in small degree, and evolves relatively easily. Thus, under most
conditions, hydrogen is not a significant factor in low-temperature properties.

Lawley et al (35) investigated the effect of hydrogen upon the yielding
behavior of molybdenum. Molybdenum wires quenched in water or in hydrogen
after exposure to hydrogen at 1 atmosphere pressure at 4620 F (2550 C) had
an upper and lower yield strength when tested at room temperature and the
lower yield was about 5000 psi higher than the yield strength prior to the
hydrogen treatment. More importantly, annealing the quenched wires in the
range of 77 to 268 F (25 to 131 C) raised the lower yield strength by a maximum of 82 percent to about 60,000 psi, and a more pronounced upper and lower
yield point developed. With longer annealing times, the yield strength decreased and the yield point drop disappeared. Wires slow cooled in hydrogen
or annealed in vacuums and quenched in helium did not develop an upper and
lower yield point, and there was no effect of subsequent annealing at temperatures up to 572 F (300 C). If the hydrogen-exposed and quenched wires
were annealed for a sufficiently long time, their stress-strain curves were
identical with those for the wires slowly cooled in hydrogen. It was deduced
from solubility data that the maximum yield drop and yield and flow stress
and minimum ductility were associated with a hydrogen concentration of less
than 6 ppm. Lawley et al concluded that the large transient increase in
yield and flow stress was due to a strong electrical interaction between
hydrogen atoms and dislocations, with the hydrogen atom actually moving
with the dislocations in the plastic region. "Over-aging" corresponded to
a decrease in the average hydrogen concentration by desorption to a level
insufficient to form effective atmospheres at the dislocations. This
concentration was of the order of 1 ppm or less.

Feuerstein (36) performed tensile tests on molybdenum at 77 F (25 C)
in 10^{-10} torr vacuum and in dry nitrogen at 1 atmosphere pressure. Prior
to tensile testing, some of the molybdenum specimens were exposed to hydrogen at 10^{-5} torr pressure and 1472 F (800 C) for 4 hours, followed by soak-
ing for 6 hours under a vacuum of 5 x 10^{-7} torr at the same temperature.
This treatment increased the difference in the strain to fracture between
that in vacuum and that in 1-atmosphere nitrogen, predominantly by reducing
the ductility at 1 atmosphere. The magnitude of the effect was a function
of grain size, being most procounced at intermediate sizes. From the
investigation of Lawley et al (35) it is difficult to see how the treatment
described could have resulted in sufficient hydrogen remaining in the
molybdenum specimens to have affected the stress-strain curves. It is more
likely, as Feuerstein suggests, that the role of hydrogen in the embrittlement
results from the interaction of the hydrogen with other impurities.

TUNGSTEN

SOLUBILITY, PERMEABILITY, AND DIFFUSION
OF HYDROGEN IN TUNGSTEN

No direct measurements of the solubility of hydrogen in tungsten are
known. It has been stated (37, 38) that the solubility of hydrogen in
tungsten is immeasurably small. From their investigations of thermal dissociation of hydrogen on tungsten, Moore and Unterwald (16) calculated a
solubility of 1 atom of hydrogen per 400 atoms of tungsten (about 14 ppm)
at 3933 F (2167 C). Aitken et al (39, 40) measured the permeability of
hydrogen through tungsten and, using the permeation coefficients they
obtained and the diffusion coefficients for hydrogen in tungsten from
Ryabchikov (27), they calculated the solubilities of hydrogen in tungsten
(in equilibrium with hydrogen gas at 1 atmosphere of pressure), as shown
in Fig. 3. Aitkin et al claim (39) that tentative, unpublished results
from solubility measurements they have made support their estimates as
opposed to the much higher value of Moore and Unterwald.

The permeability of hydrogen through tungsten is extremely low as shown
in Fig. 4, which contains data reported by Steigerwald (25), Aitkin et al
(39, 40), Webb (41), and Frauenfelder et al (42).

No direct measurements of the diffusion of hydrogen in tungsten are
known. Table 2 contains diffusion parameters calculated by Ryabchikov
(27) from data on evolution of hydrogen from tungsten and by Moore and
Unterwald (16) from data on thermal dissociation of hydrogen on tungsten.

EFFECT OF HYDROGEN UPON THE
PROPERTIES OF TUNGSTEN

As with molybdenum, hydrogen apparently has little effect upon the
mechanical properties of tungsten and tungsten-base alloys. Very few
studies have been conducted on the effects of hydrogen upon tungsten;
however, again, many fabrication and heat treating operations for tungsten
are carried out in hydrogen environments with no apparent effects.

The stress-rupture studies conducted at General Electric included tests
on pure tungsten and tungsten-25 weight percent rhenium sheet specimens in
argon and hydrogen environments (30). For high-purity powder metallurgy
tungsten, similar results were obtained in the two environments at 5072 F
(2800 C). However, commercial tungsten-25 weight percent rhenium material
of relatively high purity was stronger and more creep resistant in hydrogen
than in argon at 4352 F (2400 C) and 4712 F (2600 C), but at 3992 F (2200 C),
the results showed considerable scatter in both environments with no apparent difference between environments. These results resemble those for
powder metallurgy molybdenum which were explained on the basis of increased
sintering promoted by the hydrogen environment. To improve the quality of
tungsten-25 weight percent rhenium sheet, processing and fabrication procedures were developed at General Electric. Sheet produced using these
procedures had essentially the same properties in hydrogen as in argon at
4712 F (2600 C).

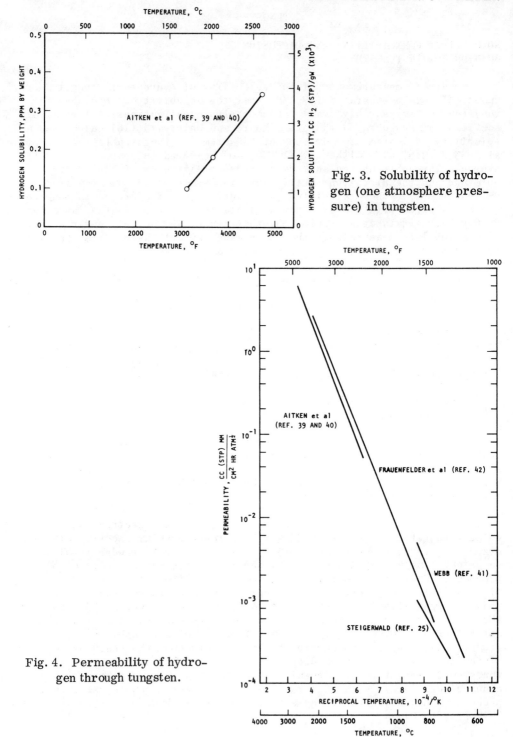

Fig. 3. Solubility of hydrogen (one atmosphere pressure) in tungsten.

Fig. 4. Permeability of hydrogen through tungsten.

As with the other bcc refractory metals, and in particular the VI A group, interstitial impurities, in general, are expected to have a dominant effect upon the low-temperature mechanical properties of tungsten (34, 43). However, the yield behavior of tungsten is somewhat anomalous compared to the other bcc refractory metals (44). Yield points have been observed but the occurrence is erratic. Usually, the specific impurities causing the yield points have not been identified. Impurities and, in particular, the segregation of impurities to grain boundaries, is believed to play an important role in the ductile-to-brittle transition of tungsten (34, 43, 45). Again, however, the relative potency of individual interstitials (or other impurities) has not been established.

Information specifically dealing with the effect of hydrogen on low-temperature properties is meager and not unambiguous. Atkinson, et al (46) found that a yield point could be produced in tungsten by annealing in hydrogen for 1-1/2 hours at 1112 F (600 C). However, the hydrogen contained water vapor, and in later work (47), it was found that: (1) the concentration of hydrogen as determined by vacuum fusion analysis could not be correlated with the yield point, (2) the magnitude of the yield point and the increase in flow stress varied with the amount of water vapor in the hydrogen, and (3) a yield point could not be developed by electrolytic charging of tungsten with hydrogen. As suggested (48), it appears that the yield point produced by annealing in the wet hydrogen may have been due to oxygen.

Sutherland and Klopp (49) found that the ductile-to-brittle transition temperatures for bend tests of five typical commercial lots of tungsten sheet increased with hydrogen content, except for one lot. The hydrogen concentrations were 1 ppm or less. However, the transition temperatures also increased with oxygen content, and carbon and nitrogen contents varied.

COLUMBIUM

In contrast to molybdenum and tungsten, columbium and tantalum are greatly affected by hydrogen, catastrophically under certain conditions at lower temperatures. However, at higher temperatures, above approximately 1500 F (816 C) for hydrogen pressures of approximately 1 atmosphere, hydrogen does not significantly affect the properties of columbium or tantalum. The reason for this can be seen from the solubility curves and phase diagram for these metals with hydrogen.

SOLUBILITY OF HYDROGEN IN COLUMBIUM AND THE
COLUMBIUM-HYDROGEN PHASE DIAGRAM

A single solid solution exists in the columbium-hydrogen system above approximately 340 F (170 C). The equilibrium solubilities of hydrogen in columbium have been determined by Albrecht et al (50), Komjathy (51), Veleckis (52), Makrides et al (53), and Oakwood and Daniels (54), and the results are in good agreement. The results from Albrecht et al are shown in Fig. 5. One can see that for hydrogen pressures near 1 atmosphere, the solubility of hydrogen in columbium is decreasing rapidly with temperature above 1000 F (538 C) to quite small values above 1500 F (816 C). In Fig. 5 are plotted points determined by Perminov (55) for the solubility of hydrogen in columbium at 1243 F (673 C) for hydrogen pressures from 1 atmosphere to 1000 atmospheres. Perminov found that for hydrogen pressures above 1 atmosphere,

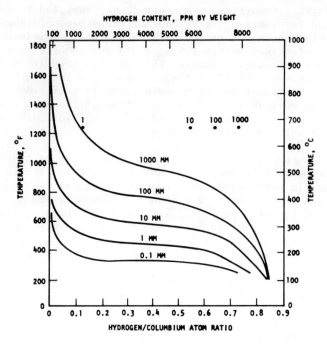

Fig. 5. Solubility of hydrogen in columbium (curves taken from Albrecht et al., Ref. 50; points taken from Perminov, Ref. 55; numbers associated with points are pressures in atmospheres).

Table 2. Diffusion of Hydrogen in Tungsten

D, cm²/sec	Temperature, F	D_0, cm²/sec	Q, cal/mole	Temperature Range, F	Reference	Method of Obtaining
5.5×10^{-8}	3456	0.081	19,800	2912 to 4172	27	Desorption of hydrogen from tungsten
3.01×10^{-8}	3222	7.25×10^{-4}	41,500	3069 to 3456	16	Thermal dissociation of hydrogen on tungsten
1.81×10^{-8}	3069					

the solubility first increased rapidly with pressure and then more slowly, according to a logarithmic relationship. Work is currently under way at Rocketdyne under Air Force Contract No. AF33(615)-2854 to determine the solubility of hydrogen in columbium and the B-66 columbium alloy for hydrogen pressures of 800 and 1500 psi at temperatures up to 1800 F (982 C).

All investigators have noted that impurities, in particular other interstitial impurities, can have a great effect upon the solubility of hydrogen in columbium. However, few quantitative correlations, none with the other interstitials, have been experimentally determined.

The solubility of hydrogen in the B-66 columbium alloy for a hydrogen pressure of 1 atmosphere was determined by Walter and Ytterhus (56) and is shown in Fig. 6. For comparison, the curve for the solubility of hydrogen in columbium for a hydrogen pressure of 1 atmosphere is also included in Fig. 6. This curve for columbium was determined by Walter and Chandler (57), and their results are in complete agreement with those of Albrecht et al (50).

As discussed in the section on molybdenum, Jones et al (17, 18, 19) have proposed that the solubility of hydrogen in bcc alloys is a function of the electron/atom ratio. Specifically, for columbium alloys, Jones and McQuillan (18) found that the heat of solution of hydrogen decreased as molybdenum or rhenium was added to columbium and became zero when the electron/atom ratio reached 5.6 or higher (assuming values of 5 for columbium, 6 for molybdenum, and 7 for rhenium). The effect of rhenium additions to columbium upon hydrogen absorption was investigated by Stephens and Garlick (58). They heated specimens at 3000 F (1649 C) for 1 hour and cooled them to room temperature in hydrogen. The amount of hydrogen absorbed dropped with increasing rhenium content from 47.4 atomic percent (H/Cb = 0.9) for pure columbium to 1.2 atomic percent (H/Cb = 0.012) for Cb-25 atomic percent Re (e/a = 5.5) and 0.08 atomic percent (H/Cb = 0.0008) for Cb-40 atomic percent Re (e/a = 5.8). At Rocketdyne, the solubility of hydrogen at 800 F (427 C) and 1 atmosphere of pressure was found to be H/M = 0.70 for the Cb-10 weight percent Hf alloy, 0.65 for pure columbium, and 0.39 for the B-66 alloy (Cb-5 weight percent Mo-5 percent V-1 percent Zr). Assuming electron atom ratios of 4 for hafnium, 5 for columbium and vanadium, and 6 for molybdenum, these solubilities again decrease as electron/atom ratios increase.

The phase diagram of the columbium-hydrogen system at lower temperatures as determined by Walter and Chandler (59) is shown in Fig. 7. The diagram was determined by hot-stage X-ray diffraction and differential thermal analysis methods. As an aid in seeing the relationship between the solubility curves and the phase diagram, the curve for the solubility of hydrogen in columbium for a hydrogen pressure of 1 atmosphere is included in Fig. 7.

The columbium hydride (β) phase is orthorhombic, with only slight deviation from bcc symmetry (60, 61). Paxton et al (62) found by X-ray diffraction examination of hydrided single crystals of columbium that the habit plane of the columbium-hydride plates is {100}.

Paxton et al (62) and Komjathy (51) suggested that the orthorhombic columbium hydride forms on cooling by an athermal martensitic-type

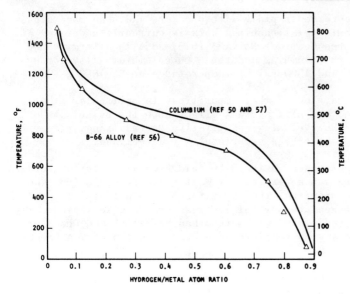

Fig. 6. Solubility of hydrogen in columbium and in the B-66 columbium alloy for hydrogen pressure of one atmosphere.

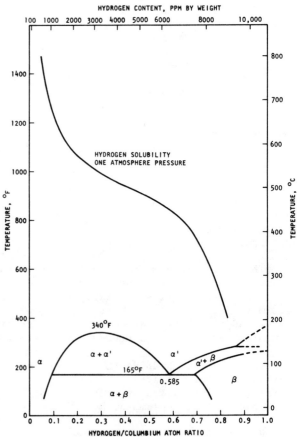

Fig. 7. Phase diagram of columbium-hydrogen system (after Walter and Chandler, Ref. 59).

transformation. However, Walter and Chandler (59) were able to form the orthorhombic hydride by charging columbium powder at room temperature.

Classical eutectoid microstructures do not form as a result of the eutectoid transformation (57), undoubtedly because of the rapid hydrogen diffusion. At hydrogen concentrations of approximately 0.2 H/Cb atom ratio and lower, the hydride precipitates as needles (Fig. 8) which form a Widmanstatten structure (57, 63). At high concentrations, e.g., 0.7 H/Cb atom ratio, the hydride appears as platelets. The dissolution of the hydride on heating and the formation of the hydride on cooling was followed by hot-stage metallography (57). The method was not useful for determining the phase diagram because of the very large hysteresis between the temperatures for dissolution and the temperatures for formation. It was found that thermal cycling through the transformation range led to considerable coarsening of the microstructure.

Recently, Benson (64) used transmission electron microscopy and electron diffraction to study columbium containing small concentrations of hydrogen at temperatures below room temperature. He reported that the first second phase to form with increasing hydrogen had a distorted bcc structure which resulted from the ordering of the interstitial hydrogen atoms.

However, it should be noted that Westlake (65) states that structure observed by transmission electron microscopy in thin foils containing hydrogen is not necessarily representative of the initial bulk material. He reports that hydrogen, originally present in a specimen of vanadium, columbium, or tantalum, can be concentrated by the thinning operation used to prepare foils from the bulk materials. These foils exhibited domain walls indicative of a tetragonal distortion of the bcc unit cells, which Westlake felt was probably caused by ordering of the hydrogen in preferred interstitial sites. Westlake believes that this ordering does not occur at room temperature in the bulk material which contains less than 1 atomic percent hydrogen (H/Cb = 0.01).

The phase diagram shown in Fig. 7 was determined using columbium rod of approximately 99.85 percent purity and powder of 99.5 percent minimum purity. Walter and Chandler (59) also determined a diagram using a more impure columbium of 99.0 percent minimum purity. The additional impurities were mainly oxygen, carbon, nitrogen, iron, and tantalum. The main effect of the impurities was to lower the $\alpha + \alpha'$ miscibility gap. The peak temperature was lowered to 275 F, the eutectoid transformation temperature to 150 F, and the eutectoid composition to 0.515 H/Cb atom ratio. The 275 F peak temperature corresponds closely to the value of 284 F obtained by Albrecht et al (50). The 340 F peak temperature found for the higher-purity columbium corresponds more closely to the 361 F peak temperature calculated by Veleckis (52).

The absorption studies of Stephens and Garlick (58), described previously, indicated that no hydride is formed down to room temperature in columbium-rhenium alloys containing 25 atomic percent rhenium or more.

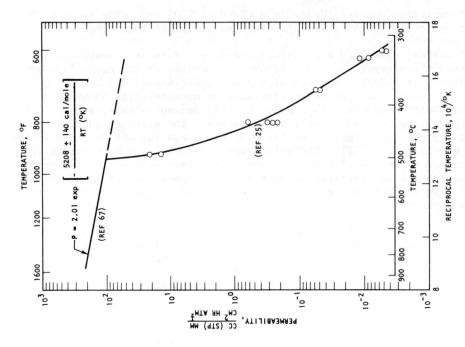

Fig. 9. Permeability of hydrogen through columbium as a function of temperature.

Fig. 8. Hydride Widmanstatten structure formed in hydrogenated columbium with H/Cb = 0.22 (1400 ×).

Upadhyaya and McQuillan (66) found that the hydrogen content of the hydride increased as titanium was added to columbium from H/Cb = 0.88 for pure columbium to H/M = 1.4 for an alloy containing 26 atomic percent titanium.

PERMEABILITY, DIFFUSION, AND KINETICS OF ABSORPTION AND DESORPTION OF HYDROGEN IN COLUMBIUM

The permeability of hydrogen through columbium has been determined by Rudd et al (67) for temperatures between 1740 F (950 C) and 1950 F (1065 C) and by Steigerwald (25) for temperatures below 900 F (482 C). These permeabilities were determined with a hydrogen pressure of 1 atmosphere on the inlet side of the specimen and essentially a zero pressure on the outlet side. The results, which are shown in Fig. 9, indicate a discontinuity in the permeability vs temperature curve at approximately 1000 F (540 C). At the higher temperatures, the permeability was essentially equal to the product of diffusivity and solubility. Steigerwald is undoubtedly correct in suggesting that surface reactions were rate controlling in his lower-temperature tests. As will be discussed shortly, surface, thermal, and mechanical pretreatments greatly affect the absorption of hydrogen in columbium at temperatures below approximately 1100 F (593 C). The authors believe that if Steigerwald had used an "activation" pretreatment, his results would have been more in line with those of Rudd et al.

Gulbransen and Andrew (68) investigated the rate of absorption of hydrogen in columbium over the temperature range of 482 F (250 C) to 1652 F (900 C) for hydrogen pressures up to 57 torr. They found no consistent rate law governing absorption. However, at 572 F (300 C), the initial rate of absorption was linear and a function of the square root of pressure. At 482 F (250 C), the rate increased as the reaction proceeded, while at 662 F (350 C), 1292 F (700 C), and 1652 F (900 C), the rate decreased as absorption proceeded. Figure 10, from Gulbransen and Andrew, shows a comparison among the rates of absorption of hydrogen in columbium, titanium, tantalum, and zirconium.

Accompanying their work on the solubility of hydrogen in columbium, Albrecht et al (50) performed absorption rate determinations. They determined the rates of absorption of hydrogen in columbium to produce hydrogen to columbium atom ratios of 0.05, 0.10, 0.50, and 0.70 in the temperature range of 572 F (300 C) to 1292 F (700 C). From 572 F (300 C) to 1022 F (550 C), the initial rate of absorption was linear, usually until about 40 to 50 percent of saturation was attained. Then, the rate decreased and stopped when equilibrium was established. Thus, these results are not in agreement with those of Gulbransen and Andrew except at 572 F (300 C). The linear rate constants found by Albrecht et al are given in Table 3. At temperatures from 1112 F (600 C) to 1292 F (700 C), Albrecht et al found that absorption could be described by a diffusion equation in which absorption was initially parabolic up to approximately 40 percent of saturation. The diffusion equation obtained by Albrecht et al was:

$$D \ (cm^2/sec) \ = \ 0.021 \ exp \left[- \frac{9370 \ \pm 600 \ cal/mole}{RT(°K)} \right] \tag{2}$$

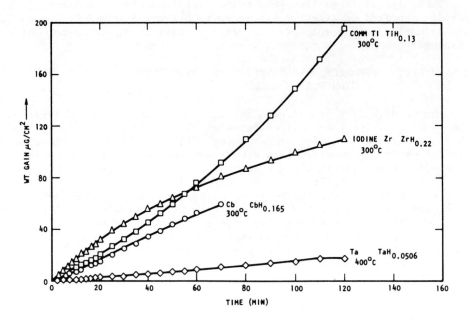

Fig. 10. Comparison of rate of absorption of hydrogen in columbium, titanium, tantalum, and zirconium (after Gulbransen and Andrew, Ref. 68).

Table 3. Linear Rate Constants for the Initial Reaction to Produce Various Hydrogen-Columbium Compositions (after Albrecht et al., Ref. 50)

Composition, H/Cb	Temperature, C	Pressure, mm Hg	Rate Constant x 10^3, ml/cm^2- sec
0.05	550	67	21
	550	69	12
0.1	450	52	1.7
	500	112	9.1
	500	114	10
	550	218	29
0.5	400	145	1.9
	425	261	6.6
	450	450	14
	450	480	14
	480	744	37
0.7	300	83	0.35
	325	210	1.7
	350	460	6.7
	366	745	11

This appears to be the most reliable diffusion information available. However, Albrecht et al also conducted some desorption experiments, and the diffusion coefficients calculated from these data were smaller than those calculated from the absorption data. It is assumed that a surface reaction must have influenced the desorption of hydrogen. Oakwood and Daniels (54) also conducted absorption experiments at 1112 F (600 C), 1202 F (650 C), and 1292 F (700 C), and the diffusion coefficients they determined from their data agree very well with those of Albrecht et al. Oakwood and Daniels obtained an activation energy of approximately 9000 cal/mol.

From desorption experiments conducted at rather high temperatures, 2912 F (1600 C) and 4172 F (2300 C), Ryabchikov (27) obtained the following diffusion equation:

$$D \ (cm^2/sec) \ = \ 0.056 \ exp \left[- \frac{19,200 \ cal/mole}{RT(°K)} \right] \qquad (3)$$

The high activation energy for such high temperatures, at which surface oxides should have been removed, is difficult to explain.

The absorption and desorption rates of hydrogen in columbium and the B-66 columbium alloy for hydrogen pressures of 800 and 1500 psi and for temperatures up to approximately 1800 F (982 C) are currently being determined at Rocketdyne under Air Force Contract No. AF33(615)-2854.

All investigators studying the columbium-hydrogen system have noted the profound effect of hydrogen gas purity and surface condition on all rate processes, e.g., absorption, desorption, and permeability, particularly at lower temperatures--below approximately 1100 F (593 C) for hydrogen pressures of 1 atmosphere. Although, in recent years, most investigators have taken considerable care in purifying the hydrogen and in surface preparation of specimens, most of the discrepancies in rate data from different investigators are still probably due to surface reactions. Surface preparation usually consists of abrasion and/or chemical cleaning followed by a thermal treatment, usually in high vacuum. Albrecht et al (50), for example, abraded their specimens with carbide papers and then annealed at 2172 F (1150 C) in 10^{-6} torr vacuum. The effect of such treatments has usually been attributed to the removal of oxide films which act as barriers to hydrogen absorption. The most comprehensive study of the effect of vacuum treatments at various temperatures on the subsequent rate of absorption of hydrogen was made by Gulbransen and Andrew (69) on zirconium. As an example, they found that the ratio of the rate of reaction with hydrogen of zirconium specimens having the room temperature oxide present to a specimen annealed at 1292 F (700 C) in a vacuum of 10^{-6} torr was 1:7700. Makrides et al found (53) that vanadium and tantalum specimens which were electrolytically etched and then covered with palladium by sputtering had greatly increased absorption of hydrogen even after exposure to air.

Oakwood and Daniels (54) found a considerable difference in rate of hydrogen absorption between columbium specimens which were etched with an HF, HNO_3, H_2O solution and the same specimens which were abraded after etching. All specimens were held in a 5×10^{-6} torr vacuum at the hydrogenation temperature for several hours prior to exposing the specimen to

the hydrogen. The etched specimens gave slow absorption rates; the calculated activation energy for diffusion was a very high 40,000 cal/mole.
Oakwood and Daniels suggested that the etching contaminated the surface
so as to retard absorption of hydrogen. The abraded specimens absorbed
hydrogen more rapidly, and for temperatures of 1112 F (600 C) and above,
the calculated activation energy for diffusion was the 9000 cal/mole mentioned above. However, at 1022 F (550 C), even the abraded specimens gave
low absorption rates. This again accentuates the problem of obtaining rapid
absorption or studying absorption rates below 1100 F (593 C). The vacuum
treatments of Oakwood and Daniels were conducted at temperatures too low
to effectively remove surface oxide layers. It should be noted that Oakwood
and Daniels found no effect of grain size on hydrogen absorption by
columbium.

At Rocketdyne (57, 70), columbium specimens to be charged with hydrogen
are chemically cleaned with a solution of 50-percent hydrofluoric acid and
50-percent nitric acid. This is a concentrated solution and gives a polish-
ing action rather than the etching action of the water-containing solution
used by Oakwood and Daniels. If, then, the specimen is to be charged at
any temperature below 1500 F (816 C), it is first given a "hydrogen activation" treatment. This treatment consists of temperature cycling the specimen in hydrogen three times between 1500 F (816 C) and the temperature at
which the subsequent tests in hydrogen are to be performed. At this point,
of course, the specimen contains considerable hydrogen. If the specimen is
to be exposed to hydrogen simply to achieve a certain concentration of hydrogen, then after activation, the specimen is brought directly to the
desired temperature in hydrogen. If the subsequent experiment is to be
one in which the rate of reaction with hydrogen is to be studied, e.g.,
an investigation of the rate of hydrogen absorption or the effect of time
of exposure to hydrogen environments on properties, then after activation,
the hydrogen is removed by outgassing the specimen in flowing high-purity
inert gas at 1500 F (816 C). The specimen should not be exposed to air
after this treatment and prior to the subsequent experiments in hydrogen.

As noted above, the effect of "hydrogen activation" treatments on
subsequent reactions with hydrogen generally has been attributed to removal
of oxide surface layers by dissolution in the metal. Observations made at
Rocketdyne (57) suggest that this is not the only effect, although undoubtedly an important one. Initially, a 1-day activation treatment consisting
of heating to 1500 F (816 C) and slow cooling was tried. This treatment
was somewhat effective but not nearly as effective as the temperature-
cycling treatment in activating a specimen for subsequent reaction with
hydrogen. Because the time at the higher temperatures was at least as
long for the slow-cooling from 1500 F (816 C) treatment, both treatments
should have had about the same effect upon oxide removal. Also, a few tests
indicated that exposure to air after activation eliminated some but not
nearly all of the effect of activation in decreasing the ductility of B-66
columbium alloy specimens tensile tested in hydrogen at 400 F (204 C), as
shown by Walter and Ytterhus (56).

Temperature gradient-induced diffusion of hydrogen in the B-66 columbium
alloy was qualitatively studied by Walter and Chandler (71, 72). In one
experiment, a room temperature to 1500 F (816 C) temperature gradient was
maintained in a 16-inch-long specimen which was exposed to flowing hydrogen

for 48 hours. In a second experiment, a series of 2- and 3-inch-long samples totaling 16 inches in length were placed in a furnace along a room
temperature to 1500 F (816 C) temperature gradient and exposed to flowing
hydrogen for 48 hours. Thus, temperature gradient-induced diffusion could
occur in the first experiment but not in the second. The results are shown
in Fig. 11, which also contains the equilibrium solubility curve for hydrogen in the B-66 alloy for a 1-atmosphere hydrogen pressure. A comparison
of the curves certainly indicates that temperature gradient-induced diffusion occurred in the first experiment. However, the results are only
qualitative because it is not possible to separate the amount of hydrogen
which was absorbed from the gas in the low-temperature region and the amount
present because of diffusion from the elevated-temperature regions.

THE EFFECT OF HYDROGEN ON THE
PROPERTIES OF COLUMBIUM

   Investigation of the effects of hydrogen on the properties of columbium
has not been extensive and there is considerable disagreement among the
results, probably because of the influence of other interstitials.

   McCoy and Douglas (73) concluded that hydrogen additions up to approximately 200 ppm (H/Cb = 0.018) raised the yield strength, although there was
considerable scatter in the data and the concentration of other interstitials,
particularly oxygen, varied somewhat.

   Wilcox and Huggins (74-76) investigated strain-aging of as-received
columbium (10 ppm hydrogen) by yield point return and of both as-received
and hydrogenated (779 ppm hydrogen) columbium by dynamic modulus measurements. The tensile tests were conducted at -58 F (-50 C) and a strain rate
of 0.005/min and an upper and lower yield point was observed. The return
of the yield point was determined for deformed specimens aged at temperatures
from 250 F (121 C) to 500 F (260 C). The return of the modulus of elasticity
at room temperature was determined for deformed specimens aged at temperatures from 75 F (24 C) to 201 F (94 C). The following activation energies
were calculated: 10,500 cal/mole for the yield point return, 7830 to 8280
cal/mole for the return of the modulus for as-received columbium, and 8080
to 9920 cal/mole for the return of the modulus for hydrogenated columbium.
These activation energies are close to that for the diffusion of hydrogen
in columbium and, thus, Wilcox and Huggins concluded that the yield point
and modulus behavior was due to pinning of dislocations by hydrogen.

   Wilcox and Huggins (74, 76) also measured the lower yield stress at
75 F (24 C) as a function of grain size for as-received (7 ppm hydrogen)
and hydrogenated (15 and 61 ppm hydrogen) columbium. They interpreted
their results in terms of the Petch equation (77):

$$\sigma_{\ell y} = \sigma_0 + k_y \, d^{-1/2} \tag{4}$$

where $\sigma_{\ell y}$ is the lower yield stress, 2 d is the grain diameter, and σ_0 and
k_y are experimental constants initially associated with lattice friction
stress and dislocation locking stress, respectively. It should be noted
however, that the exact significance of σ_0 and k_y is still under discussion

(e.g., by 78–80). In any case, Wilcox and Huggins determined that k_y increased with increasing hydrogen content. From their investigations, Wilcox and Huggins concluded that the observed upper and lower yield point, the strain-aging effects, and variation of k_y with hydrogen content were due to the pinning of mobile dislocations by hydrogen.

Wilcox et al (81, 82) determined the effect of strain rate (0.005, 0.10, and 6.0/min) and temperature, from –320 F (–196 C) to 77 F (25 C), on the tensile properties of arc-melted columbium containing 1 ppm hydrogen. From these data, they calculated an activation energy for the early stages of deformation of 8300 cal/mole, which is close to the activation energy for the diffusion of hydrogen in columbium. Wilcox et al also determined the tensile properties of columbium containing 1, 9, and 30 ppm hydrogen over the temperature range indicated above, for a strain rate of 0.005/min. They found that the specimens containing the higher hydrogen contents had higher ultimate tensile and yield strengths. They also observed a strain-aging peak at –58 F (–50 C) for the specimens containing 30 ppm hydrogen, but not for the others, and attributed the strain-aging peak to hydrogen. At –58 F (–50 C) for the 30 ppm hydrogen specimen, there was only a slight waviness of the stress–strain curve to indicate a tendency toward the serrated stress–strain curve which normally accompanies strain aging. However, a well-defined, serrated, stress–strain curve was observed in coarse-grained columbium which contained 89 ppm hydrogen and was tensile tested at 77 F (25 C) at a strain rate of 0.005/min.

In contrast to Wilcox and coworkers, Imgram et al (83), Wood and Daniel (84), Oakwood and Daniels (54), Longson (85), and Walter and Chandler (57, 71) have concluded that hydrogen content has little or no effect upon the yield or ultimate strengths of columbium regardless of test temperature, except insofar as severe embrittlement by higher hydrogen contents can lead to low ultimate strength values. Longson actually found that serrated yielding and yield point drop, which he attributed to oxygen and/or nitrogen was suppressed by increasing hydrogen contents, but still higher hydrogen concentrations were necessary before embrittlement resulted. Thus, most of the evidence appears to indicate that hydrogen does not pin dislocations in columbium. Oakwood and Daniels suggested that the strain-aging effects reported by Wilcox and coworkers were small enough to be within experimental error. The increases in strengths and serrated stress–strain curves reported by Wilcox and coworkers were likely due to other interstitials, probably oxygen.

One concludes then, as did Oakwood and Daniels (54), that hydrogen does not embrittle columbium by pinning of dislocations, leading to strengthening of the lattice or dislocation pile-ups. In this regard, dislocation pile-ups have never been observed in any bcc metal (86) and, specifically, Van Torne and Thomas (87) did not find them in columbium.

There is no disagreement as to the fact that hydrogen embrittles columbium, and severely. Eustice and Carlson (88) found that the ductile-to-brittle transition temperature for a strain rate of 0.008/min was below –320 F (–196 C) for columbium containing 1 ppm hydrogen and was approximately –94 F (–70 C) for columbium containing 20 ppm hydrogen (H/Cb = 0.0018). Wilcox et al (81, 82) found a somewhat smaller increase in the ductile-to-brittle transition temperature with hydrogen content. With a strain rate

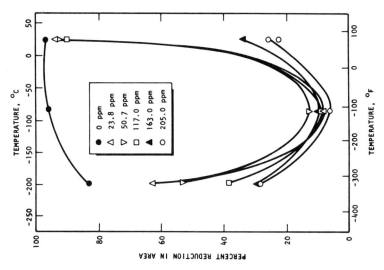

Fig. 12. Reduction in area vs temperature for several recrystallized columbium-hydrogen alloys (after Oakwood and Daniels, Ref. 54).

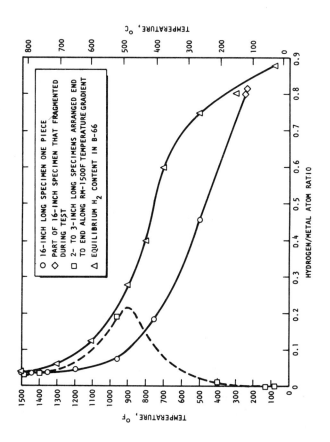

Fig. 11. Absorption of hydrogen by B-66 columbium alloy exposed to temperature gradient in flowing hydrogen for 48 hrs (after Walter and Chandler, Ref. 71 and 72).

of 0.005/min, the ductile-to-brittle transition temperatures were approximately -283 F (-175 C) for 1 ppm hydrogen and -184 F (-120 C) for 30 ppm
hydrogen. They found that increasing the strain rate lowered the ductile-
to-brittle transition temperature of the columbium containing 1 ppm hydrogen
and attributed this to slow strain-rate hydrogen embrittlement which was
controlled by hydrogen diffusion. Wilcox et al observed voids near the
fractured surface of both shear- and cleavage-type fractures, and the void
density was higher with higher hydrogen contents. They noted the possible
role of such voids in embrittlement, i.e., rupturing due to hydrogen pressure buildup in voids or lowering of surface energy by hydrogen leading to
crack propagation.

Imgram et al (83) performed tests on notched and unnotched, wrought
and recrystallized columbium specimens containing different hydrogen contents. With unnotched specimens tested at a constant crosshead speed of
0.02 in./min, the ductile-to-brittle transition temperatures were approximately -390 F (-234 C) for pure columbium (unspecified hydrogen content),
110 F (43 C) for columbium containing 200 ppm hydrogen, and 140 F (60 C)
for columbium containing 390 ppm hydrogen. There was no significant difference between the behavior of the wrought and recrystallized hydrogenated
specimens, and hydrogen had a marked effect upon notch sensitivity. Imgram
et al thus concluded that hydrogen promotes cleavage fracture. Imgram
et al (89) performed tensile tests at -104 F (-76 C) on columbium containing
200 ppm hydrogen (H/Cb = 0.018) at four strain rates between 0.00015 per
minute to 4.0 per minute and found that both strength and ductility increase
with an increase in strain rate.

None of the investigators just discussed found the ductile-brittle-ductile
transition behavior, i.e., a ductility minimum at a certain temperature,
that has been observed for vanadium by Roberts and Rogers (90) and Eustice
and Carlson (91), and for tantalum by Nunes et al (92) and Imgram et al (83)
However, Wood and Daniels (84) and Oakwood and Daniels (54) reported pronounced ductility minima at a temperature of approximately -116 F (-82 C).
Results from Oakwood and Daniels are shown in Fig. 12. Wood and Daniels
found that strain rate had little effect upon ductility. They also observed that the fracture behavior at the temperature of minimum ductility
varied from ductile cup and cone-type fractures in hydrogen-free specimens
to completely brittle cleavages in specimens with large amounts of hydrogen.
At the lower hydrogen concentrations which exhibited the ductility minimum,
the fractures at the minimum temperature were brittle but uniform elongation
was large compared to that observed in ductile specimens because numerous
cracks developed in the gage section prior to final fracture. Oakwood and
Daniels found that coarse-grained columbium was more sensitive to hydrogen
than wrought or fine-grained material, but this was attributed to the lower
initial ductility of the coarse-grained structure. They observed extensive
porosity near the fracture surface in fine-grained recrystallized and wrough
(but not coarse-grained) columbium specimens tensile tested at -320 F (-196
and 81 F (27 C) but not in specimens tested at -116 F (-82 C). The porosity
increased with increasing hydrogen content. They concluded that many of
the voids which formed during plastic straining at -320 F (-196 C) and 81 F
(27 C) were blunted, while at -116 F (-82 C), a few voids expanded rapidly
into cracks of sufficient length to propagate to failure. Thus, the formation and subsequent growth of voids was believed to be dependent upon some
ductility in the metal.

Longson (85) found that ductility of columbium was not affected by hydrogen additions until a critical amount of hydrogen for a given temperature was reached; thereafter, the ductility was drastically reduced over a small range of composition. The critical hydrogen content required for embrittlement was not a function of strain rate (from slow bend to impact) but increased with temperature and appeared to be close to but less than the solubility limits for hydrogen in columbium. Longson found no recovery of ductility at low temperatures.

Tensile tests were performed at Rocketdyne (57) at temperatures above 200 F (93 C) with a constant crosshead rate of 0.002 in./min on columbium which contained various quantities of hydrogen. The results indicated that severe loss in ductility occurred at temperatures as high as 800 F (427 C), the highest temperature tested, for a specimen containing 31 atomic percent hydrogen (H/Cb = 0.45). Tests cannot be conducted on hydrogenated columbium specimens at temperatures much above 800 F (427 C) in inert atmospheres without desorption of hydrogen. Tests can be conducted in hydrogen environments at pressures commensurate with the hydrogen content to be tested although additional effects can then occur, as will be discussed later.

The ductile-to-brittle transition data obtained by the various investigators are given in Fig. 13 (from 56). Figure 13 also contains the columbium-hydrogen phase diagram of Walter and Chandler (Fig. 7). Although the data at temperatures above room temperature are very sparse, it appears that a discontinuity occurs in the ductile-to-brittle transition temperature vs hydrogen content curve at about the peak of the miscibility gap. Considerably lower hydrogen contents are needed to cause embrittlement in the temperature region in which a second phase is stable, than at higher temperatures.

The abrupt onset of embrittlement of columbium at lower temperatures at a critical hydrogen content which is near the boundary of the two-phase (α + hydride) region suggests that the embrittlement is associated with the precipitation of the brittle hydride phase. Daniels and coworkers have suggested that the brittle hydride particles serve as nuclei for microcracks. As indicated by Fig. 13 and by Longson, it appears that the critical hydrogen content is below the solubility limit for hydrogen in the α phase. To explain this, it has been proposed by Oakwood and Daniels (54) and at Rocketdyne (56, 94) that a stress-induced hydride precipitation may occur with hydrogen concentrations near the solubility limit but in the one-phase region. Benson (64) studied crack propagation in hydrogenated columbium at temperatures below room temperature. He concluded that the critical propagation step in embrittled material was the formation of a microcrack of a critical size, which acts as a stress concentrator to nucleate microcracks at a second phase structure which was present. The main crack was then purported to propagate by the joining of microcracks to each other, and to the main crack front. As noted earlier, Benson reported that the second phase, the first to form as hydrogen was added to columbium, had a distorted bcc structure which resulted from the ordering of the hydrogen atoms.

To explain the ductile-brittle-ductile behavior they observed with hydrogenated columbium, Daniels and coworkers proposed that although a brittle second phase may be necessary for initiation of microcracks, the growth of the microcracks to critical size for embrittlement depends upon

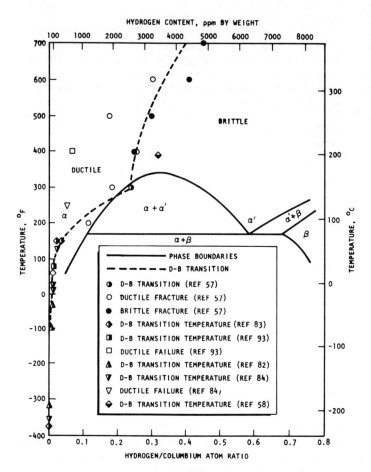

Fig. 13. Ductile-to-brittle transition temperature for columbium-hydrogen alloys.

diffusion of hydrogen. They speculated that the mechanism by which hydrogen in solution affects crack propagation in columbium is similar to that described for hydrogen embrittlement of ferrous materials. Thus, the ductile behavior at -320 F (-196 C) of hydrogenated columbium specimens was attributed to the low mobility of hydrogen at that temperature. At -116 F (-82 C), the temperature of minimum ductility, it was presumed that the low hydrogen solubility, high yield strength, and adequate hydrogen mobility combined to provide optimum conditions for crack propagation and brittle fracture.

Recently, Westlake (95) investigated columbium containing approximately 110 ppm hydrogen (H/Cb = 0.01). He found that specimens quenched from room temperature to -320 F (-196 C) and tested at that temperature were ductile while specimens quenched to and tested at -243 F (-153 C) were quite brittle. However, when the specimens were slow cooled to the test temperature, interestingly, the specimen tested at -320 F (-196 C) was brittle. The specimen tested at -243 F (-153 C) was more brittle than the quenched specimen tested at the same temperature. Thus, according to Westlake, the ductile-brittle-ductile behavior in columbium, and also vanadium, is a nonequilibrium phenomenon resulting from the quenching treatment. He proposes that slow cooling leads to the formation of an embrittling, rather coarse hydride structure while quenching to -320 F (-196 C) leads to the precipitation of a very fine dispersion of hydride particles. The ductility is then attributed to the glide of prismatic dislocation loops punched out by the hydride particles.

Although the brittle hydride may play an important role in the embrittlement of hydrogenated columbium at low temperatures, it, of course, cannot at higher temperatures. From the limited higher temperature data (Fig. 13), it is difficult to say whether in the miscibility gap temperature region the formation of α' has a role in embrittlement or that the proximity of the ductile-to-brittle transition curve to the phase field boundary is simply fortuitous. No hydride or other second phase is present, of course, at temperatures above the miscibility gap and yet columbium is embrittled by sufficiently high hydrogen contents. The shape of the ductile-to-brittle transition curve, as drawn in Fig. 13, does suggest the possibility that a different mechanism may be operative at temperatures above and below the peak of the miscibility gap.

The {100} columbium planes are the hydride habit planes (62) and the {100} planes tend to lie at 45 degrees to the working direction in wrought material. Furthermore, the most reasonable locations for hydrogen atoms are in the tetrahedral sites on the {100} planes. The presence of hydrogen in, or hydride formation on, the {100} planes can be particularly embrittling because cleavage fracture of bcc metals occurs predominantly on the {100} planes in the ⟨011⟩ directions.

It has been suggested (56) that hydrogen segregation to the cleavage planes, and the growth of the hydride from these cleavage planes at lower temperatures, could lower the cleavage strength of the metal. In other words, preferred segregation of hydrogen atoms to the tetrahedral sites may be the cause of the embrittlement by hydrogen at temperatures above which the hydride is stable. Investigation of the fractures occurring in columbium specimens at the higher temperatures is made difficult by the precipitation of hydride when the metal is cooled to room temperature.

Because alloying elements affect the phase diagram, it may be possible to use alloys in which hydride formation is suppressed to below room temperature to study the "high-temperature" embrittlement which would then be occurring at room temperature.

The effect of alloying elements on hydrogen solubility and hydride formation in columbium was discussed above, and it would be deduced that alloying could have quite an effect upon the hydrogen embrittlement of columbium. However, few direct experiments have been conducted to corroborate this. Eustice and Carlson (88) investigated the hydrogen embrittlement at temperatures below room temperature of a series of alloys of columbium and vanadium, which form a complete series of solid solutions. They found that for a given hydrogen content the ductile-to-brittle transition temperature was higher for vanadium than columbium and higher for the alloys than for either pure metal. For example, with a 10-ppm hydrogen content, the highest ductile-to-brittle transition temperature was exhibited by a V-40 weight percent Cb alloy. A ductile-brittle-ductile behavior was found for pure vanadium and all the alloys, which contained up to 70 weight percent columbium, but not for pure columbium.

Thompson et al (96) investigated delayed cracking in B-66 (Cb-5 weight percent Mo-5 percent V-1 percent Zr) and D-43 (Cb-10 weight percent W-1 percent Zr-0.1 C) alloys by conducting static load tests at room temperature on internally notched weld specimens containing up to 100 ppm hydrogen. They found that the D-43 welds were susceptible to delayed cracking while B-66 welds were not.

All of the investigations of the effect of hydrogen upon the properties of columbium discussed thus far involved the charging of specimens with hydrogen and then testing these specimens in air or inert environments. A few studies have also been conducted in which columbium or columbium alloy specimens were tested in a hydrogen environment.

McCoy and Douglas (73) determined the creep properties of columbium in various environments, including hydrogen, at 1 atmosphere of pressure at 1800 F with an applied stress of 3500 psi. The creep rate was greatly accelerated in hydrogen compared to argon, but the ductility was not affected. Of course, at 1800 F, the solubility of hydrogen in columbium is very low. Wet argon and wet hydrogen also increased the creep rate but not as much as the dry hydrogen.

Kieger et al (97) conducted tensile tests on columbium specimens at room temperature in air and in hydrogen at 1 atmosphere of pressure. For as-received columbium, no effect due to the hydrogen atmosphere was observed for strain rates of approximately 0.1 per min to 7.3 per min. However, for columbium specimens, to which oxygen (up to 2.5 atomic percent) or nitrogen (up to 0.77 atomic percent) had been added, the ductility (percent of elongation) was less in hydrogen than in air, and the difference appeared to depend upon both strain rate and oxygen or nitrogen content.

The ductility of the B-66 columbium alloy was determined in 1 atmosphere hydrogen at temperatures from room temperature to 1000 F (538 C) by Walter and Chandler (71). The tensile tests were run as follows. The specimen was heated in flowing purified argon to the test temperature and the temperature stabilized. The furnace was flushed with purified hydrogen, and after 15 minutes in flowing hydrogen, the specimen was loaded to approximately 50 percent of the room-temperature yield strength. The load was held constant for 30 minutes, then the specimen was pulled to failure at a strain rate of approximately 0.001 per min. Both unactivated and hydrogen-activated (as described earlier) specimens were tested. The results are shown in Fig. 14. Tests conducted in an approximately 10^{-5} torr vacuum are included for comparison. It can be seen that at room temperature, there was severe loss of ductility for all specimens tested in hydrogen compared to tests in vacuum. However, above room temperature, unactivated specimens had the same ductility in hydrogen as in vacuum. The activated specimens were less ductile in hydrogen than in vacuum at temperatures up to 700 F (371 C), but at 800 F (427 C) and 1000 F (538 C), the ductilities were the same in hydrogen and in vacuum.

Tensile and bend tests were used (56, 71) to evaluate the effectiveness of various coatings on B-66 alloy specimens as protection against a 1-atmosphere pressure hydrogen environment at temperatures up to 1500 F (816 C). The Sylcor R-505 tin-aluminum coating was found to be a very effective hydrogen barrier at temperatures between 800 and 1500 F (427 and 816 C) but not at temperatures below 400 F (204 C). The Sylcor silver-silicon coating showed promise over the whole temperature range but testing was limited.

Tensile tests on columbium and the B-66 columbium alloy in hydrogen at pressures of 800 and 1500 psi at temperatures up to approximately 1900 F (1038 C) are in progress at Rocketdyne under Air Force Contract AF33(615)-2854. The test procedure is similar to that described above for the tests in hydrogen at 1 atmosphere of pressure. The specimen is put in place in the tensile test furnace and heated in purified helium to 1450 F (788 C) or to the final test temperature if it is higher than 1450 F. Purified hydrogen is introduced and the pressure brought to the desired level, i.e., 800 or 1500 psi. The specimen temperature is then stabilized at the desired test temperature, and the tensile test is conducted at a strain rate of approximately 0.001 per min. The results indicate (98) that the ductile-to-brittle transition temperature for columbium in both 800- and 1500-psi hydrogen is between 1100 and 1200 F (593 and 649 C); i.e., the failure is of a distinctly brittle nature below this temperature and of a ductile nature above. The ductile-to-brittle transition temperature is considerably higher for the B-66 columbium alloy, 1800 F (982 C) to 1900 F (1038 C) in 800-psi hydrogen.

Tensile tests have also been conducted (70) on columbium and B-66 columbium alloy specimens in 1-atmosphere pressure water vapor/hydrogen environments with H_2O/H_2 mixture ratios of 3 and 1 by volume. The test procedure was the same as that described above for the tests in hydrogen at 1-atmosphere pressure. Tests were run at temperatures from 400 F (204 C) to 1500 F (816 C). The ultimate tensile strengths for both columbium and the B-66 alloy were about the same in the water vapor/hydrogen environments as in argon. Ductility minima occurred at 400 F (204 C) and 1050 F (566 C) for columbium tested in either water vapor/hydrogen environment. However, even at the minima, the ductilities were not too low, 16-percent elongation at 400 F (204 C), which was only slightly less than for tests in argon, and 20-percent elongation at 1050 F (566 C). The 1050 F (566 C) minimum was attributed to hydrogen. The oxide formed in the water vapor/hydrogen

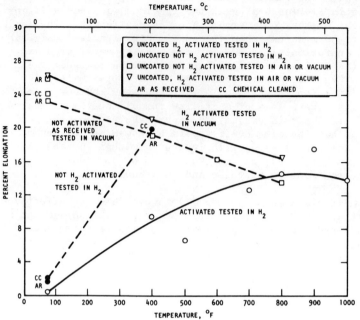

Fig. 14. Ductility of
B-66 columbium alloy
in vacuum and in hy-
drogen at 1 atmosphere
pressure as a function
of temperature (after
Walter and Chandler,
Ref. 71).

Fig. 15. Solubili-
ty of hydrogen in
tantalum (curves
from Veleckis,
Ref. 52; points
from Perminov,
Ref. 55; numbers
associated with
points are pres-
sures in atmo-
(spheres).

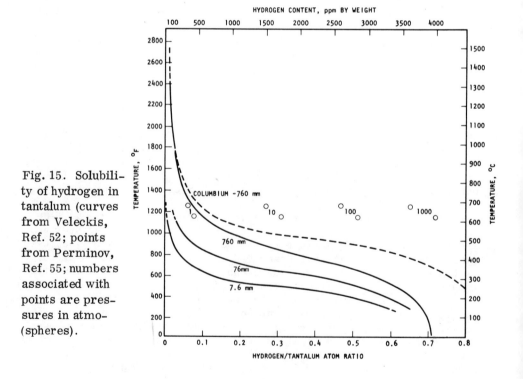

environments was thin and much more adherent than the oxide formed on columbium in air at the same temperatures. The ductility of the B-66 alloy was essentially the same in the water vapor/hydrogen environment as in argon at lower temperature but was considerably lower in the water vapor/hydrogen environments at 1500 F (816 C) for H_2O/H_2 = 1 and at 1200 F (649 C) for H_2O/H_2 = 3. The embrittlement of B-66 alloy at the higher temperatures was attributed to oxygen.

A catastrophic embrittlement of columbium or columbium alloys can occur in hydrogen environments under certain conditions. It is associated with hydride formation and we shall call it fragmentation. It has long been known that heating columbium to a high temperature, say 1500 F (816 C) in hydrogen (1 atmosphere pressure), and then slow cooling in hydrogen will cause fragmentation when the specimen reaches low temperatures at which hydride can form. Rapid cooling or cooling from the high temperature in an inert gas may prevent fragmentation. However, Stephens and Garlick (58) conducted temperature gradient tests on columbium specimens in flowing hydrogen at a 1-atmosphere pressure and found that even very rapid cooling did not prevent fragmentation. Their tests were made as follows. While in hydrogen, one end of the unloaded specimen was heated to 3000 F while the other was maintained at room temperature. The specimen was held under these conditions usually for 1 hour and then cooled in hydrogen. Fragmentation occurred with cooling rates (for the hot end) of up to 1500 F (816 C)/min, the fastest rate investigated. Cooling in argon or helium at any rate after holding in hydrogen resulted in low hydrogen contents and no fragmentation. However, if the cooling in helium was interrupted when the maximum temperature was reduced to 800 F (427 C) and hydrogen was introduced into the furnace, then fragmentation would occur on cooling from the 800 F (427 C) temperature in hydrogen.

It was found by Walter and Chandler (56, 71, 72) that progressive fragmentation of columbium can occur in hydrogen at room temperature with or without prior heating in hydrogen. For example, it was found that if the surface of a columbium or B-66 columbium alloy specimen, not previously exposed to hydrogen, was filed while the specimen was exposed to high-purity flowing hydrogen (1-atmosphere pressure) at room temperature, then fragmentation would begin at the file mark and continue until the specimen had completely fragmented. Fragmentation occurs at a surprisingly rapid rate at room temperature and with substantial release of energy as evidenced by the velocity of the fragments leaving the specimen and striking the enclosure.

The fragmentation rate is believed too rapid to be accounted for by volume diffusion which indicates that hydrogen entry is by way of grain boundaries, subgrain boundaries, and/or dislocation networks. The energy release indicates the introduction of large internal stresses from hydrogen absorption and, undoubtedly, hydride formation. For the B-66 alloy, fragmentation resulted from filing in hydrogen at temperatures up to 400 F (204 C), but not 500 F (260 C), and the rate of fragmentation increased as the temperature increased from room temperature to 400 F (204 C), as shown by Walter and Ytterhus (56). From these data, an activation energy for fragmentation of approximately 1600 cal/mole was calculated. This very low value again suggests that volume diffusion of hydrogen is not rate determining. If a columbium specimen which was fragmenting was exposed to air, the fragmentation would, of course, stop. Re-exposure to hydrogen would cause restart

of fragmentation unless the time of exposure to air was longer than approximately 1 hour. It was also observed that fragmentation would initiate in
B–66 alloy specimens in hydrogen at room temperature at locations where
chromel–alumel thermocouples had been welded onto the specimen. The specimens had no prior exposure to hydrogen. Also, when specimens were tensile
tested to failure in hydrogen at room temperature, fragmentation would
initiate from the fracture surface and proceed until the specimen was
completely fragmented.

TANTALUM

Hydrogen effects have not been investigated as extensively for tantalum as for columbium. However, investigations that have been conducted
show great similarities between hydrogen effects in tantalum and in columbium and this would, of course, be expected because of the great similarity
between these two metals in most respects. Thus, in a general sense, most
or all of what has been written about hydrogen effects in columbium is
expected to be true for tantalum, the only differences being those of degree
and not of kind.

SOLUBILITY OF HYDROGEN IN TANTALUM AND THE
TANTALUM–HYDROGEN PHASE DIAGRAM

As with columbium, a single solid solution exists in the tantalum–
hydrogen system at higher temperatures. The equilibrium solubilities of
hydrogen in tantalum have been determined by Sieverts and Bergner (99),
Kofstad et al (100), Veleckis (52), Mallett and Koehl (101), and Makrides
et al (53). The results of these investigators are in good agreement. The
results of Veleckis as plotted by Elliott (102) are shown in Fig. 15. For
comparison purposes, the curve for the solubility of hydrogen in columbium
for 1–atmosphere hydrogen pressure is included in Fig. 15 as the dashed
curve. The curves for tantalum and columbium are very similar in shape,
but the one for tantalum is displaced to lower hydrogen solubilities at
all temperatures except those above approximately 1650 F (900 C). In Fig.
15 are plotted results from Perminov (55) for the solubility of hydrogen in
tantalum at 1152 (622 C) and 1254 F (679 C) for hydrogen pressures from 1
to 1000 atmospheres. Work is currently under way at Rocketdyne under Air
Force Contract No. AF33(615)–2854 to determine the solubility of hydrogen
in tantalum and the tantalum–10 weight percent tungsten alloy for hydrogen
pressures of 800 and 1500 psi at temperatures up to 1800 F (982 C).

The solubility of hydrogen in the tantalum–10 weight percent tungsten
alloy for a hydrogen pressure of 1 atmosphere was determined (56) and is
shown in Fig. 16 with the curve for pure tantalum from Fig. 15 for comparison. The lower solubility of hydrogen in Ta–10W than in pure tantalum is
consistent with the electron/atom ratio effect discussed in the section on
columbium.

In experiments of the type discussed in the section on columbium,
Stephens and Garlick (58) found that rhenium additions to tantalum reduced
the amount of hydrogen absorbed. The amount of hydrogen absorbed dropped
from 40.8 atomic percent (H/Ta = 0.69) for pure tantalum to 0.2 atomic
percent (H/Ta = 0.002) for Ta–25 atomic percent Re (e/a = 5.5) and 0.018
atomic percent (H/Ta = 0.00018) for Ta–37.5 atomic percent Re (e/a = 5.75).

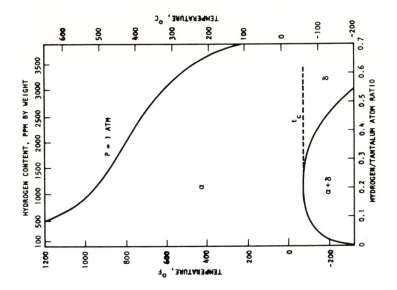

Fig. 17. Phase diagram of tantalum–hydrogen system (after Veleckis, Ref. 52).

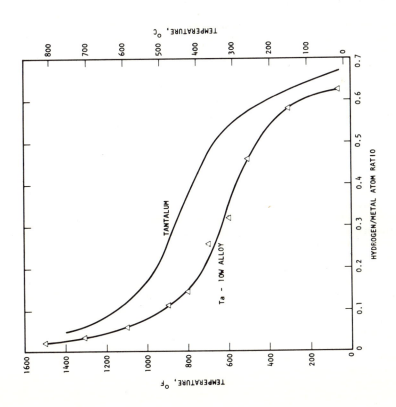

Fig. 16. Solubility of hydrogen in tantalum and in tantalum–10 w/o tungsten alloy for hydrogen pressure of 1 atmosphere (after Walter and Ytterhus, Ref. 56).

Although there have been a number of investigations of the tantalum–hydrogen system at lower temperatures, the tantalum–hydrogen phase diagram has not been established. Figure 17 shows a miscibility gap as predicted by Veleckis (52) and, for comparison, the solubility curve for 1-atmosphere pressure is included. The peak of the miscibility gap was calculated to occur at –73 F (–58.5 C) and H/Ta $\simeq$ 0.2. Veleckis actually conducted equilibrium experiments over the temperature range from 662 F (350 C) to 1135 F (613 C) and obtained the lower–temperature isotherms from which he determined the miscibility gap by extrapolating using van't Hoff plots. Veleckis noted that the accuracy of the miscibility gap is in question because of the rather narrow composition range for which he obtained experimental data and the extreme extrapolation to low temperatures that is required.

Other pressure–temperature–composition equilibria data indicate the existence of a miscibility gap but at higher temperatures. The results of Mallet and Koehl (101) suggested that a two–phase region exists, but they could only estimate that it would appear at some temperature below 212 F (100 C). Veleckis notes that Sieverts and Bergner's data (99) would indicate higher temperatures for the miscibility gap.

K. K. Kelly (103) found that hydrogenated tantalum exhibited anomalous heat capacity above –280 F (–173 C) for hydrogen contents up to H/Cb = 0.1. Waite et al (104) used Kelly's heat capacity data and the results of X–ray and electrical resistivity measurements they made to construct the phase diagram shown in Fig. 18. The α phase was bcc and β was reported to be Ta_2H with a body-centered-tetragonal structure. It was also found that as the hydrogen concentration in the β phase was increased beyond that of Ta_2H, the lattice appeared to distort gradually into a face–centered–orthorhombic structure.

A number of X–ray diffraction investigations of the tantalum–hydrogen system have been made in addition to the one by Waite et al, and have yielded highly conflicting results. Hägg (105) observed three phases at room temperature: a bcc phase from 0 to 12 atomic percent hydrogen (H/Ta = 0.14), a hexagonal–close–packed phase at about 33 atomic percent hydrogen (H/Ta = 0.5), and a face–centered–orthorhombic phase at 50 atomic percent hydrogen (H/Ta = 1). Pietsch and Lehl (106) found two bcc phases and a miscibility gap at room temperature, but no compositions were determined. Horn and Ziegler (107) investigated the tantalum–hydrogen system at room temperature for hydrogen contents up to 34 atomic percent hydrogen (H/Ta = 0.5) and found no second phase, only the bcc terminal solid solution phase whose lattice parameter increased continuously as the hydrogen content increased. Brauer and Hermann (60) observed two phases at room temperature, a bcc phase from 0 to 17 atomic percent hydrogen (H/Ta = 0.2), and a face–centered–orthorhombic phase above 37 atomic percent hydrogen (H/Ta = 0.6) with the two phases existing together for compositions between 17 and 37 atomic percent hydrogen. Brauer and Hermann proposed that the hexagonal–close–packed phase reported by Hägg was actually a nitride. Stalinski (108) also reported two phases at room temperature with a bcc phase existing from 0 to 27 atomic percent hydrogen (H/Ta = 0.37), a two–phase field existing from 27 to 32 atomic percent hydrogen, and an orthorhombic (probably face–centered orthorhombic) phase above 32 atomic percent hydrogen (H/Ta = 0.47) which extended up to 43 atomic percent hydrogen (H/Ta = 0.75).

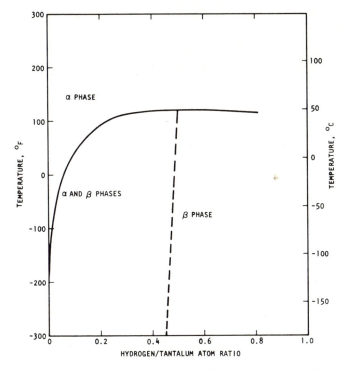

Fig. 18. Phase diagram of tantalum-hydrogen system
(after Waite et al., Ref. 104).

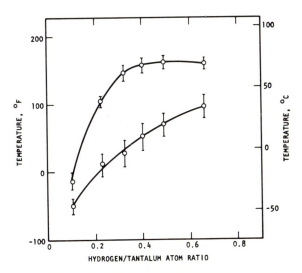

Fig. 19. Phase diagram of tantalum-hydrogen
system (after Pedersen et al., Ref. 111).

Bakish (109) found that the habit plane for the precipitation of the β –tantalum hydride is $\lfloor 100 \rfloor$ as is the case for the β columbium hydride.

Saba et al (110) measured the heat capacity of a composition corresponding to Ta_2H and found two anomalies which were interpreted as showing that several polymorphic modifications of Ta_2H exist: β_1 below 90 F (32 C), β_2 between 90 F (32 C) and 138 F (59 C), β_3 between 138 F (59 C) and 140 F (60 C), and α above 140 F (60 C).

Pedersen et al (111) conducted a proton magnetic resonance study of the temperature dependence of the spin–lattice relaxation time, T_1, as a function of hydrogen content in tantalum. Discontinuities in the T_1 curves were observed at temperatures close to the temperatures at which Saba et al observed the heat capacity anomalies. From their results, Pedersen et al constructed a phase diagram, as shown in Fig. 19.

There appears to be considerable disagreement among the investigations of the tantalum–hydrogen system at the lower temperatures and among the proposed phase diagrams. It is suggested by the authors that the tantalum–hydrogen phase diagram is in reality much like the columbium–hydrogen phase diagram shown in Fig. 7 with similar α, α', and β hydride phases. Apparently, in comparison to the columbium–hydrogen diagram, the tantalum–hydrogen diagram is shifted to lower temperatures and also to lower hydrogen concentrations. Some of the discrepancies in results from different investigators is probably because of impurities.

As was pointed out in the section on columbium, Walter and Chandler found that using a 99.0–percent, minimum–purity columbium rather than a 99.5–percent, minimum–purity columbium lowered the $\alpha + \alpha'$ miscibility gap, with the peak temperature being lowered by approximately 65 F (36 C). Thus, it is suggested that in the tantalum–hydrogen system, the peak of the miscibility gap is just above room temperature for pure tantalum and just below for somewhat less–pure tantalum. The tantalum used by Horn and Ziegler may have been impure enough to lower the miscibility gap below room temperature, thus accounting for their finding only one phase up to approximately H/Ta = 0.5. The body–centered–tetragonal phase observed by Waite et al may have been the α' phase. It would appear from the investigations of Brauer and Herman and of Stalinski that the β hydride phase must extend to lower hydrogen concentrations in the tantalum–hydrogen system than in the columbium hydrogen system, probably down to H/Ta = 0.5. The anomalous heat capacity results reported by Saba et al for a composition corresponding to Ta_2H may be associated with a peritectoid reaction such as is indicated for the columbium–hydrogen system.

PERMEABILITY, DIFFUSION, AND KINETICS OF ABSORPTION
AND DESORPTION OF HYDROGEN IN TANTALUM

The permeation rate of hydrogen through tantalum was investigated by Makrides et al (53) and Wright et al (112) over the temperature range of 752 F (400 C) to 1292 F (700 C) for pressures up to 160 psig (less than 50 μ on the low–pressure side). The tantalum membranes were prepared by electrolytic etching in HF or chemical etching in an $HF–H_2SO_4–HNO_3–H_2O$ solution and then coating with a 1000 Å thick layer of palladium by sputtering. Between

752 F (400 C) and 977 F (525 C), the permeability was reported to be diffusion controlled with the diffusion coefficients given by:

$$D(cm^2/sec) \; = \; 0.075 \; exp \left[- \frac{14,400 \; cal/mole}{RT(^\circ K)} \right] \; for \; H/Ta = 0.065 \quad (5)$$

$$D(cm^2/sec) \; = \; 0.17 \; exp \left[- \frac{14,400 \; cal/mole}{RT(^\circ K)} \right] \; for \; H/Ta = 0.16 \quad (6)$$

At temperatures below 752 F (400 C), the tantalum membranes failed mechanically when the H/Ta ratio approached 0.6. At temperatures above 977 F (525 C), the permeability rates dropped rapidly with time, which was attributed to breaking up of the palladium coating and poisoning of the tantalum surface. Wright et al (112) also determined the effect of adding approximately 5 percent CO_2 or CO to the hydrogen. The decrease in the rate of permeation of hydrogen through tantalum was approximately 15 percent, i.e., it was greater than one would predict from dilution alone.

Gulbransen and Andrew (68) investigated the rate of absorption of hydrogen in tantalum over the temperature range of 392 F (200 C) to 1472 F (800 C) for hydrogen pressures up to 38 torr. They found that tantalum absorbed hydrogen at a much slower rate than did columbium, as shown in Fig. 10. When a tantalum specimen was slowly heated in hydrogen, absorption did not begin at any measurable rate until the temperature reached approximately 644 F (340 C). Gulbransen and Andrew found for tantalum, as they did for columbium, that there was no consistent rate law governing absorption of hydrogen at 662 F (350 C) to 752 F (400 C), the rate of absorption was almost linear with a slight tendency for the rate to increase with time. At 842 F (450 C) and 932 F (500 C), there was a rapid initial absorption and the maximum amount absorbed was reached in a relatively short time. At 1112 F (600 C) to 1472 F (800 C), absorption of hydrogen by tantalum followed a parabolic rate law.

Mallett and Koehl (113) studied the absorption of hydrogen by tantalum to produce hydrogen atom fractions of 0.05 (H/Ta = 0.053) at 932 F (500 C) to 1292 F (700 C) and 0.10 (H/Ta = 0.11) at 842 F (450 C) to 1112 F (600 C). For temperatures at and above 932 F (500 C), the absorption of hydrogen was initially parabolic, but at lower temperatures, the nature of the reaction was not as clear although the absorption was not linear. Mallett and Koehl interpreted their data in terms of diffusion behavior and obtained the following diffusion coefficients:

$$D(cm^2/sec) = 1560 \; exp \left[- \frac{32,230 \pm 3,140 \; cal/mole}{RT(^\circ K)} \right] \; for \; H/Ta = 0.053 \; (7)$$

$$D(cm^2/sec) = 13,960 \; exp \left[- \frac{33,620 \pm 1,370 \; cal/mole}{RT(^\circ K)} \right] \; for \; H/Ta = 0.11 \quad (8)$$

These values for D_0 and the activation energy are very high when compared to those for diffusion of hydrogen in similar metals. For example, the activation energies in cal/mole for the diffusion of hydrogen are 9370 in columbium, 12,380 in α-titanium (114), 6640 in β-titanium (114), and 7060 in α-zirconium (115). Thus, the diffusion equations developed by Makrides et al (53)(Eq. 5, 6) appear more reasonable. As Wright et al (112) noted, it appears that surface reactions rather than volume diffusion were rate determining in Mallett and Koehl's experiments.

Zubler (116) measured the absorption of hydrogen by tantalum at low pressures, 0 to 0.2 torr. He found that the hydrogen was absorbed reversibly from 392 F (200 C) to 1346 F (730 C). Between 392 F (200 C) and 932 F (500 C), the reaction obeyed a parabolic rate law. At higher temperatures, the reaction was too rapid to measure by the method used (measurement of the decrease of gas pressure with a micromanometer). For a tantalum specimen which was degassed at a temperature of 3272 F (1800 C) or higher, the minimum temperature at which absorption was observed was 698 F (370 C). However, if the specimen had absorbed hydrogen at a higher temperature, e.g., 932 F (500 C) to 1112 F (600 C), then absorption was observed at 392 F (200 C).

From measurements of the kinetics of hydrogen desorption from tantalum, Klyachko (117) determined the following diffusion equation:

$$D(cm^2/sec) \; = \; 0.000339 \; exp \left[- \frac{14,400 \; cal/mole}{RT(°K)} \right] \qquad (9)$$

Merisov (118) determined a diffusion coefficient for hydrogen in tantalum over the temperature range of 32 F (0 C) to 320 F (160 C) by measuring the increase in the electrical resistivity of a section of wire, depending upon the amount of hydrogen diffused to it from an adjoining electrolytically charged section. The diffusion coefficient obtained was:

$$D(cm^2/sec) \; = \; 0.00061 \; exp \left[- \frac{3500 \; cal/mole}{RT(°K)} \right] \qquad (10)$$

The values of D_o and the activation energy appear low.

At this time, the most reasonable values for the diffusion parameters appear to be those of Makrides et al (53) given earlier (Eq. 5, 6).

The absorption and desorption rates of hydrogen in tantalum and the Ta-10W alloy for hydrogen pressures of 800 and 1500 psi and for temperatures up to approximately 1800 F (982 C) are currently being determined at Rocketdyne under Air Force Contract No. AF33(615)-2854.

Again, for tantalum, as discussed earlier for columbium, the great effect of surface condition upon hydrogen absorption is apparent and the conflicting absorption data are undoubtedly caused by differences in surface preparation. Activation treatments are required for tantalum, as well as for columbium, prior to absorption studies or to obtain reasonably rapid absorption rates for any reason at lower temperatures. The temperature below which prior activation treatments are required to obtain rapid absorption is not as well established for tantalum as for columbium. The work of Zubler, described previously, indicates that for rapid absorption at low temperatures, previous absorption of hydrogen at higher temperatures is more effective than simply heating to a high temperature in vacuum. At Rocketdyne (71, 72), activation of tantalum for hydrogen absorption is accomplished by temperature cycling in hydrogen in much the same manner as described earlier for columbium. Thus, to introduce hydrogen at any temperature below 1500 F (816 C), the tantalum specimen is first temperature cycled three times in hydrogen between 1500 F (816 C) and the final test temperature. For introduction of hydrogen into tantalum at 1500 F (816 C) and higher, prior activation is not believed necessary although no studies have been made of the effect of prior activation by temperature cycling in hydrogen on absorption at the higher temperatures.

EFFECT OF HYDROGEN ON THE PROPERTIES OF TANTALUM

That tantalum is embrittled by hydrogen is well known, but the effects of hydrogen on properties have been even less extensively investigated for tantalum than for columbium and, again, data are conflicting.

Wilcox and Huggins (76) investigated strain-aging of arc-melted tantalum by yield point return and dynamic modulus measurements. The tantalum contained 5-ppm hydrogen (H/Ta = 0.0009) and 150-ppm carbon, 90-ppm oxygen and 20-ppm nitrogen. The return of the yield point at -108 F (-78 C) was determined on deformed specimens which had been aged at temperatures from 150 F (65 C) to 275 F (135 C). The return of the modulus of elasticity at room temperature was determined for deformed specimens aged at temperatures from 350 F (177 C) to 550 F (288 C). Activation energies calculated were: 16,900 cal/mole for the return of the yield point and 13,100 - 14,400 cal/mole for the return of the modulus. These activation energies are in good agreement with the 14,400 cal/mole activation energy for the diffusion of hydrogen in tantalum determined by Makrides (Eq. 5 and 6).

Wilcox and Huggins investigated the effect of hydrogen content on the deformation of tantalum at 32 F (0 C) and -108 F (-78 C) for hydrogen concentrations from 4 to 196 ppm (H/Ta = 0.0007 to 0.035). The hydrogenating treatments lowered the concentration of other interstitials. It was found that the hydrogen content had no effect upon the upper or lower yield point at either temperature. However, the specimen containing 196 ppm hydrogen had a drastically reduced ductility. Microscopic examination at a magnification of 1000 did not reveal any hydride. It was noted in the specimen containing 196 ppm hydrogen and tested at 32 F (0 C) that numerous surface cracks formed perpendicularly to the stress axis. Even with the formation of these cracks, the specimen still had a final reduction of area of 46 percent (compared to 99 percent for a specimen containing 4 ppm hydrogen).

As they did for columbium, Imgram et al (83) also performed tests on notched and unnotched, wrought and recrystallized tantalum specimens containing different hydrogen contents. The specimens were tested at a constant cross-head speed of 0.02 in./min for unnotched specimens and 0.005 in./min for notched specimens. They found for tantalum, as they had for columbium, that hydrogen had essentially no effect upon the strength. However, the ductility and notch-unnotch strength ratios were significantly affected by hydrogen and a ductile-brittle-ductile behavior was observed with the ductility minimum occurring at -105 F (-76 C). The results of Imgram et al are presented in Fig. 20. There was no significant difference between the behavior of wrought and recrystallized hydrogenated specimens. Imgram et al concluded that hydrogen promotes cleavage fracture. Imgram et al (89) performed tensile tests at -105 F (-76 C) on tantalum containing 135 ppm hydrogen (H/Ta = 0.024) at four strain rates from 0.00015 per minute to 4.0 per minute and found that both strength and ductility increased with strain rate.

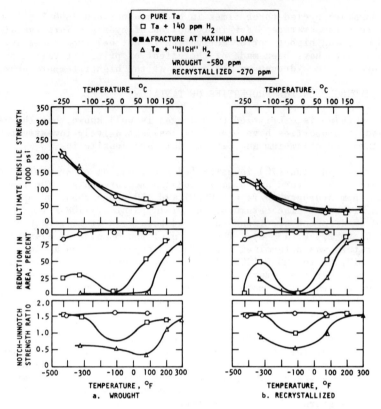

Fig. 20. Low-temperature tensile and notch-tensile properties of tantalum with different hydrogen contents (after Im-
gam et al., Ref. 83).

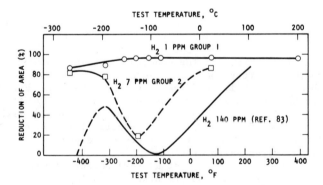

Fig. 21. Ductility of tantalum as influenced by
hydrogen content and test temperature (after
Nunes, et al., Ref. 92).

Nunes et al (92) obtained true stress-true strain tensile properties of commercially pure tantalum at temperatures from 392 F (200 C) to -452 F (-269 C) using an initial strain rate of 0.01 per minute. Two groups of specimens were tested, one containing 1 ppm hydrogen and the other 7 ppm hydrogen. Surprisingly, the second group (7 ppm hydrogen) exhibited severe embrittlement, as measured by reduction of area, between room temperature and -320 F (-196 C) and a ductile-brittle-ductile behavior with the ductility minimum occurring at approximately -202 F (-130 C). The results are shown in Fig. 21 with a curve taken from Imgram et al (83). Nunes et al also noted that for the specimens with 7 ppm hydrogen, there was a definite suppression of the upper yield point with some evidence of increased strain hardening and decreased elongation. Also, the specimen tested at -202 F (-130 C) showed no evidence of strain hardening and failed by cleavage. The difference in hydrogen content in the two groups of specimens resulted from different recrystallization annealing treatments at 2372 F (1300 C), one conducted in a furnace at 10^{-5} torr and one with specimens sealed in a quartz tube. Nunes et al stated that although the evidence indicating hydrogen embrittlement was strong, other possibilities could not be ruled out. The surprising fact is the large degree of embrittlement from such a small hydrogen content.

In a number of instances, the effect of hydrogen on the mechanical behavior of tantalum has been interpreted by utilizing the Petch equation (see Eq. 4).

Abowitz and Burn (119) conducted tensile tests at temperatures from 77 F (25 C) to -320 F (-196 C) with a strain rate of 10^{-3} per second on tantalum containing less than 1 to 394 ppm hydrogen. The ultimate tensile strength increased as the hydrogen content increased, but it varied in the same way with the total interstitial content. Abowitz and Burn used the Petch analysis and concluded that hydrogen did not affect σ_i, the lattice friction stress, but did raise k_y, a measure of the pinning strength. However, there was considerable scatter in the data.

Gazza (120) conducted tensile tests at temperatures from 77 F (25 C) to -320 F (-196 C) with a platen speed of 0.05 in./min on pure tantalum specimens which contained 2 to 5 ppm hydrogen, and on hydrogenated tantalum specimens which contained 20 to 30 ppm hydrogen. Gazza observed that the hydrogenated tantalum exhibited ductile-brittle-ductile behavior, as shown in Fig. 22. Gazza used the Petch analysis and concluded that hydrogen did not affect k_y but did influence σ_i, thus disagreeing with Abowitz and Burn. Gazza's σ_i values were quite consistent but the k_y values were quite erratic.

Owen et al (121) conducted tests at room temperature with a strain rate of 2.3 x 10^{-4} per second on tantalum specimens which were charged with 99 ppm hydrogen and then aged in vacuum for 1 hour at temperatures from 1112 F (600 C) to 1832 F (1000 C). They also conducted a thin-film electron microscopy investigation of foils treated in the same way as the tensile specimens. Some strengthening of the hydrogen charged material was observed. From a Petch analysis, Owen et al concluded that there was no effect of aging or of introduction of hydrogen on σ_i but that k_y was increased. k_y increased only slightly with hydrogen charging, increased considerably with aging at 1112 F (600 C), 1292 F (700 C), and 1472 F (800 C), and then decreased with aging at 1652 F (900 C) and 1832 F (1000 C) to values near that for the hydrogen charged but unaged material. The electron microscopy revealed no evidence of precipitates in the as-hydrogen charged foils, but precipitates were observed in the aged foils and Owen et al correlated k_y with precipitate size. Owen et al also reported observing an order-disorder transformation directly in the electron microscope.

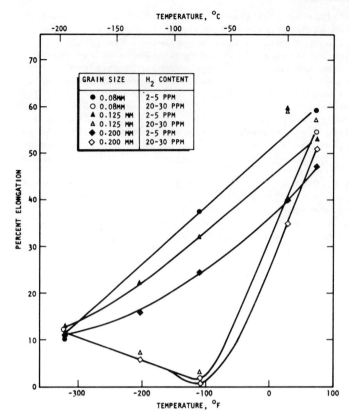

Fig. 22. Effect of hydrogen, temperature, and grain size on the percent elongation to fracture of tantalum (after Gazza, Ref. 120).

Fig. 23. S-N curves for uncharged and hydrogen-charged, wrought, stress-relieved tantalum tested in tension-compression (2000 cycles/minute) at –108°F (–78°C) (after Wilcox, Ref. 122).

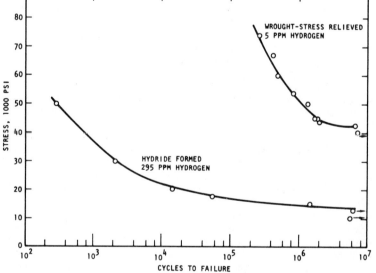

There are some aspects of the investigation by Owen et al which appear to require clarification. It would be surprising if a considerable amount of the hydrogen was not desorbed from the specimens (and particularly the electron microscopy foils), during the aging (vacuum annealing) treatments, particularly at the higher temperatures. No hydrogen analyses were reported for aged specimens. The available information on the tantalum-hydrogen phase diagram, as discussed earlier, would indicate that no second phase would form at room temperature and above in tantalum which contains only 90 ppm hydrogen. As noted earlier, it has been pointed out by Westlake (65) that it is difficult to study hydrogen effects in thin films (by transmission electron microscopy) because of the considerable effect of thinning operations on hydrogen content.

Wilcox (122) conducted a limited investigation of the effect of hydrogen on the fatigue strength of tantalum. Tantalum specimens containing 5 ppm and 295 ppm hydrogen were tested in tension-compression at -108 F (-78 C). The results are shown in Fig. 23. Wilcox reported a uniform dispersion of a hydride precipitate across the 295 ppm hydrogen specimens. In a few tests, Wilcox also found that 35 ppm hydrogen (no precipitate formed) reduced the fatigue strength somewhat.

Stephens and Garlick (58) determined the effect of hydrogen on the ductile-to-brittle bend transition temperature of tantalum for temperatures from -320 F (-196 C) to 1500 F (816 C). The ductile-to-brittle bend transition temperature was defined as the minimum temperature at which specimens with the dimensions 0.025 by 0.5 by 1.25 inches could be bent 120 degrees over a 4 T bend radius using a crosshead speed of 1 in./min without fracture. Their results are shown in Fig. 24.

A few investigations have been made of the properties of tantalum in hydrogen environments. Studies were conducted at General Electric (30) on the stress-rupture of tantalum in argon and in hydrogen at 3992 F (2200 C) and 4712 F (2600 C). For arc-cast tantalum, it was found that the stress-rupture properties were the same in hydrogen as in argon for both temperatures. However, powder metallurgy tantalum, which was tested at 4712 F (2600 C), had a higher rupture strength in hydrogen than in argon. The results for tantalum resemble those for molybdenum for which the higher rupture strength in hydrogen was attributed to additional sintering, promoted by the hydrogen environment during the test.

The ductility of the tantalum-10 weight percent tungsten alloy was determined in 1-atmosphere pressure hydrogen at temperatures from room temperature to 1000 F (538 C) by Walter and Chandler (71). The tensile tests were run in the same way as for the tests on the B-66 columbium alloy as described earlier. The results are shown in Fig. 25. Tests conducted in vacuum, approximately 10^{-5} torr, are included in the figure for comparison. Loss of ductility, as measured by percent of elongation, of Ta-10W because of hydrogen was observed only at room temperature and 400 F (204 C). At room temperature, specimens in the as-received condition, i.e., not chemically cleaned and not hydrogen activated, were not embrittled, and, surprisingly, hydrogen-activated specimens were not embrittled at 400 F (204 C), although others were. It should be noted that these "activated" specimens were exposed to air before testing after activation by temperature cycling in hydrogen. Tests on activated specimens which were not exposed to air after activation showed a considerable loss of ductility but still not as much as did the unactivated specimens.

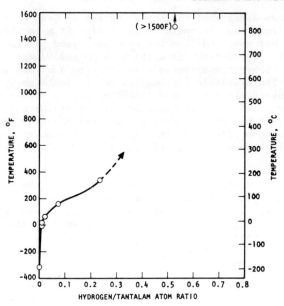

Fig. 24. Effect of hydrogen on the ductile-to-brittle bend transition temperature of tantalum (after Stephens and Garlick, Ref. 58).

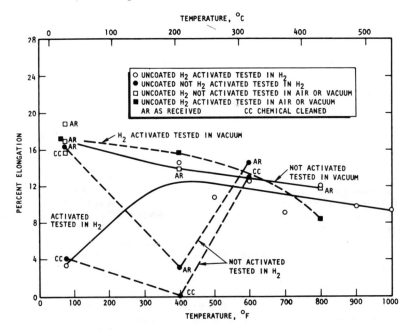

Fig. 25. Ductility of tantalum–10 tungsten alloy in vaccum and in hydrogen at one atmosphere pressure as a function of temperature (after Walter and Chandler, Ref. 71).

The effectiveness of various coatings as hydrogen barriers was evaluated at Rocketdyne (56, 71) for Ta-10W as well as for the B-66 columbium alloy as described earlier. Again, the Sylcor R-505 tin-aluminum coating was found to be an effective hydrogen barrier at temperatures between 800 F (427 C) and 1500 F (816 C) but not at lower temperatures.

Tensile tests on tantalum and Ta-10W in hydrogen at pressures of 800 and 1500 psi at temperature up to approximately 1900 F (1038 C) are in progress at Rocketdyne under Air Force Contract No. AF33(615)-2854. The test procedure has been described under the section on columbium. The results indicate (123) that the ductile-to-brittle transition temperature for tantalum is approximately 1200 F (649 C) in both 800- and 1500-psi hydrogen. The investigation of Ta-10W is still in progress.

Tensile tests have also been conducted by Bentle and Chandler (98) on tantalum and Ta-10W in 1-atmosphere pressure water vapor/hydrogen environments with H_2O/H_2 mixture ratios of 3 and 1 by volume. Both tantalum and Ta-10W showed lower ductilities in the water vapor/hydrogen environments than in argon at temperatures above approximately 800 F (427 C), but not below; however the major effect is probably due to oxidation and oxygen absorption.

The catastrophic low-temperature fragmentation phenomenon described earlier for columbium also occurs with tantalum and tantalum alloys. The conditions for occurrence are essentially the same for tantalum as for columbium. Cooling tantalum from elevated temperatures in hydrogen will lead to fragmentation. Stephens and Garlick (58) conducted temperature gradient (3000 F to 72 F) tests on tantalum and tantalum alloy specimens in flowing hydrogen at a 1-atmosphere pressure and found that even rapid cooling (1500 F per minute) did not prevent fragmentation for some distance along the specimen from the hot end. These experiments were described in the section on columbium.

As described earlier for columbium, it was found at Rocketdyne (56, 71, 72) that fragmentation of Ta-10W will initiate at a file mark made in a specimen exposed to high-purity, flowing (1-atmosphere pressure) hydrogen at room temperature or at a location in a similarly exposed specimen, where a chromel-alumel thermocouple had previously been welded. Also, with Ta-10W, but not columbium, bending in hydrogen at room temperature led to the initiation of fragmentation.

SUMMARY

In summary, the solubility and permeability of hydrogen is small in molybdenum and very small in tungsten. The effect of hydrogen on the mechanical properties of these metals is minimal, the only reported effects coming as the result of secondary effects such as removal of other interstitials or reaction with other alloying elements or promotion of sintering. Suggestions have been made that molybdenum and tungsten are subject to hydrogen embrittlement similar to that occurring in steels but no such embrittlement has been established.

On the other hand, columbium and tantalum can be highly embrittled by hydrogen, catastrophically under certain conditions. However, there are somewhat alleviating circumstances. Low-temperature hydrogen embrittlement

of columbium and tantalum is generally associated with hydride formation and
the hydrides are stable at only relatively low temperatures. The solubility
of hydrogen in these metals decreases as the temperature increases, rapidly
above approximately 900 F (482 C), and is quite low above 1600 F (871 C).
Because the rate of hydrogen absorption decreases as the temperature decreases,
under many circumstances exposure to hydrogen in the intermediate temperature
range near 900 F (480 C) will result in the most severe embrittlement because
the solubility and rate of absorption of hydrogen are both high enough to
result in considerable absorption. Absorption of hydrogen at temperatures
below approximately 1100 F (593 C) is quite slow unless the metal is previously
"activated" for hydrogen absorption. Also, the effect of a given amount of
hydrogen on properties decreases as the temperature increases. Because of
these factors, there may be practical conditions of exposure to hydrogen that
will not embrittle columbium or tantalum. A catastrophic embrittlement that
must be avoided in any case is the fragmentation which results from cooling
from elevated temperatures to room temperature in hydrogen or from exposure of
a fresh surface to hydrogen at room temperature. Alloying and hydrogen barrier
coatings give promise of alleviating the hydrogen embrittlement problem with
columbium and tantalum.

As this review indicated, a very great deal remains to be done in the
area of hydrogen effects in columbium and tantalum. The tantalum-hydrogen
phase diagram must be determined; there are many aspects of hydrogen embrittlement in this system that are difficult to discuss without it. Hydrogen
embrittlement at low temperatures, e.g., stress-induced hydride formation
needs further investigation, and there have been no investigations of the
mechanism of the hydrogen embrittlement which occurs in columbium and tantalum under conditions such that no hydride can form. The reasons for the
effectiveness of surface activation treatments in increasing the rates of
subsequent absorption of hydrogen and the mechanism of fragmentation require
study. Finally, work on practical solutions to the hydrogen embrittlement
problem in columbium and tantalum, such as alloying and hydrogen barrier
coatings, has hardly begun.

ACKNOWLEDGMENTS

Information used in this report from work accomplished at Rocketdyne, a
Division of North American Rockwell Corporation, was conducted under Independent
Research and Development funds, National Aeronautics and Space Adminstration
Contract No. NAS8-19, and Air Force Contract No. AF33(615)-2854. Mr. L. D.
Parsons and Lt. L. D. Blackburn served as project engineers on the Air Force
program. Valuable contributions on these programs have been made by Dr. G.
Bentle and Dr. J. A Ytterhus. The assistance of J. Pero, R. D. Lloyd, J.
Mosher, P. E. Miller, and G. V. Sneesby in experimental work is gratefully
acknowledged. The authors also wish to thank Dr. R. P. Jewett for helpful
comments on the manuscript.

REFERENCES

1. Smith, D. P., Hydrogen in Metals, University of Chicago Press, Chicago, Illinois (1948).

2. Cotterill, P., "The Hydrogen Embrittlement of Metals," in Progress in Materials Science, Vol. 9, B. Chalmers, ed., Pergamon Press, New York (1961), p. 205.

3. Rylski, O. Z., Department of Mines and Technical Surveys, Mines Branch, Ottawa, Canada PM 203 (CAN).

4. Muehlenkamp, G. T. and A. D. Schwope, "Effect of Hydrogen on Mechanical Properties of Zirconium and Its Tin Alloys," USAEC Rept. No. BMI 845 (1953).

5. Jaffee, R. I., G. A. Lenning, and C. M. Craighead, J. Metals, 8, 907 (1956).

6. Williams, D. N., F. R. Schwartzberg, and R. I. Jaffee, Trans. ASM, 51, 802 (1959).

7. Forscher, F., Trans. AIME, 206, 536 (1956).

8. Magnusson, A. W. and W. M. Baldwin, J. Mech. Phys. of Solids, 5, 172 (1957).

9. Johnson, H. H., "On Hydrogen Brittleness in High Strength Steel," Paper to be published in Proceedings of the International Conference on Fundamental Aspects of Stress-Corrosion Cracking held at Ohio State University, Columbus, Ohio, from 11 to 15 September 1967, R. W. Staehle, ed.

10. Oriani, R. A., "Hydrogen in Metals," ibid.

11. Gibb, T. R. P., Jr., "Primary Solid Hydrides," in Progress in Inorganic Chemistry, Vol. 3, F. A. Cotton, ed., Interscience Publishers, New York (1962), p. 315.

12. Sieverts, A. and K. Bruning, Arch. Eisenhüttnw., 7, 641 (1934).

13. Hill, M. L., J. Metals, 12, 725 (1960).

14. Martin, E., Arch. Eisenhüttenw, 3, 407 (1929).

15. Jones, P. M. S., R. Gibson and J. A. Evans, AWRE Report No. 0-16/66, Atomic Weapons Research Establishment, Aldermaston, England, February 1966.

16. Moore, G. E. and F. C. Unterwald, J. Chem. Phys., 40 (9), 2639 (1964).

17. Jones, D. W., N. Pessall, and A. D. McQuillan, Phil. Mag., 6, 455 (1961).

18. Jones, D. W. and A. D. McQuillan, J. Phys. Chem. Solids, 23, 1441 (1962).

19. Jones, D. W., Phil. Mag., 9, 709 (1964).

20. Booth, J. G., "Effect of Electron Concentration on Mechanical Properties
     of Alloys of Refractory Metals," Annual Topical Report on Office of
     Naval Research Contract No. Nonr-3589-(00), 17 October 1963, ASTIA
     AD No. 421178.

21. Smithells, C. J. and C. E. Ransley, Proc. Roy. Soc., A, 150, 172 (1935).

22. Dushman, S., Scientific Foundations of Vacuum Technique, John Wiley and
     Sons, Inc., New York, 608 (1949).

23. Huffine, C. L. and J. M. Williams, Corrosion, 16, 432t (September 1960).

24. Norton, F. J., in Trans. Eighth Vacuum Symp., American Vacuum Soc.,
     L. E. Preuss, ed., Pergamon Press, New York (1962), Vol. 1, p. 8.

25. Steigerwald, E. A., "The Permeation of Hydrogen Through Materials for
     the Sunflower System," Engineering Report ER-5623, Materials Technology,
     Tapco, A Division of Thompson Ramo Wooldridge Inc., Cleveland, Ohio
     (15 November 1963).

26. McCracken, G. M. and J. H. C. Maple, Brit. J. Appl. Phys., 18, 919 (1967).

27. Ryabchikov, L. M., Ukr. Fiz. Zhur., 9 (3), 293 (1964).

28. "High Temperature Materials and Reactor Component Development Programs,"
     GEMP-177A, Second Annual Report, Volume I - Materials, General Electric
     Company, Contract No. AT(40-1)-2847 (28 February 1963).

29. Conway, J. B., D. G. Salyards, W. L. McCullough, and P. N. Flagella,
     Paper presented at American Nuclear Society Meeting, Cincinnati, Ohio,
     17-19 April 1963. NASA Doc. N64-11984 (1963).

30. "High Temperature Materials and Reactor Component Development Programs,"
     GEMP-270A, Third Annual Report, Volume I - Materials, General Electric
     Company, Cincinnati, Ohio, Contract No. AT(40-1)-2847 (28 February 1964).

31. Flagella, P. N., in Refractory Metals and Alloys III; Applied Aspects
     Part 2, AIME Metallurgical Society Conferences Vol. 30, R. I. Jaffee, ed.,
     Gordon and Breach Science Publishers, New York (1966), p. 917.

32. Flagella, P. N., J. AIAA, 5, (2), 281 (1967).

33. Perkins, R. A. and J. L. Lytton, "Effect of Processing Variables on the
     Structure and Properties of Refractory Metals," Technical Report No.
     AFML-TR-65-234, Part I, Air Force Materials Laboratory, Wright-Patterson
     Air Force Base, Ohio (July 1965).

34. Hahn, G. T., A. Gilbert, and R. I. Jaffee, "The Effects of Solutes on the
     Ductile-to-Brittle Transition in Refractory Metals," DMIC Memorandum
     No. 155, 28 June 1962, and in Refractory Metals and Alloys, II, AIME
     Metallurgical Society Conferences Vol. 17, M. Semchyshen and I. Perlmutter,
     eds., Interscience Publishers, New York (1963), p. 23.

35. Lawley, A., W. Liebmann, and R. Maddin, Acta Met., 9 (9), 841 (1961).

36. Feuerstein, S., in the Effects of Space Environment on Materials, Proceedings of National Symposium of Society of Aerospace Material and Process Engineers, St. Louis, Missouri, 19 to 21 April 1967, p. 299.

37. Hansen, M., Constitution of Binary Alloys, 807, McGraw-Hill, New York (1958).

38. Cupp, C. R., Progress in Metal Physics, 4, 105, Pergamon Press (1953).

39. Aitken, E. A., P. K. Conn, E. C. Duderstadt, and R. E. Fryxell, "Measurement of the Permeability of Tungsten to Hydrogen and to Oxygen," Final Report, NASA Contract NAS3-6216, NASA Report CR-54918 (May 1966) AEC Accession No. 27493.

40. Aitken, E. A., H. C. Brassfield, P. K. Conn, E. C. Duderstadt, and R. E. Fryxell, Trans. Met. Soc. AIME, 239 (10), 1565 (1967).

41. Webb, R. W., "Permeation of Hydrogen Through Metals," NAA-SR-10462, Atomics International, a Division of North American Aviation, Inc., 25 July 1965.

42. Frauenfelder, R., W. J. Lange, and J. H. Singleton, Semi-Annual Report - Ultrahigh Vacuum Techniques, Report WERL-2823-24, Westinghouse Research Laboratories, Pittsburgh, Pa., U.S.A.E.C. Contract No. AT(30-1)-2823 for period 1 June to 30 November 1966.

43. Meyers, C. L., Jr., G. Y. Onoda, Jr., A. V. Levy, and R. J. Kotfila, Trans. Met. Soc. AIME, 233 (4), 720 (1965).

44. Farrell, K., A. C. Schaffhauser, and J. O. Steigler, J. Less-Common Metals, 13 (5), 548 (1967).

45. Farrel, K., A. C. Schaffhauser, and J. O. Steigler, J. Less-Common Metals, 13 (2), 141 (1967).

46. Atkinson, R. H. and Staff of Metals Research Group, Westinghouse Lamp Division, WADD Technical Report 60-37, May 1960 (ASTIA AD240981).

47. Sell, H. G., G. H. Keith, R. C. Koo, R. H. Schnitzel, and R. Corth, Westinghouse Lamp Division, WADD Technical Report 60-37, Part II, May 1961 (ASTIA AD266330).

48. Atkinson, R. H., G. H. Keith, and R. C. Koo, "Tungsten and Tungsten-Base Alloys" in Refractory Metals and Alloys, AIME Metallurgical Society Conferences Vol. 11, M. Semchyshen and J. J. Harwood, eds., Interscience Publishers, New York (1961), p. 319.

49. Sutherland, E. C. and W. D. Klopp, "Observations of Properties of Sintered Wrought Tungsten Sheet at Very High Temperatures," NASA Technical Note D-1310, February 1963.

50. Albrecht, W. M., W. D. Goode, and M. W. Mallett, J. Electrochem. Soc., 106 (11), 981 (1959).

51. Komjathy, S., J. Less-Common Metals, 2, 466 (1960).

52. Veleckis, E., Ph.D. Thesis, Illinois Institute of Technology, ASTIA AD282433, January 1960.

53. Makrides, A. C., M. Wright, and R. McNeill, "Hydrogen Permeation Through Group Vb Metals," Final and Summary Report for Harry Diamond Laboratories Contract DA-49-186-AMC-136(d), Vol. I, Tyco Laboratories, Inc., Waltham, Massachusetts, October 1965.

54. Oakwood, T. G. and R. D. Daniels, "The Sorption Kinetics and Mechanical Properties of Dilute Columbium-Hydrogen Solutions," University of Oklahoma, Report No. ORC-2570-10, AEC Contract No. AT(40-1)-2570, February 1967.

55. Perminov, P. S., Doklady Akad. Nauk SSSR (Proceedings of the Academy of Sciences USSR) 121, 1041 (1958).

56. Walter, R. J. and J. A. Ytterhus, "Behavior of Columbium and Tantalum in Hydrogen Environments," Research Report RR 67-7, Rocketdyne, a Division of North American Aviation, Inc., Canoga Park, California, September 1967.

57. Walter, R. J., "The Columbium-Hydrogen System and Hydrogen Embrittlement of Columbium," Research Report RR 64-6, Rocketdyne, a Division of North American Aviation, Inc., Canoga Park, California, February 1964.

58. Stephens, J. R. and R. G. Garlick, "Compatibility of Tantalum, Columbium and Their Alloys with Hydrogen in the Presence of Temperature Gradient," NASA Technical Note D-3546, August 1966.

59. Walter, R. J. and W. T. Chandler, Trans. Met. Soc. AIME, 233, 762 (1965).

60. Brauer, G. and R. Herman, Z. Anorg. Chem., 274, 11 (1953).

61. Wainwright, C., A. J. Cook, and B. E. Hopkins, J. Less-Common Metals, 6, 362 (1964).

62. Paxton, H. W., J. M. Sheehan, and W. J. Babyak, Trans. AIME, 215, 725 (1959).

63. Rauch, G. C., R. M. Rose, and J. Wulff, J. Less-Common Metals, 8, 99 (1965).

64. Benson, R. B., Jr., "The Ductile-Brittle Transition in the Niobium-Hydrogen System," Ph.D. Thesis, University of California, Berkeley, Diss. Abstr. B 27 (8), 2729 (1967) Order No. 66-15,345.

65. Westlake, D. G., Trans. Met. Soc. AIME, 239, 1106 (1967).

66. Upadhyaya, G. S. and A. D. McQuillan, Trans. Met. Soc., AIME, 224, 1290 (1962).

67. Rudd, D. W., D. W. Vose, and S. Johnson, J. Phys. Chem., 66, 351 (1962).

68. Gulbransen, E. A. and K. F. Andrew, Trans. AIME, 188, 586 (1950).

69. Gulbransen, E. A. and K. F. Andrew, J. Electrochem. Soc., 101, 348 (1954).

70. Walter, R. J., J. A. Ytterhus, R. D. Lloyd, and W. T. Chandler, "Effect of Water Vapor/Hydrogen Environments on Columbium Alloys," Technical Report No. AFML-TR-66-322, Air Force Materials Laboratory, Wright-Patterson Air Force Base, Ohio, December 1966.

71. Walter, R. J. and W. T. Chandler, J. AIAA, 4 (2), 302 (1966).

72. Walter, R. J., "Compatiblity of Tantalum and Columbium Alloys with Hydrogen," Research Report RR 65-3, Rocketdyne, a Division of North American Aviation, Inc., Canoga Park, California, February 1965.

73. McCoy, H. E. and D. A. Douglas, in Columbium Metallurgy, D. L. Douglass and F. W. Kunz, eds., Interscience Publishers, New York (1961), p. 85.

74. Wilcox, B. A. and R. A. Huggins, J. Less-Common Metals, 2, 292 (1960).

75. Wilcox, B. A. and R. A. Huggins, in Columbium Metallurgy, D. L. Douglass and F. W. Kunz, eds., Interscience Publishers, New York (1961), p. 257.

76. Wilcox, B. A. and R. A. Huggins, "The Effect of Interstitial Atom-Dislocation Interactions on the Deformation Behavior of Columbium, Tantalum, and 1020 Steel," Technical Documentary Report No. ASD-TR-61-351, Aeronautical Systems Division, Wright-Patterson Air Force Base, Ohio, February 1962.

77. Petch, N. J., J. Iron Steel Inst., 174, 25 (1953).

78. Christian, J. W. and B. C. Masters, Proc. Roy. Soc., A, 281, 223 (1964).

79. Armstrong, R. W., "Strengthening Mechanisms and Brittleness in Metals," Paper Presented at Symposium on Materials - Key to Effective Use of the Sea, New York, September 1967.

80. Armstrong, R. W., "Grain Boundary Strengthening and the Polycrystal Deformation Rate," in Dislocation Dynamics, McGraw-Hill Book Co., New York (1968), p. 293.

81. Wilcox, G. A., A. W. Brisbane, and R. F. Klinger, "The Effects of Strain Rate and Hydrogen Content on the Low Temperature Deformation Behavior of Columbium," Technical Report No. WADD-TR-61-44, Aeronautical Systems Division, Wright-Patterson Air Force Base, Ohio, May 1961.

82. Wilcox, B. A., A. W. Brisbane, and R. F. Klinger, Trans. ASM, 55, 179 (1962).

83. Imgram, A. G., E. S. Bartlett, and H. R. Ogden, Trans. Met. Soc. AIME, 227, 131 (1963).

84. Wood, T. W. and R. D. Daniels, Trans. Met. Soc. AIME, 233, 898 (1965).

85. Longson, B., United Kingdom Atomic Energy Authority, TRG Report No. 1035 (c), (1966).

86. Keh, A. S. and S. Weissman, in Electron Microscopy and the Strength of Crystals, G. Thomas and J. Washburn, eds., Interscience Publishers, New York (1963), p. 881.

87. VanTorne, L. I. and G. Thomas, Acta Met. 11, 881 (1963).

88. Eustice, A. L. and O. H. Carlson, Trans. ASM, 53, 501 (1961).

89. Imgram, A. G., E. S. Bartlett, H. R. Ogden et al, "Further Investigation of Notch Sensitivity of Refractory Metals," Technical Documentary Report No. ML-TDR-64-35, Air Force Materials Laboratory, Wright-Patterson Air Force Base, Ohio, February 1964.

90. Roberts, B. W. and M. C. Rogers, J. Metals, 8, 1213 (1956).

91. Eustice, A. L. and O. H. Carlson, Trans. Met. Soc. AIME, 221, 238 (1961).

92. Nunes, J., A. A. Anctil, and E. B. Kula, "Low Temperature Flow and Fracture Behavior of Tantalum," Technical Report AMRA TR-64-22, U.S. Army Materials Research Agency, Watertown, Massachusetts, August 1964.

93. Wood, T. W. and R. D. Daniels, J. Met., 13, 83 (1961).

94. Ytterhus, J. A. and R. J. Walter, "Mechanisms of Hydrogen Embrittlement of Columbium," Report R-6134P, Rocketdyne, a Division of North American Aviation, Inc., Canoga Park, California, April 1965.

95. Westlake, D. G., "The Ductile-Brittle-Ductile Transition in Nb-H and V-H Alloys," paper presented at AIME Annual Meeting, New York, 29 February 1968, abstract in J. Metals, 20 (1), 114A (1968).

96. Thompson, S. R., W. R. Young, and W. H. Kearns, "Investigation of the Structural Stability of Welds in Columbium Alloys," Technical Report ML-TDR-64-210, Part II, Air Force Materials Laboratory, Wright-Patterson Air Force Base, Ohio, April 1966.

97. Kieger, R., A. Clauss, and H. Forestier, Mem. Sci. Rev. Met. 64 (2), 195 (1967).

98. Bentle, G. and W. T. Chandler, "Effects of Hydrogen Environments on Columbium and Tantalum Alloys," Quarterly Report for period 1 August 1967 to 31 October 1967 for Air Force Contract AF33(615)-2854, Rocketdyne, a Division of North American Rockwell Corporation, Report No. R-7076-3, December 1967.

99. Sieverts, A. and E. Bergner, Ber. deut. chem. Ges., 44, 2394 (1911).

100. Kofstad, P., W. E. Wallace, and L. J. Hyvönen, J. Am. Chem. Soc., 81, 5015 (1959).

101. Mallett, M. W. and B. G. Koehl, J. Electrochem. Soc., 109, 611 (1962).

102. Elliott, R. P., Constitution of Binary Alloys, First Supplement, McGraw-Hill Book Company, New York, 1965.

103. Kelly, K. K., J. Chem. Phys., 8, 316 (1940).

104. Waite, T. R., W. E. Wallace, and R. S. Craig, J. Chem. Phys., 24, 634 (1956).

105. Hägg, G., Z. Physik. Chem., B11, 433 (1931).

106. Pietsch, E. and H. Lehl, Kolloidz., 68, 226 (1934).

107. Horn, F. H. and W. T. Ziegler, J. Am. Chem. Soc., 69, 2762 (1947).

108. Stalinski, B., Bull. Acad. Polon. Sci. 2, 245 (1954).

109. Bakish, R., J. Electrochem. Soc., 105, 574 (1958).

110. Saba, W. G., W. E. Wallace, H. Sandmo, and R. S. Craig, J. Chem. Phys., 35, 2148 (1961).

111. Pedersen, B., T. Krogdahl, and O. E. Stokkeland, J. Chem. Phys., 42 (1), 72 (1965).

112. Wright, M., D. Jewett, and A. C. Makrides, "Research Studies on Solid Hydrogen Purification Membranes," Report Covering Period from 15 June 1965 to 15 April 1966, on U.S. Army Engineer Contract No. DA44-009-AMC-1183(T), Tyco Laboratories, Inc., Waltham, Massachusetts, April 1966.

113. Mallett, M. W. and B. G. Koehl, J. Electrochem. Soc., 109, 968 (1962).

114. Wasilewski, R. J. and G. L. Kehl, Metallurgia Manchester, 50, 225 (1954).

115. Mallett, M. W. and W. M. Albrecht, J. Electrochem. Soc., 104, 142 (1957).

116. Zubler, E. G., J. Electrochem. Soc., 110, 1072 (1963).

117. Klyachko, Yu. A., L. L. Kurin, S. P. Fedorov, and I. N. Larionov, Trudy Komissii Anal. Khim., Akad. Nauk S.S.S.R., Inst. Geokhim. i Anal. Khim. im. V. I. Vernadskogo 10, 49 (1960).

118. Merisov, B. A., V. I. Khotkevich, and A. I. Karnus, Phys. Metal. Metallography, 22 (2), 308 (1966).

119. Abowitz, G. and R. A. Burn, "The Mechanical Properties of Tantalum with Special Reference to the Ductile-Brittle Transition," Technical Report No. ASD-TR-61-203, Part III, Air Force Materials Laboratory, Wright-Patterson Air Force Base, Ohio, May 1963.

120. Gazza, G. E., "Petch Analysis of Hydrogenated Columbium Sheet," Technical Report AMRA TR-66-10, U.S. Army Materials Research Agency, Watertown, Massachusetts, May 1966.

121. Owen, W. S., D. C. Hull, J. Bryson, and C. L. Formby, Technical Report AFML-TR-66-369, Vol. I, Air Force Materials Laboratory, Wright-Patterson Air Force Base, Ohio, February 1967.

122. Wilcox, B. A., Trans. Met. Soc. AIME, 230, 1199 (1964).

123. Bentle, G. and W. T. Chandler, "Effects of Hydrogen Environments on Columbium and Tantalum Alloys," Quarterly Report for period 1 November 1967 to 31 January 1968 for Air Force Contract AF33(615)-2854, Rocketdyne, a Division of North American Rockwell Corporation Report No. R-7067-4, February 1968.

THE COMPATIBILITY OF REFRACTORY METALS WITH LIQUID METALS

E. E. Hoffman and R. W. Harrison

Abstract

Both basic investigations and engineering experiments conducted for the most part during the last ten years have demonstrated that refractory metals generally have excellent compatibility with liquid metals. The temperature range of usefulness of the refractory metals begins at temperatures where conventional superalloys still have nominal mechanical strength but do not have the required resistance to corrosion by the liquid metals. The excellent compatibility of refractory metals with liquid metals can be degraded by impurity elements, and the controls required to assure compatibility will be presented.

The most significant investigations of refractory metal - alkali metal compatibility have been performed to establish the required technology for the development of Rankine cycle nuclear electric space power systems. This compatibility review is limited to the most significant results obtained in flowing systems where careful control was exercised to control test variables. The discussion will be limited to tests of refractory metals in contact with mercury, cesium, potassium, sodium, and lithium. A detailed review is presented describing the compatibility results obtained in a 5000-hour test of a two-loop Cb-1Zr facility in which sodium was circulated in the heater circuit and potassium was boiled and circulated in a two-phase secondary loop which contained turbine simulator test sections of Mo-TZM alloy.

E. E. Hoffman and R. W. Harrison are metallurgists with Nuclear Systems Programs, General Electric Company, Cincinnati, Ohio.

251

Introduction

Interest in applying the refractory metals as containment and
structural materials for liquid metals systems designed to operate at
elevated temperatures has existed since the late 1940's when the first
studies directed at the development of aircraft nuclear propulsion systems were initiated. Although many potential applications involving
the combination of these classes of materials have been considered during the last twenty years, by far the most significant advances in this
technology have been associated with nuclear power devices and primarily those related to power systems for long-time space applications.
No attempt will be made in this paper to describe the industrial applications of refractory metals as liquid metal containment materials;
information on this subject was recently reviewed (1). Most of the
refractory metal - liquid metal compatibility testing has been performed in order to make optimum material selections for Rankine cycle
turbogenerator electric power systems (2,3). However, the compatibility
information generated by these studies has been extremely useful in
selecting the most promising candidate materials for other applications
which utilize liquid metals including thermoelectric, thermionic, and
magnetohydrodynamic systems.

Extensive investigations of containment material - liquid metal
compatibility have been conducted using conventional materials, primarily iron- (4,5), nickel- (4,5), and cobalt-base alloys (5), but
problems such as evaporation, strength limitations, and corrosion prevent or discourage the use of these materials for long-time space
applications at temperatures in excess of 1600°-1700°F. In the case
of boiling mercury containment, the best of the conventional alloys
have only limited corrosion resistance at temperatures of 1200°-1300°F
(6), and refractory metals hold promise of solving this difficult containment problem.

A number of recently published papers contain very complete reviews of the results of refractory metal - alkali metal corrosion investigations (7,8,9,10). It is the intent of the authors to summarize
as succinctly as possible the status of containment technology for the
various refractory metal - liquid metal combinations of interest and
to cite several examples of interactions which have been noted in those
systems. The last portion of this paper will include a more detailed
review of the results of a 5000-hour refractory metal - alkali metal
Rankine system compatibility experiment which serves to illustrate the
inertness of refractory metals to liquid metals which is generally
observed when sufficient attention is directed to the control of variables, i.e., materials and testing environment purity.

Liquid Metals and Refractory Metals of Interest

Prior to discussing the test methods used to evaluate the various refractory metals in contact with liquid metals and reviewing the current status of the compatibility technology for the various combinations, it is in order to discuss briefly several of the more important physical properties of the liquid metals, the status of liquid metal analytical and purification technology, and the refractory metals and alloys of principal current interest as liquid metal containment materials.

Liquid Metals Properties

The liquid metals, mercury, cesium, potassium, sodium, and lithium, have many properties which make them candidate energy transfer fluids for both all-liquid and two-phase applications. Of the many important properties of liquid metals which must be considered for a particular application, only the melting points and the vapor pressure of these liquid metals will be discussed here. The advantages of using a liquid metal which has a melting point near that of the ambient environment in which the system is to operate are fairly obvious. In general, those who have worked with liquid metal systems have found that it is easier to transfer, sample, etc. liquid metals such as potassium (m.p. $146°F$) or sodium (m.p. $208°F$), where a modest amount of external heating is required, than it is to handle either those liquid metals with relatively high or subambient melting points. Vapor pressure as a function of temperature is of particular importance in selecting the working fluid for Rankine systems. The pressure and temperature regimes of the Rankine systems which have received the most attention in the last ten years (11) are indicated in Figure 1. The shaded region on the right is the region of principal interest to refractory metal technology, since at the high temperatures and the stress levels associated with advanced alkali metal systems only the refractory metals can be considered as potential containment materials.

Liquid Metal Purity

Most of the confusion and negative information regarding the limitations of various potential containment materials for liquid metals has resulted from inattention of the investigators to the problem of obtaining and maintaining the purity of the liquid metal under study. This is particularly true of the alkali metals, more so than mercury, because of the tendency of the former to react readily with the atmosphere when exposed. The deleterious consequences of such contamination, particularly by oxygen, on the corrosion resistance of austenitic and ferritic stainless steels to sodium have been described in the literature (4,12). The effects of liquid metal contamination are much more serious in the case of the refractory metals than for conventional alloys due to the increased reaction rates which may result in such gross effects as the loss up to 33 mils per month of columbium observed in pumped sodium systems operating at $1100°F$ (13).

During the last ten years, substantial progress has been made in developing methods of purifying liquid metals, in developing the analytical methods required to monitor the purity of these materials, and

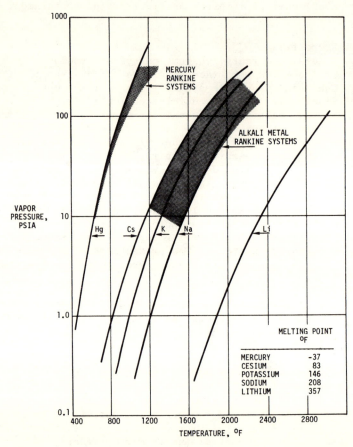

Fig. 1. Melting points and vapor pressures of alkali metals and mercury shaded regions denote pressure–temperature regions for Rankine systems.

in defining the mechanisms responsible for the deleterious effects of liquid metal impurities on refractory metals. Goldman and Minushkin (14) have written an excellent summary of the analytical methods for determining sodium purity and the varied methods of purifying sodium. Purification methods for removal of such elements as oxygen, nitrogen, and carbon from alkali metals include: 1) low-temperature (just above the melting point) filtration, 2) diffusion cold traps, 3) chemical reaction or hot-gettering methods, and 4) vacuum distillation. In general, these techniques are useful in purifying all the alkali metals. A detailed description of the equipment, procedures, and specifications used to purchase, purify, analyze, and transfer sodium and potassium has been given in a recent report by Dotson and Hand (15).

One of the more difficult alkali metal purification problems has been the removal of oxygen from lithium. First, because of the thermodynamic stability of lithium oxide, it is very difficult to "getter" oxygen from lithium with more stable oxide formers and secondly, analytical methods for determining oxygen in lithium have been limited. However, in recent studies at General Electric - Nuclear Systems Programs, approximately 20 pounds of lithium was purified by hot gettering followed by vacuum distillation using a condenser constructed of Cb-1Zr alloy. The product was analyzed both by thermal neutron activation (16) and fast neutron activation (17) analytical methods and was found to contain less than 50 ppm oxygen. In summary, the technology of liquid metal analysis and purification is well established and can be applied to minimize or eliminate deleterious interactions between the refractory metals and the liquid metals resulting from impurities in the liquid metal.

Refractory Metals
The refractory metals which will be discussed in this paper include columbium, tantalum, molybdenum, tungsten, and the alloys of these metals. Rhenium is not included although limited corrosion evaluations indicate that its compatibility with liquid metals is similar to that of tungsten.

The various refractory alloys which have been evaluated in liquid metals to any significant extent are listed in Table I. This table also lists the most useful references to compatibility investigations for the various combinations of liquid metal and refractory metal. Comments in this section will be confined to a discussion of the refractory metals only. It may be noted that with the exception of the alloy, 90Ta-10W, all the columbium and tantalum alloys listed contain small amounts of elements, such as zirconium, hafnium, and yttrium, which form oxides that have high negative free energies of formation, i.e., they are thermodynamically stable relative to most other metal oxides. Extensive alkali metal corrosion investigations have indicated the advantages of the "gettered" alloys for alkali metal containment applications, and more will be presented on this aspect of liquid metal compatibility in the portion of this paper entitled "Refractory Metal - Alkali Metal Corrosion Phenomena."

The detrimental effect of high oxygen concentrations in refractory materials, particularly the pure metals, has necessitated the requirement for low-oxygen products from the producers. Substantial advances have been made in this technology in recent years as a result of the development and utilization of high-vacuum processing equipment.

TABLE 1. LIST OF REFRACTORY METALS AND ALLOYS EVALUATED IN CONTACT WITH LIQUID METALS

MATERIAL	COMPOSITION	LIQUID METAL CORROSION TEST REFERENCES				
		Hg	Cs	K	Na	Li
COLUMBIUM						
Cb-1Zr	Cb-1Zr	6,18,19	8,20,21,22	25,26	13,44,45,46,47,48	9,45,48,58,59,60
D-43	Cb-10W-1Zr-0.1C		8	8,10,22,25,27,28	8,21,30,40,47,49,50,51	8,9,60,61,62,63,64,65
AS-55	Cb-5W-1Zr-0.3Y			8,10,27,31,34,35,36,37	8	8,60,66
Cb-752	Cb-10W-2.5Zr			10,34,35	8	66
SCb-291	Cb-10Ta-10W			10	8	
FS-85	Cb-27Ta-10W-1Zr		22	10	8	7,66
B-66	Cb-5Mo-5V-1Zr			10	8	66
TANTALUM						
T-111	Ta-8W-2Hf	6,18	23	38	48,52,53,54	45,48,58,59
T-222	Ta-10W-2.5Hf-0.1C			10,36	8	7,66,67
90Ta-10W	Ta-10W		22,24	10	8	67
MOLYBDENUM						
Mo-0.5Ti	Mo-0.5Ti		22,24		8	48,58
Mo-TZM	Mo-0.5Ti-0.08Zr		8	27,29,30,31,36,39,40,41,42,43	48,55,56,57	45
Mo-TZC	Mo-1.25Ti-0.15Zr-0.12C		8	41,42	8	9,68
Mo-50Re	Mo-50Re					68
TUNGSTEN						
W-10Re	W-10Re			43	48,52	48,58,68
W-25Re	W-25Re					68
W-30Re-30Mo	W-30Re-30Mo				8	7,68,69

(ATOMIC PERCENT)

Compatibility Evaluation Techniques

The first factor to be considered in the selection of containment materials for liquid metals is the equilibrium solubility of the
particular solid metal in the liquid metal of interest. Solubility
data have been very useful in indicating the superiority of refractory
metal containment materials for mercury over conventional materials
such as iron, nickel, and chromium (70). As illustrated in Figure 2,
columbium and tantalum are orders of magnitude less soluble in mercury
at elevated temperatures than the conventional metals. In the review
of test results given in the following section of this paper, the compatibility of tantalum with mercury suggested by these solubility data
were subsequently confirmed in a two-phase, natural-circulation loop
test.

The utilization of equilibrium solubility data for refractory
metals in various alkali metals as a guide in selecting the containment
metal or alloy which might be most resistant to dissolution and temperature gradient mass transfer is complicated by several factors. First
of all, only a very limited number of worthwhile investigations have
been conducted and secondly, the results to date suggest that such
problems as fine particle formation, subtle impurity effects, complex
compound formation (26), or other unknown phenomena are contributing
to high "apparent solubility" values and great scatter in the data.
DiStefano and Litman (71) have reviewed the substantial effect of
oxygen and nitrogen in alkali metals on increasing the extent of dissolution of columbium. A study conducted by McKisson and co-workers
(72) to determine the solubility of various refractory metals and
alloys in potassium utilized very elaborate test methods to minimize
the possibility of introducing impurities which might influence the
results. The solubility results obtained at 1000°C and 1200°C for
ungettered columbium and tantalum are listed in Table 2. The minimum
and maximum values obtained at each temperature are given and indicate
that the values are quite high for ungettered columbium and tantalum
and that there was considerable scatter in the results at a given
temperature. These results and the temperature coefficients determined in the referenced study would suggest that ungettered columbium
and tantalum would experience rapid and substantial temperature-
gradient mass transfer in flowing nonisothermal systems. Additional
more recent solubility investigations by McKisson (73) to determine
the extent of dissolution of columbium from Cb-1Zr and tantalum from
T-111 (Ta-8W-2Hf) have yielded solubility values of generally less
than 25 ppm for tantalum and less than 10 ppm for columbium over the
temperature range from 1000° to 1600°C. The cause of this marked reduction in solubility from the alloys compared to the values given in
Table 2 for the pure metals is not understood at present.

For the reasons cited above, the selection of alkali metal containment materials based on solubility data is limited to nonflowing
systems, and compatibility evaluation techniques are required which
simulate the application conditions as nearly as is practical.

A number of types of tests have been utilized to determine the
compatibility of various materials in liquid metals. The most common
testing methods are listed below and are arranged in order of increasing

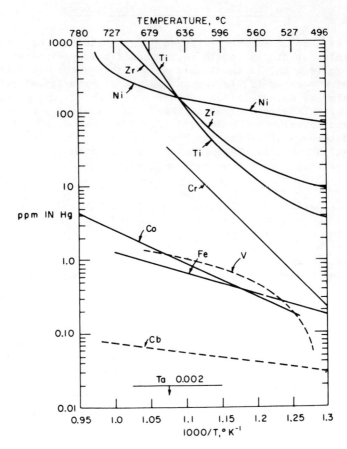

Fig. 2. Solubility of various metals in mercury.

Table 2. "Solubility" of Various Metals in Potassium[a]

Metal	Temperature °C	°F	Concentration in Potassium, ppm Minimum	Maximum
Cb	1000	1832	15	171
	1200	2192	33	72
Ta	1000	1832	371	2730
	1200	2192	1520	2880

[a] Reference 72: McKisson, et al., NASA CR-610, October 1966.

complexity and cost.

Static Capsule Tests
Seesaw- or Rocking-Furnace Capsule Tests
Refluxing Capsule Tests
Natural-Circulation, Single-Phase Loop Tests
Natural-Circulation, Two-Phase Loop Tests
Forced-Circulation, Single-Phase Loop Tests
Forced-Circulation, Two-Phase Loop Tests

Static Capsule Tests

Static capsule tests are most often used for preliminary screening
of metals for liquid alkali metal containment. Although general compatibility can be determined by evaluation of the metal capsule itself,
specimens suspended within the capsule can, in addition, provide weight
change and mechanical property data. Specimens of different composition from that of the capsule can be evaluated in this manner providing
no interaction between the dissimilar metal specimens confuses the
evaluation.

Seesaw- or Rocking-Furnace Capsule Tests

Seesaw- or rocking-furnace capsule tests have been employed to
evaluate temperature-gradient mass transfer in refractory metal -
liquid metal systems. Capsules partially filled with alkali metal are
heated in a manner to maintain a temperature differential between the
two ends of the capsule. The capsules are rocked slowly back and forth
causing the alkali metal flow alternately to the hot end and the cold
end.

Refluxing Capsule Tests

The compatibility of refractory alloys under two-phase, vapor-
liquid alkali metal exposure is most easily evaluated by refluxing
capsule tests similar to the potassium test system shown in Figure 3.
Potassium in the lower half of the capsule is heated, vaporized, and
condensed on the tight-fitting insert specimens in the cooled upper
half of the capsule or on the tube wall itself if tubular inserts are
not used. The potassium condensate flows down the wall into the liquid
pool, and the process is repeated. The insert specimens are desirable
as they provide a means of obtaining weight change data. A water-
cooled heat sink is used in the system illustrated in Figure 3. By
measuring the water flow rate and temperature rise and knowing the
heat of condensation for the metal, it is possible to determine the
condensation rate per unit area on the capsule wall.

Natural-Circulation Loop Tests

Natural-circulation loops are employed to evaluate compatibility
under dynamic, nonisothermal conditions. Single-phase (all-liquid) or
two-phase (liquid-vapor) loop tests of the type shown in Figure 4 have
been widely used to study the resistance of various materials to
temperature-gradient mass transfer. The system illustrated is the type
used in testing at the Oak Ridge National Laboratory and incorporates
a continuous string of interlocking sheet specimens which are utilized
to obtain weight and chemistry change information. Although very useful data can be obtained in loops of this type, the extrapolation to

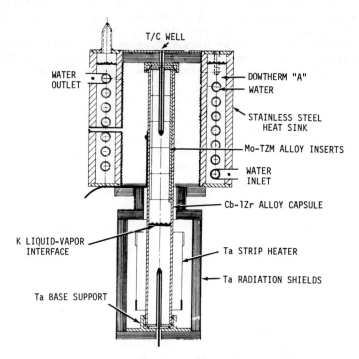

Fig. 3. Refluxing liquid metal capsule test system.

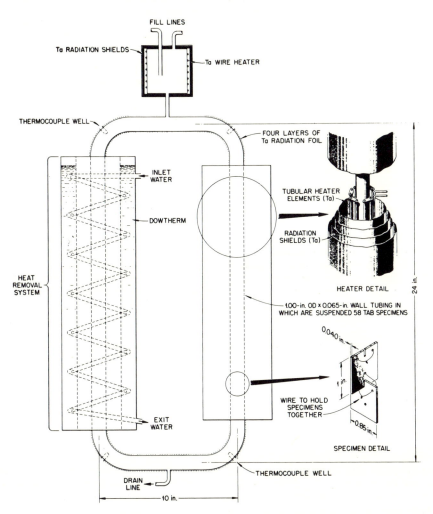

Fig. 4. Natural-circulation, single-phase compatibility test loop
(ORNL design).

high-velocity pumped systems is sometimes difficult because of the
inherent low-flow velocities (2-10 ft/min) of the natural-circulation
test systems.

Forced-Circulation Loop Tests

Forced-circulation loops, both single-phase and two-phase, are
used to determine containment material compatibility under high-
velocity, temperature-gradient conditions which closely duplicate the
environmental conditions in the actual application. A rather complete
description of one such system containing both a single-phase circuit
and a two-phase circuit will be given in a latter portion of this
paper, "Cb-1Zr 5000-Hour Rankine System Corrosion Loop Test."

Review of Compatibility Test Results

The extent of compatibility testing of the various refractory
metals in liquid metals was indicated in Table 1, which lists the various refractory metals and alloys and references to what the authors
consider to be the most useful sources of compatibility information for
the various combinations. No attempt will be made in this review to do
any more than briefly describe the most significant results reported in
these studies. It is readily apparent on examining Table 1 that the
great bulk of the data which exists relates to studies involving the
liquid metals, potassium, sodium, and lithium. In addition, it may be
noted in Table 1 that over one-half the sources referenced involved
studies on just four of the refractory materials, columbium, Cb-1Zr,
D-43, and Mo-TZM. Some combinations of liquid metal and refractory
metals which have been evaluated only in a cursory manner are not included in the review which follows.

No illustrations of the various types of corrosion which have
been observed with refractory metals will be included in this section
of the paper. Several of the principal corrosion phenomena which have
been observed in a number of test systems will be presented later in
this paper, including a detailed report of the results of a 5000-hour
alkali metal loop test.

Mercury

Only very limited data can be found in the literature regarding
the corrosion resistance of refractory metals and alloys in mercury at
elevated temperatures. The results of solubility determinations of a
variety of metals and alloys (70) were shown previously in Figure 2
and indicate that the refractory materials are orders of magnitude less
soluble in mercury than the conventional metals. The limited amount of
corrosion testing of refractory metals in mercury summarized below,
expecially the results for tantalum, are consistent with the solubility
information.

Columbium. Refluxing mercury capsule tests at 1100°F indicated
no measurable penetration (6).

Cb-1Zr. Refluxing mercury Cb-1Zr capsule tests at 1100°-1200°F
and 1300°F showed no mercury attack (6). A Cb-1Zr loop tested for
7700 hours with a mercury boiling temperature of 1200°F and a superheat
temperature of 1400°F failed in the superheat region (19). Subsequent

capsule experiments indicated cracks in bent Cb-1Zr specimens exposed to mercury at 1400°F. The cracks were attributed to preferential leaching of zirconium from the Cb-1Zr matrix (70). More recent experiments have not indicated this difficulty. In a recently completed study, stressed Cb-1Zr specimens, exposed for 1000 hours to 1200°F mercury, showed no evidence of degradation or cracking in preliminary posttest evaluation (74).

A Cb-1Zr mercury loop tested for 1000 hours with a mercury boiling temperature of 1100°F, superheat temperature of 1300°F, and condensing temperature of 900°F operated stably without difficulty (75).

Tantalum. Refluxing capsule tests have shown no attack of tantalum up to 1400°F (6). The corrosion resistance of tantalum to mercury was further documented in a two-phase, natural-circulation loop test which ran for 19,975 hours with a boiling temperature of 1200°F and superheat temperature of 1300°F. Posttest evaluation of the loop revealed no corrosion (18). As a result of the inertness of tantalum to mercury attack demonstrated in this long-time experiment, tantalum is currently being evaluated as a replacement material for Croloy 9M steel in the SNAP-8 mercury boiler (76).

Cesium
Columbium. Refluxing capsule tests indicate surface dissolution and severe attack of columbium after 720 hours at 1800° and 2500°F (20) (Based on good corrosion resistance of Cb-1Zr cited below, oxygen contamination of cesium suspected).

Cb-1Zr. Inserts removed from the condenser section of a Cb-1Zr refluxing cesium capsule which was tested for 5000 hours at 1220°C (2228°F) showed no evidence of penetration and only slight weight gains resulting from oxygen pick-up (21).

The excellent corrosion resistance of Cb-1Zr to high-velocity cesium is best documented by the results of a two-phase, forced-circulation loop test which ran 1100 hours with a boiling temperature of 1760°F. The Cb-1Zr loop contained nozzle and blade specimens of Cb-1Zr, Mo-TZM, and Mo-TZC. Posttest metallographic examination of the blades indicated no evidence of penetration or dissolution; however, a 1-mil thick adherent deposit on the blades was analyzed as columbium (8).

Other Columbium Alloys. Refluxing capsule tests indicate no corrosion of D-43 following exposure to cesium at 2200°F for 9437 hours (8) and FS-85 exposed to cesium at 2100°F for 818 hours (22).

Tantalum. Similar lack of corrosion resistance to cesium was found for tantalum as reported above for columbium. Refluxing capsule tests indicated surface dissolution and severe attack after 720 hours at 1800° and 2500°F (20).

T-111. A two-phase, natural-circulation T-111 cesium loop was tested for 246 hours at 2400°F and 175 hours at 2000°F before a high-temperature excursion terminated the test by causing a failure in the boiler. Posttest evaluation indicated slight increases in oxygen in the T-111 and a decrease in oxygen concentration in the cesium. Although no gross corrosive attack was reported, photomicrographs of the ID of the boiler indicates some grain boundary area degradation to a depth of 2.5 mils (23).

90Ta-10W. Refluxing capsule tests of 90Ta-10W exposed to cesium at 2100°F for 528 hours showed no mass transfer or attack.

Molybdenum and Mo-0.5Ti. Severe dissolutive corrosion of molybdenum (20) and Mo-0.5Ti (22,24) was observed in refluxing capsule tests at 1800° and 2500°F for periods up to 1000 hours.

Mo-TZM and Mo-TZC. Mo-TZM and Mo-TZC blade specimens tested in the Cb-1Zr forced-circulation loop described above showed no evidence of corrosion after 1100 hours exposure to wet cesium vapor (80% quality) at 1542°F (8).

Potassium

Columbium. Columbium has excellent compatibility with potassium liquid and vapor up to 2400°F providing the oxygen concentration in the columbium and potassium is maintained at a low level (25).

Cb-1Zr. The compatibility of Cb-1Zr to liquid and boiling potassium has been studied extensively. Results of refluxing capsule tests, natural-circulation, and forced-circulation loop tests indicate weight changes resulting from interstitial element migration occurring, but no evidence of dissolution and mass transfer of metallic constituents was observed (9). The excellent corrosion resistance of Cb-1Zr to potassium will be described in detail during the discussion of the results of a two-phase, forced-circulation loop test later in the paper.

D-43. No corrosion was observed in D-43 capsules biaxially stressed in refluxing potassium for 1000 hours at 2100°F and 2000 hours at 2050°F (37). Single-phase and two-phase, natural-circulation loop tests at 1100°C (2012°F) for 2000 hours and 1200°C (2192°F) for 3000 hours, respectively, indicated no attack (8).

AS-55. No corrosion was observed in an AS-55 natural-circulation loop which ran 2000 hours at 2000°F (35).

Tantalum. In capsule tests at 600°C (1112°F) tantalum specimens with 50 ppm oxygen exposed to potassium containing 100-3300 ppm oxygen lost weight proportionately to the oxygen concentration in the potassium. Tantalum specimens with oxygen concentration in excess of 200 ppm were attacked intergranularly at 600°C (1112°F) (38).

T-111. No attack was observed in a two-phase, natural-circulation T-111 loop test which operated for 3000 hours at 2280°F (36).

Mo-TZM. Mo-TZM blade and nozzle specimens were exposed to 2000°F potassium vapor for 5000 hours in the two-phase, forced-circulation loop test described later in this paper. No attack, erosion, or migration of interstitial impurities was observed.

Sodium

Columbium. While pure columbium has excellent corrosion resistance to static sodium up to 1800°F, the presence of oxygen in the sodium may lead to penetration rates as high as 33 mils per month in flowing systems at 600°C (1112°F) (13).

Cb-1Zr. The corrosion resistance of Cb-1Zr to flowing sodium was best demonstrated in a natural-circulation loop test and a forced-circulation loop test that were completed prior to the initiation of the Cb-1Zr two-phase, forced-circulation loop test (30) which is described in detail in this paper.

The natural-circulation, single-phase loop was operated for 1000 hours at a flow rate of 11 lb/hr with maximum and minimum loop temperatures of 2380° and 1350°F, respectively (49). Maximum corrosion observed was less than 1 mil. The forced-circulation, single-phase loop test was operated 2650 hours with a sodium flow rate of 400 lb/hr

and maximum and minimum loop temperatures of 2065° and 650°F, respectively (50). No significant corrosion or interstitial elements contamination was observed.

Other Columbium Alloys. The compatibility of columbium alloys,
FS-85 and D-43, with static sodium was evaluated in capsules tested
at 2400°F for 6271 hours. No corrosion was observed for D-43; however,
2 mils of grain boundary corrosion was found in FS-85 (8).

Tantalum. The compatibility of tantalum with sodium is similar to
that of columbium, in that the presence of oxygen in the sodium leads
to weight loss in flowing systems. In addition, extensive intergranular and transgranular attack of tantalum by sodium has been observed.
This attack is attributed to the high (390 ppm) oxygen concentration
of the tantalum prior to exposure to the sodium (53).

Tantalum Alloys. The compatibility of tantalum alloys, T-111 and
90Ta-10W, with static sodium was demonstrated in capsules tested at
2400°F for 6271 hours and 3000 hours, respectively; no corrosion was
found in either alloy (8).

Molybdenum. Molybdenum exposed to static sodium at 1500°C (2732°F)
for 100 hours showed less than 1 mil of attack (56).

Recent data regarding the corrosion resistance of molybdenum in
flowing sodium was presented by Russian investigators. In loop tests
conducted at 600°C (1112°F) and 900°C (1652°F) very small weight gains
occurred in 360-hour tests at 900°C in sodium containing up to 100 ppm
oxygen flowing at 2 m/sec. No corrosion was found in similar 3600-hour
tests at 600°C in sodium with 20 to 40 ppm oxygen flowing at 5 m/sec
(77).

Tungsten. Limited sodium compatibility results exist for tungsten.
A 400-hour, 1000°C (1832°F) capsule test indicated no attack (48).

Lithium

Columbium. Columbium exhibits excellent corrosion resistance to
lithium up to 2400°F (60) providing the oxygen concentration of the
columbium does not exceed the threshold level of 200-300 ppm (71). The
"oxygen effect" on the corrosion resistance of columbium in lithium is
covered in more detail later in the next section of this paper which
describes several of the most important refractory metal - alkali metal
corrosion phenomena.

Cb-1Zr. The excellent compatibility of Cb-1Zr with lithium was
demonstrated in a 10,000-hour, 2000°F forced-circulation loop test.
Mass transfer was limited to zirconium depletion in the hot regions
and zirconium-rich layer formations in the colder regions (63). Minor
grain boundary attack was found in a large Cb-1Zr lithium loop tested
for 10,000 hours at 2000°F and a similar loop test for 500 hours at
2400°F (64). The rate of mass transfer of Cb-1Zr in lithium has been
obtained in a thermal convection loop test operated for 3000 hours at
2400°F hot-leg temperature (9). The design of this loop was shown
earlier in Figure 4. Chemical analysis indicated loss of oxygen in all
sections of the loop and transfer of zirconium, nitrogen, and carbon
from the hot leg to the cold leg. Metallographic examination indicated
no attack although some zirconium nitride was found in specimens from
the colder portions of the loop.

Other Columbium Alloys. The rate of mass transfer of D-43 in
lithium was evaluated in a similar thermal convection loop test to

that described above for Cb-1Zr.   Chemical analysis indicated loss of
both oxygen and carbon in all sections of the loop.   Zirconium was
transferred from the hot leg to the cold leg, and a gain of nitrogen
was noted in the cold leg.   Metallographic examination showed no attack,
films, or deposits (9).

     Columbium-base alloys, such as B-66, FS-85, D-43, and Cb-752, have
demonstrated good corrosion resistance to lithium at temperatures to
1800°F (66).

     Tantalum.   Tantalum compatibility with lithium is similar to co-
lumbium in that the corrosion resistance is dependent on oxygen concentration (45).   Tantalum will exhibit good corrosion resistance to
lithium as long as the oxygen concentration of the tantalum is maintained below 100 to 200 ppm.

     Tantalum Alloys.   T-111 and T-222 alloy specimens, oxygen contami-
nated to 500 ppm and welded in argon, were exposed to lithium at 750°C
(1382°F) and 1200°C (2192°F) for 100 hours.   Evaluation indicated no
attack in the weld areas; however, intergranular penetration was observed in the base metal of both alloys.   Heat treatment at 2400°F
eliminated the attack (67).

     Mo-0.5Ti.   Mo-0.5Ti in contact with lithium in seesaw capsule tests
cycled between 1100°F and 1800°F for 150 hours showed no attack or mass
transfer deposits (45).

     Other Molybdenum Alloys.   The molybdenum alloys, Mo-TZM and Mo-50Re,
showed no attack after 1000 hours exposure to static lithium at 2500°F,
2800°F, and 3000°F in Mo-TZM capsules (68).

     Mo-TZM insert specimens in a Cb-1Zr thermal convection loop were
exposed to lithium at a hot-leg temperature of 1200°C (2192°F) for 3000
hours (9).   Chemical analysis results indicate the Mo-TZM specimens
lost carbon and picked up columbium and zirconium.   Cb-1Zr specimens
in the same loop gained nitrogen, carbon, and molybdenum.   Metallographic
examination indicated zirconium carbide and nitride deposits on the
Cb-1Zr and Mo-TZM specimens in the cold leg.   Columbium crystals were
found on the Mo-TZM specimens in the hot leg.

     Tungsten.   Tungsten specimens exposed to lithium in Mo-TZM capsules
at 2500°F for 100 and 1000 hours showed no evidence of corrosion.   Similar specimens exposed at 2800°F and 3000°F showed evidence of dissolution and some grain boundary attack.   Chemical analysis of the lithium
indicated the tungsten concentration had increased from 10 ppm (before
test) to 4300 ppm after 2800°F exposure (68).

     Tungsten Alloys.   W-10Re and W-25Re specimens demonstrated excel-
lent corrosion resistance to lithium and exhibited similar behavior to
that previously described for tungsten for lithium exposures up to
3000°F.   W-30Re-30Mo (atomic percent) specimens exposed to lithium at
2600°F in a T-111 capsule for 100 hours showed no evidence of attack.
Both arc cast and powder metallurgy products were tested (69).

                Refractory Metal - Alkali Metal Corrosion Phenomena

     A limited discussion and several illustrations of the more impor-
tant types of corrosion phenomena which have been observed in refractory
metal - alkali metal systems, including impurity reactions, dissimilar
metal mass transfer, and high-velocity effects, is given below.

Impurity Reactions

The presence of oxygen as an impurity in either the refractory metal or alkali metal has been responsible for most of the dramatic and catastrophic corrosion which has been observed in refractory metal - alkali metal systems.

The initial investigation which revealed the partitioning of oxygen between a refractory metal (columbium) and an alkali metal (sodium) was conducted by Raines and co-workers (53). Other refractory metal - alkali metal combinations exhibit similar oxygen partitioning behavior as shown in Table 3 for Li-Cb (71), Li-Ta (71), and K-Cb (26). The data indicate the efficiency of the alkali metals in reducing the oxygen concentration of the refractory metals.

In many cases the removal of oxygen from the refractory metal by the alkali metal is accompanied by an intergranular and transgranular attack similar to that shown in Figure 5. This attack of oxygen contaminated tantalum by potassium observed by Klueh (38) is similar to attack by lithium observed in earlier studies (45).

The presence of oxygen as an impurity in sodium, potassium, and cesium can also destroy the inherent corrosion resistance of the refractory metals. Oxygen can react with the refractory metal to form compounds which, if soluble in these alkali metals or removed mechanically by the flowing metal in dynamic systems, can lead to an accelerated rate of attack (13,71).

The effects of oxygen on the corrosion resistance of refractory metals can be modified by alloying additions such as zirconium and hafnium. These elements form more stable oxides than the refractory-base metals. As alloy constituents, zirconium and hafnium remove oxygen from solid solution by precipitation of ZrO_2 and HfO_2, thereby increasing the threshold oxygen concentration associated with attack. Heat treatment is necessary to develop the corrosion resistant condition by enhancing the precipitation of the oxide phases.

The effect of oxygen concentration, zirconium concentration, and heat treatment on the lithium corrosion resistance of Cb-Zr alloys is best illustrated by the published work of DiStefano and Litman (71). The results of this study, summarized graphically in Figure 6, indicated columbium alloys containing zirconium (0.05 to 1.3 percent) and oxygen (0.09 to 0.23 percent) which were not heat treated were attacked by lithium at 1500°F. Alloys containing sufficient zirconium to combine with the oxygen to form ZrO_2 (atomic ratio 1/2) were not attacked when heat treated for two hours at 1300°C (2372°F) before exposure to lithium. As a result of studies such as these, gettered refractory alloys (those containing elements such as zirconium and hafnium) are preferred for alkali metal applications.

Dissimilar Metal Mass Transfer

Capsule test (45) and thermal convection loop test (33) investigations in which both columbium or Cb-1Zr and Type 316 stainless steel were in contact with alkali metal revealed substantial transfer of carbon and nitrogen from the stainless steel to the refractory metal. In tests conducted at 1700°F (45) gross increases in the carbon (240 to 680 ppm) and nitrogen (660 to 1920 ppm) concentration of columbium were noted. These contaminations resulted in substantial increases in the room temperature tensile strength and concomitant decreases in tensile

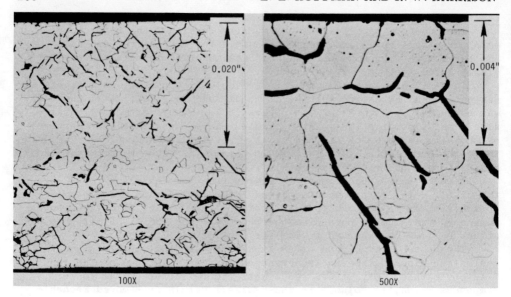

Fig. 5. Tantalum sheet specimen following 500 hours exposure to potassium at 600°C (1112°F). The pretest oxygen concentration of the tantalum was 1000 ppm–(test performed by R. L. Klueh, Oak Ridge National Laboratory).

Table 3. Oxygen Concentrations in Refractory Metals Before and After Exposure to Alkali Metals for 100 Hours at 1500°F

Alkali Metal	Refractory Metal	Oxygen Concentration of Refractory Metal	
		Before Test, ppm	After Test, ppm
Lithium[a]	Columbium	150	90
		500	210
		1000	460
		1700	660
Lithium[a]	Tantalum	220	80
		450	100
		1100	150
		2000	250
Potassium[b]	Columbium	410	160
		1300	110
		1500	130
		2400	130

[a] Reference 71: Corrosion, DiStefano and Litman, December 1964.

[b] Reference 26: ORNL-3751, Litman, July 1965.

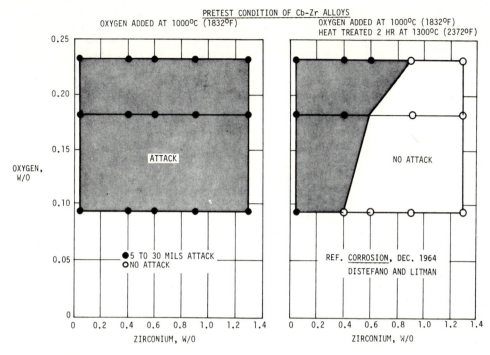

Fig. 6. Effect of oxygen concentration, zirconium concentration, and heat treatment on the corrosion resistance of Cb-1Zr alloys exposed to lithium for 100 hrs at 816°C (1500°F).

Table 4. Effect of Exposure for 500 Hours at 1700°F on Mo-TZM Alloy[a] in Potassium - Type 316 Stainless Steel System
Surface Area: SS/Mo = 5/1

Specimen Condition	Posttest Chemical Analysis Concentration, ppm				Room-Temperature Tensile Properties	
	O	N	H	C	Tensile Strength psi	Elongation in 2 in. Percent
Annealed 2 hr - 2550°F	40	15	4	270	76,200	29.0
Annealed and heated in inert gas 500 hr - 1700°F					76,000	32.0
Annealed and tested in potassium 500 hr - 1700°F	25	20	6	260	76,500	27.0

[a] Specimen thickness - 0.040 inch.

elongation. Tests of Mo-TZM specimens in contact with potassium in
Type 316 stainless steel containers under very similar conditions to
those described above for columbium revealed no carbon or nitrogen contamination of the Mo-TZM (40). The test conditions, chemical analyses,
and tensile properties of the Mo-TZM specimens are given in Table 4 and
indicate no significant effect of the potassium exposure on the properties of the material. This immunity of molybdenum to interstitial contamination while in contact with alkali metals in stainless steel systems
makes it a prime candidate for applications of this type.

High-Velocity Effects

Refractory metals which exhibit excellent corrosion resistance to
alkali metals can be degraded under the influence of high liquid metal
flow velocities. In a recent investigation by Hays (65) the corrosion
resistance of Cb-1Zr to lithium flowing at high velocities was evaluated
for potential application in a magnetohydrodynamic space power system.
The depth of material removed after 500 hours of exposure to liquid
lithium at 1073°-1143°C (1960°-2090°F), flowing at 50 m/sec (164 ft/sec),
was measured to be 0.3 mils. Extensive grain boundary grooving occurred
in the lithium flow channel.

The compatibility of Mo-TZM with potassium vapor flowing at a
velocity of 535 ft/sec was determined in a 3000-hour turbine test (42).
Mo-TZM turbine blades, shown in Figure 7, showed negligible corrosion
after exposure to the 1385°F potassium vapor. Tests have indicated negligible corrosion providing the vapor quality is high. Wet vapor can
cause material loss from turbine blades predominately along the trailing edges.

Cb-1Zr 5000-Hour Rankine System Corrosion Loop Test

A detailed description is given below of the compatibility results
obtained in a 5000-hour test of a two-loop Cb-1Zr facility in which
sodium was circulated in the heater circuit and potassium was boiled
and circulated in a two-phase secondary loop which contained turbine
simulator test sections of Mo-TZM alloy (30).

Test System Description

The two-loop system, which is shown in Figure 8, consisted of a
primary sodium heater loop and a secondary potassium boiling and condensing loop. In the primary circuit, the sodium was discharged from
a helical-induction EM pump and flowed through an electrical resistance
heater to a tube-in-tube counterflow boiler where heat was transferred
to the potassium in the secondary circuit.

In the boiling and condensing secondary loop, the potassium was
discharged from an EM pump through an electrical resistance preheater
where the temperature of the potassium was increased to near the saturation temperature before flowing into the tube-in-tube counterflow
boiler. The potassium boiler tube was 0.25 inch inside diameter and
240 inches long. A plug that consisted of a 1/16-inch diameter wire
wound with a 1-inch pitch on a 1/8-inch rod was located in the first
12 inches of the boiler. In traveling through the boiler, the potassium was converted from liquid to superheater vapor and then passed

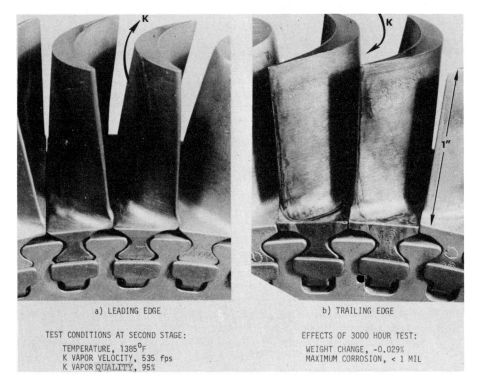

a) LEADING EDGE

b) TRAILING EDGE

TEST CONDITIONS AT SECOND STAGE:

TEMPERATURE, 1385°F
K VAPOR VELOCITY, 535 fps
K VAPOR QUALITY, 95%

EFFECTS OF 3000 HOUR TEST:

WEIGHT CHANGE, -0.029%
MAXIMUM CORROSION, < 1 MIL

Fig. 7. Mo-TZM alloy turbine blades in the second stage of potassium turbine test system following completion of a 3000-hr endurance test.

Table 5.   Vacuum Chamber Total Pressure and the Partial Pressures
of Major Gases* During the 5000-Hour Cb-1Zr Corrosion Loop Test

| Time Period | Total Pressure | Pressure, Torr | | |
		N_2, CO	Ar	H_2
0-200 hr	2×10^{-7}	4×10^{-8}	3×10^{-8}	3×10^{-8}
200-1000 hr	4×10^{-8}	2×10^{-8}	1×10^{-8}	1×10^{-8}
1000-5000 hr	3×10^{-8}	7×10^{-9}	6×10^{-9}	7×10^{-9}

* Other gases measured included H_2O, CO_2, O_2, CH_3, and He; Pressure
of each of these was less than 1×10^{-9} torr after 800 hours of
loop operation.

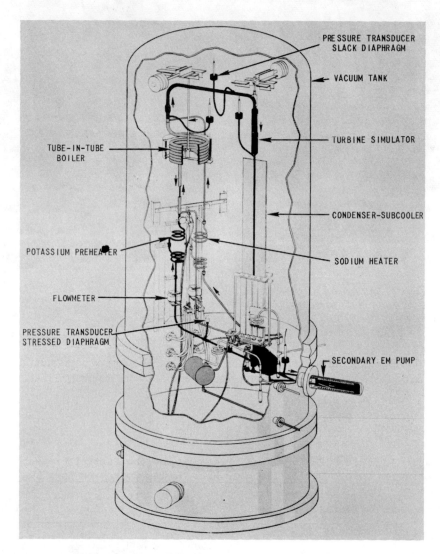

Fig. 8. Isometric drawing of the Cb-1Zr 5000-hr rankine system corrosion test loop.

through the first nozzle and impinged upon the test blade specimen. The superheated vapor passed through a radiation-type heat rejector where the vapor quality was reduced to approximately 88%. The wet vapor then passed through the turbine simulator, containing Stages 2-10, to a finned-tube, radiation-type condenser. The fins of the condenser were coated with iron titanate to increase the emittance to > 0.8. The potassium condensate was returned through a subcooler reservoir to the EM pump.

The loop was constructed of Cb-1Zr with the following exceptions: Mo-TZM turbine simulator nozzles and blade specimens, T-222 pressure transducer diaphragm, and Mo-TZM plugs in the valves in the potassium circuit of the loop. The loop following fabrication is shown in the vacuum chamber spool piece in Figure 9. The loop was tested in a 4-ft diameter by 11-ft high vacuum chamber, equipped with a 2400-liter/sec getter-ion pump.

Test Operation

The test conditions measured during operation of the corrosion test loop are shown in Figure 10. The temperatures shown on the boiler are the sodium temperatures, and it may be noted that of the total sodium temperature drop across the boiler (approximately 130°), more than two-thirds of this drop occurred in the 12-inch long plug section (straight portion) of the boiler. The sharp temperature drop of the sodium in this region associated with the high heat flux to the boiling potassium, resulted in oxygen pickup by the Cb-1Zr potassium containment tube, and this phenomena will be described in more detail below. The 150°F super-heat put into the potassium vapor was achieved in the remaining 230 inches of the boiler. The test system was extremely stable at the endurance test operating conditions with no temperature fluctuations in excess of 5°F or pressure fluctuations in excess of 0.5 lb/in^2.

Typical total pressures and partial pressures of the major residual gases measured in the vacuum chamber for time periods during the 5000-hour test are presented in Table 5. The pressures were maintained low enough to prevent significant contamination of the loop by the test chamber environment.

Test Evaluation

The extensive posttest evaluation performed on the Cb-1Zr and Mo-TZM components following draining and distillation of the alkali metals from the loop circuits included approximately 100 chemical analyses and the examination of 85 metallographic specimens, as well as weight change measurements, dimensional analysis, electron beam microprobe analysis, room temperature and elevated temperature tensile testing, bend testing, and hardness measurements.

Mo-TZM Turbine Simulator Components. The outstanding corrosion resistance of the Mo-TZM turbine simulator nozzles and blades is illustrated in Figure 11. These test components are shown as they appeared when removed from the turbine simulator casing following distillation of residual potassium from the loop. No cleaning of the nozzles and blades was performed prior to photographing them. The test conditions for two turbine simulator stages and the weight changes observed are also given in Figure 11. The changes noted were so slight as to be considered insignificant, and this appraisal was further substantiated

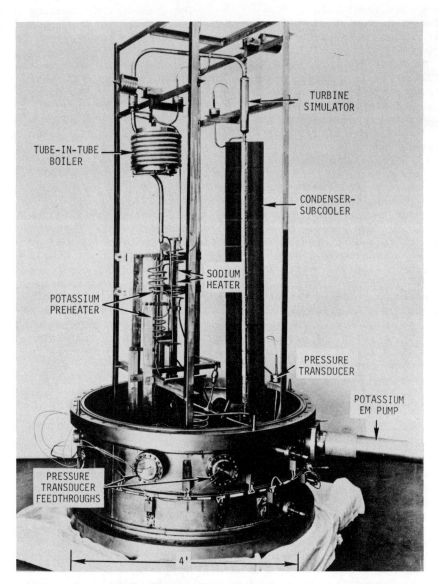

Fig. 9. Cb-1Zr corrosion test loop installed in the vacuum chamber
spool piece following completion of fabrication.

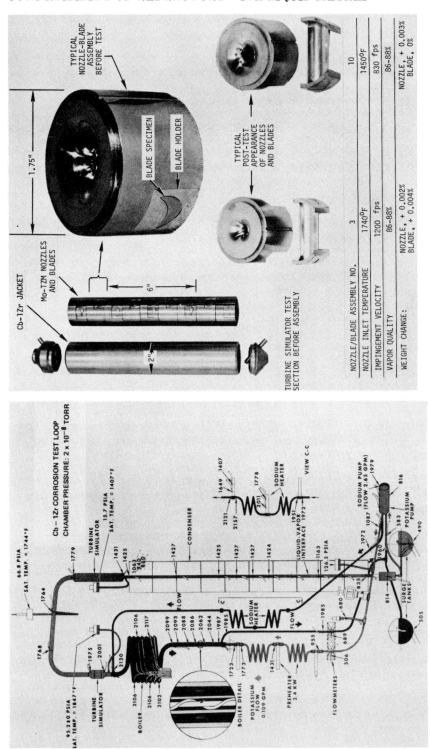

Fig. 11. Mo–TZM alloy turbine simulator components before and following 5000-hr exposure to potassium in the Cb–1Zr corrosion test loop.

NOZZLE/BLADE ASSEMBLY NO.	3	10
NOZZLE INLET TEMPERATURE	1740°F	1450°F
IMPINGEMENT VELOCITY	1200 fps	830 fps
VAPOR QUALITY	86–88%	86–88%
WEIGHT CHANGE:	NOZZLE, + 0.002% BLADE, + 0.004%	NOZZLE, + 0.003% BLADE, 0%

Fig. 10. Operating conditions during the 5000-hr Rankine system corrosion loop test. Temperatures given in °F.

by the extensive metallographic examination and chemical analysis of
the turbine simulator components.

Photomicrographs of blade specimens, shown in Figure 12, do not
show any changes in microstructure which could be attributed to the
high-temperature, high-velocity potassium exposure. Dimensional anal-
sis of the ten nozzles did not indicate significant changes in the
throat diameters. The excellent condition of the nozzles is exemplified in the cross sections and photomicrographs shown in Figure 13.

Cb-1Zr Loop Components. The results of extensive evaluation of
the various regions of the loop revealed no significant corrosion in
any portion of either the potassium or the sodium circuit. However,
the sodium circuit in general and in particular, the boiler portions of
this circuit, did reveal extensive migration of oxygen from the hotter
to the cooler regions of the test loop. This phenomena was most obvious
in the plug section of the boiler, where a thin gray deposit was observed on the sodium side of the potassium boiler tube in the high-heat
flux region. Figure 14 illustrates the appearance of the components in
this region following disassembly and the metallographic cross sections
of specimens taken from this region. The extent of oxygen migration
in the boiler region of the loop may be noted in Figure 15 which includes oxygen analyses of the Cb-1Zr tubing for both whole wall and
gradient samples. Most interesting is the marked oxygen depletion
(260 ppm to 36 ppm, 245 ppm to 40 ppm) noted in the highest temperature
regions and the very large oxygen increases on the sodium side of the
potassium containment tube in the highest heat flux region. An extremely steep oxygen gradient may be noted in the potassium containment
tube; 2445 ppm, 538 ppm, 176 ppm; in the three sections of the wall.

The very high oxygen concentration observed in the various regions
of the loop suggested possible deleterious effects on the ductility of
the "oxygen contaminated" tubing. Reduced gauge-length tube tensile
specimens were prepared from various high and low oxygen regions of the
loop, and duplicate specimens were tested both at room temperature and
2000°F. All results indicated no significant effects on ductility or
tensile strength. The results of the room temperature tests are given
in Table 6. In addition, Cb-1Zr tube specimen rings cut from the
section of the boiler shown in Figure 14 which were covered with oxide
crystals and contained 1240 ppm oxygen were flattened at room temperature and showed no signs of cracking.

The small gray particles which covered the sodium side of the
potassium boiler tube in the high oxygen region were studied in an
attempt to determine their composition. Metallographic specimen cross
sections of this region containing the 1/2-mil thick, ceramic-like
deposit were analyzed by electron microprobe techniques. An enlarged
view of the deposit is shown in the photomicrograph in Figure 16. An
X-ray wavelength scan of the deposit indicated the presence of zirconium. The zirconium X-ray image shown in Figure 16 confirmed the
existence of a high concentration of zirconium in the deposit. Subsequent quantitative analysis scans across the deposit revealed approximately 85 weight percent zirconium and greater than 10 weight percent
oxygen. The deposit is believed to be ZrO which has a stoichiometric
composition of 85 weight percent Zr and 15 weight percent oxygen.

In summary, evaluation of the corrosion test loop after 5000 hours
of operation under two-phase potassium flow and single-phase sodium

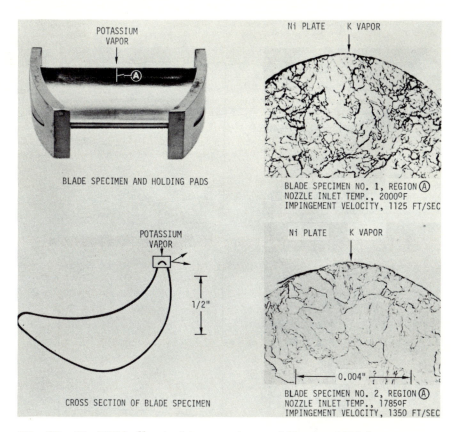

Fig. 12. Mo–TZM alloy turbine specimens following 5000–hrs exposure to potassium in the Cb–1 Zr corrosion test loop.

Table 6. Results of Room Temperature Tensile Tests of Cb–1 Zr Tube Specimens[a] from the Sodium Circuit of the Cb–1 Zr Corrosion Test Loop

Specimen	Oxygen Concentration ppm	Yield Strength psi	Ultimate Tensile Strength psi	Elongation % in 1-inch
Before Test	260	18,700	35,700	37
Boiler Inlet Tube (2130°F)	36	18,500	31,350	37
Boiler Outlet Tube (1985°F)	730	16,250	33,350	35

[a] Reduced gauge length tube tensile specimen.

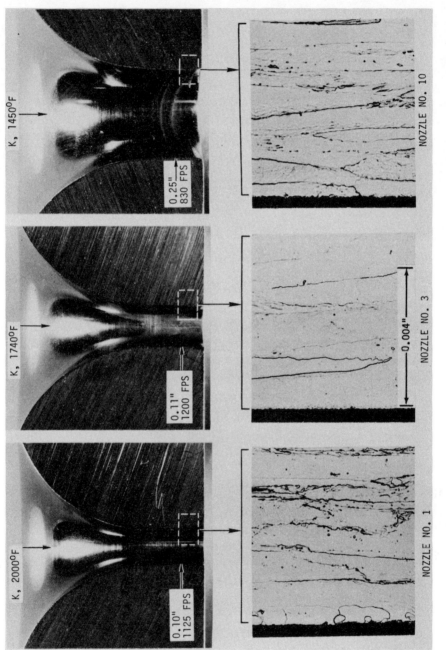

Fig. 13. Cross section views and photomicrographs of turbine simulator nozzles following exposure to potassium for 5000-hrs in the Cb-1Zr corrosion test loop.

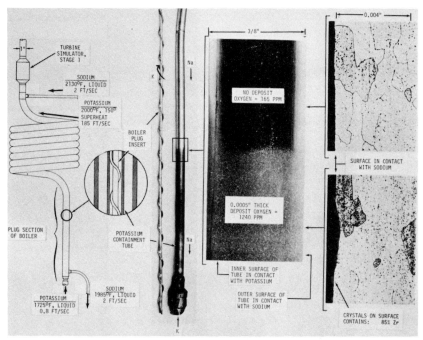

Fig. 14. Boiler plug section components following 5000-hrs of operation of the Cb-1Zr corrosion test loop.

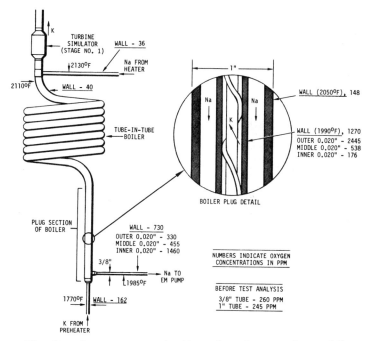

Fig. 15. Oxygen concentration of various regions of the boiler of the Cb-1Zr loop following 5000-hrs operation.

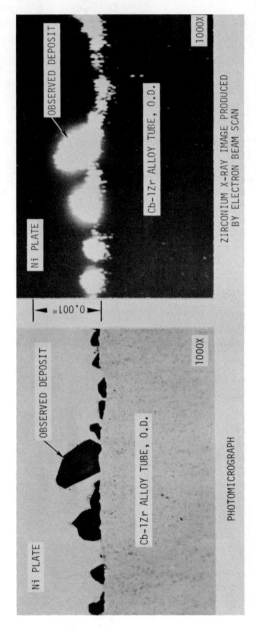

Fig. 16. Deposit found on the sodium side of the potassium boiler tube of the Cb–1Zr loop in the high heat flux region.

flow conditions revealed interesting metallurgical phenomena but no
significant corrosion or deleterious alterations of the properties of
the system materials.

Summary

Refractory Metals in Mercury

Equilibrium solubility results and a limited amount of corrosion
testing indicates that tantalum and Cb-1Zr have excellent resistance
to attack by mercury.

Columbium, Tantalum, and Their Alloys in Alkali Metals

Pure columbium and tantalum have excellent corrosion resistance to
pure alkali metals.

The inherent corrosion resistance of these metals to alkali metals
is adversely affected by oxygen.

Columbium- and tantalum-base alloys containing metals which form
stable oxides are resistant to oxygen-associated attack.

Oxygen-contaminated alkali metals will attack columbium, tantalum,
and their alloys.

No significant temperature-gradient mass transfer has been observed
up to 2000°F.

At temperatures of 2200°F and above, temperature-gradient mass
transfer of zirconium, nitrogen, oxygen, and carbon has been observed.

Transfer of elements such as carbon and nitrogen to these metals
and alloys from dissimilar materials such as stainless steel, in a
common alkali metal system has been noted.

Molybdenum, Tungsten, and Their Alloys in Alkali Metals

Corrosion evaluations of molybdenum and its alloys have been primarily limited to tests in potassium.

Corrosion evaluations of tungsten and its alloys have been limited
to a few capsule investigations.

Although the testing has been limited, these metals have demonstrated excellent corrosion resistance at temperatures to 3000°F.

Molybdenum has excellent resistance to carbon and nitrogen mass
transfer in stainless steel - alkali metal systems.

Acknowledgment

The section of this paper entitled, "Cb-1Zr 5000-Hour Rankine System Corrosion Loop Test," is a description of results obtained under
Contract NAS 3-2547, NASA - Lewis Research Center.

REFERENCES

1. Barto, R. L. and Hurd, D. T., "Refractory Metals in Liquid Metals Handling," Research/Development, No. 11, 26-9, November 1966.

2. Zipkin, M. A., "Alkali Metal Rankine Cycle Power Systems for Electric Propulsion," J. Spacecraft, Vol. 4, No. 7, pp. 852-858.

3. Hodgson, J. N., Geimer, R. G., and Kruger, A. H., "SNAP-8 - A Technical Assessment," Advances in Energy Conversion Engineering, ASME 1967, p. 135.

4. DeVan, J. H., "Corrosion of Iron- and Nickel-Base Alloys in High-Temperature Sodium and NaK," Alkali Metal Coolants, International Atomic Energy Agency, Vienna, 1967, p. 643.

5. Jansen, D. H. and Hoffman, E. E., Type 316 Stainless Steel, Inconel, and Haynes Alloy No. 25 Natural-Circulation Boiling-Potassium Corrosion Test Loops, ORNL-3790, June 1965.

6. Scheuermann, C. M., Barrett, C. A., Lowdermilk, W. H., and Rosenblum, L., "SNAP-8 Materials," Astronautics and Aerospace Engineering, Vol. 1, No. 11, December 1963, p. 40.

7. Harms, W. O. and Litman, A. P., Compatibility of Materials for Advanced Space Nuclear Power Systems, ASME Paper 67-WA/AV-1, 1967.

8. Romano, A., Fleitman, A., and Klamut, C., "Behavior of Refractory Metals and Alloys in Boiling Sodium and Other Boiling Alkali Metals," Alkali Metal Coolants, International Atomic Energy Agency, Vienna, 1967, p. 663.

9. DeVan, J. H., Litman, A. P., DiStefano, J. R., and Sessions, C. E., "Lithium and Potassium Corrosion Studies with Refractory Metals," Alkali Metal Coolants, International Atomic Energy Agency, Vienna, 1967, p. 675.

10. Rosenblum, L., Scheuermann,C. M., and Moss, T. A., "Space-Power-System Material Compatibility Tests of Selected Refractory Metal Alloys with Boiling Potassium," Alkali Metal Coolants, International Atomic Energy Agency, Vienna, 1967, p. 699.

11. Rosenblum, L., et al., "Potassium Rankine System Materials Technology," Space Power Systems Advanced Technology Conference, NASA SP-131, 1966, p. 185.

12. Goldman, K., Minushkin, B., "Sodium Technology," Reactor Technology, Selected Reviews, USAEC Report No. TID-8541, 1966, p. 321.

13. Davis, M. and Draycott, A., "Compatibility of Reactor Materials in Flowing Sodium," Proceedings of the Second United Nations International Conference on the Peaceful Uses of Atomic Energy, 7, 1958, pp. 95-110.

14. Goldman, op. cit.

15. Dotson, L. E. and Hand, R. B., Purification, Analysis, and Handling of Sodium and Potassium - Potassium Corrosion Test Loop Development Topical Report No. 4, R66SD3012, General Electric Company, Cincinnati, Ohio, June 13, 1966.

16. Bates, L. C., "Nuclear Methods of Oxygen Analysis," Nucleonics, 21 (7), 1963, p. 72.

17. Litman, A. P. and Strain, J. E., "Determination of Oxygen in Alkali Metals," Metals and Ceramics Division Annual Progress Report, ORNL-3970, October 1966, p. 82.

18. Nuclear Engineering Department Annual Report, December 31, 1965, Brookhaven National Laboratory Report No. BNL 954.

19. Progress Report of the Nuclear Engineering Department, Brookhaven National Laboratory Report No. BNL-759, May 1 - August 31, 1962, p. 57.

20. Chandler, W. T. and Hoffman, N. J., Effects of Liquid and Vapor Cesium on Container Metals, Report No. ASD-TDR-965, North American Aviation, Inc., March, 1963.

21. Metals and Ceramics Division Annual Progress Report for Period Ending June 30, 1966, USAEC Report ORNL-3970, October 1966, p. 77.

22. Tepper, F. and Greer, J., Factors Affecting the Compatibility of Liquid Cesium with Containment Materials, ASD-TDR-63-824, Part I, Mines Safety Appliances Research Corporation, September 1963.

23. Ammon, R. L., Begley, R. T., and Eichinger, R. L., "T-111 Cesium Natural Connection Loop," AEC-NASA Liquid Metals Information Meeting, USAEC Report CONF-650411, 1966, p. 359.

24. Tepper, F. and Greer, J., "Factors Affecting the Compatibility of Liquid Cesium with Containment Metals," AEC-NASA Liquid Metal Information Meeting, USAEC Report CONF-650411, 1966, p. 323.

25. Chandler, W. T., "Alkali Metal Corrosion Studies at Rocketdyne," NASA-AEC Liquid Metals Corrosion Meeting, USAEC Report TID-7626 (Part 1), April 1962, pp. 52-53.

26. Litman, A. P., The Effect of Oxygen on the Corrosion of Niobium by Liquid Potassium, USAEC Report ORNL-3751 (1965).

27. DeVan, J. H., DiStefano, J. R. and Jansen, D. H., "Compatibility of Refractory Metals with Boiling Alkali Metals," Transactions American Nuclear Society, Vol. 8, No. 2, 1965, pp. 390-391.

28.   Jansen, D. H. and Hoffman, E. E., Niobium - 1% Zirconium, Natural-Circulation, Boiling-Potassium Corrosion Loop Test, USAEC Report ORNL-3603, 1964.

29.   Blecherman, S. S. and Hodel, J., The Compatibility of Structural and Turbo-Machinery Alloys in Boiling Potassium, USAEC Report PWAC-501, 1965.

30.   Hoffman, E. E. and Holowach, J., "Cb-1Zr Rankine System Corrosion Test Loop," Corrosion Test Loop Development Topical Report No. 7, R67SD3016, General Electric Company, Cincinnati, Ohio, May 1, 1968.

31.   Harms, W. O., "High-Temperature Materials," Metals and Ceramics Division Annual Report for Period Ending June 30, 1967, USAEC Report ORNL-4170, Part II, 1967.

32.   Semmel, Jr., J. W., Engel, Jr., L. B., Frank, R. G., and Harrison, R. W., "Carbon Mass Transfer in Multimetallic Systems Containing Potassium," Alkali Metal Coolants, International Atomic Energy Agency, Vienna, 1967, p. 181.

33.   Goldman, K., Hyman, N., Kostman, S., and McKee, J., "Carbon and Nitrogen Transfer in a Type 316 Stainless Steel, Cb-1%Zr, Liquid-Potassium System," Transactions American Nuclear Society, Vol. 8, No. 2, 1965, p. 117.

34.   Engel, L. B. and Frank, Jr., R. F., Evaluation of High-Strength Columbium Alloys for Alkali Metal Containment, NASA-CR-54226, 1964.

35.   NASA-AEC Liquid-Metals Corrosion Meeting, Lewis Research Center October 2-3, 1963, NASA Report SP-41, Vol. I, 1964, pp. 189-197.

36.   DeVan, J. H., DiStefano, J. R., and Jansen, D. H., "Compatibility of Boiling Potassium with Refractory Alloys," Metals and Ceramics Division Annual Progress Report for Period Ending June 30, 1965, USAEC Report ORNL-3870, pp. 124-129.

37.   Harrison, R. W., Compatibility of Biaxially Stressed D-43 Alloy with Refluxing Potassium, NASA Report NASA CR-807, June 1967.

38.   Personal Communication, J. H. DeVan, Oak Ridge National Laboratory.

39.   Simons, E. M. and Lagerdrost, J. F., "Mass Transfer of TZM Alloy by Potassium in Boiling-Refluxing Capsules," AEC-NASA Liquid Metals Information Meeting, USAEC Report CONF-650411, 1966, p 237.

40.   DiStefano, J. R. and Hoffman, E. E., A Survey of the Compatibility of Some Potential Refractory-Metal Fuel Cladding Materials with Potassium, USAEC Report ORNL-TM-708, December 13, 1963.

41. Zimmerman, W. F., Hand, R. B., Engleby, D. S., and Semmel, Jr., J. W., Two-Stage Potassium Test Turbine, Vol. IV, Material Support, NASA Report CR-925, February 1968.

42. Schnetzer, E. (Ed.), 3000-Hour Test - Two-Stage Potassium Turbine, NASA Report NASA CR-72273, July 1967.

43. Kovacevich, E., "Potassium Corrosion Studies," NASA-AEC Liquid Metals Corrosion Meeting, USAEC Report TID-7626 (Part 1), April 1962, p. 65.

44. Annual Report for 1963, Metallurgy Division, USAEC Report ANL-6677, Argonne National Laboratory.

45. DiStefano, J. R. and Hoffman, E. E., "Corrosion Mechanism in Refractory Metal-Alkali Metal Systems," The Science and Technology of Tungsten, Tantalum, Molybdenum, Niobium and Their Alloys, Edited by N. E. Promisel, Pergamon Press, London, 1964, p. 257.

46. Goldman, K. and Minushkin, B., "Sodium Technology," Reactor Technology, Selected Reviews, USAEC Report No. TID-8541, p. 352.

47. DiStefano, J. R., Mass-Transfer Effects in Some Refractory Metal-Alkali Metal-Stainless Steel Systems, USAEC Report ORNL-4028, November 1966.

48. Brasunas, A. de S., Interim Report on Static Liquid Metal Corrosion, USAEC Report ORNL-1647, May 11, 1954.

49. Hoffman, E. E. and Holowach, J., Cb-1Zr Thermal Convection Loop, Potassium Corrosion Test Loop Development Topical Report No. 5, R67SD3014, General Electric Company, Cincinnati, Ohio, June 15, 1967.

50. Hoffman, E. E. and Holowach, J., Cb-1Zr Pumped Sodium Loop, Potassium Corrosion Test Loop Development Topical Report No. 6, R67SD3015, General Electric Company, Cincinnati, Ohio, December 20, 1967.

51. Yaggee, F. L. and Gilbert, E. R., "Effect of Sodium Exposure on the Mechanical Properties of Potential Fuel Jacket Alloys at 550° to 700°C," Alkali Metal Coolants, International Atomic Energy Agency, Vienna, 1967, p. 215.

52. Kelman, L. R., Wilkinson, W. D., and Yaggee, F. L., Resistance of Materials to Attack by Liquid Metals, USAEC Report ANL-4417, July 1950.

53. Raines, G. E., Weaver, C. V., and Stang, J. H., Corrosion and Creep Behavior of Tantalum in Flowing Sodium, Battelle Memorial Institute Report BMI-1284, August 1958.

54.   Bowers, H. I. and Ferguson, W. E., "Structural Materials in LASL
      Liquid Sodium Systems," IMD Special Report Series No. 12, Nuclear
      Metallurgy, AIME, IX, 1963.

55.   Carter, R. L., Eichelberger, R. L., and Siegel, S., "Recent
      Developments in the Technology of Sodium-Graphite Reactor
      Materials," Proceedings of the United Nations International
      Conference on the Peaceful Uses of Atomic Energy, 2nd Geneva 7,
      1958, p. 74.

56.   Cygan, R. and Reed, E., Molybdenum Corrosion by Sodium, USAEC Report
      NAA-SR-161, November 20, 1951.

57.   Neserov, B. A., et al., "Corrosion Resistance of Constructional
      Materials in Alkali Metals, A/Conf. 28F/343, Third United Nations
      International Conference on the Peaceful Uses of Atomic Energy,
      Geneva, 1964.

58.   Hoffman, E. E., Corrosion of Materials by Lithium at Elevated
      Temperatures, USAEC Report ORNL-2674, March 1959.

59.   DiStefano, J. R., Corrosion of Refractory Metals by Lithium, USAEC
      Report ORNL-3551, March 1964.

60.   Freed, M. S. and Kelly, K. J., Corrosion of Columbium-Base and
      Other Structural Alloys in High-Temperature Lithium, Pratt and
      Whitney Aircraft Report No. PWAC-355, June 1961.

61.   DeVan, J. H. and Sessions, C. E., "Mass Transfer of Niobium-Base
      Alloys in Flowing Nonisothermal Lithium," Nuclear Applications,
      Vol. 3, No. 2, February 1967, p. 102.

62.   Sessions, C. E. and DeVan, J. H., Thermal Convection Loop Tests of
      Refractory Alloys in Lithium, USAEC Report ORNL-

63.   Bourdon, C. J., Lymperes, C. J., and Schenck, G. F., Columbium-
      1 Zirconium Alloy Lithium Corrosion Loop Tests, LCCBK-5, 6, 7, 8,
      USAEC Report TIM-823, Pratt and Whitney Aircraft, 1965.

64.   SNAP-50/SPUR Final Summary Report-Coolants and Working Fluids,
      USAEC Report PWAC-491, Pratt and Whitney Aircraft, 1965.

65.   Hays, L. G., Corrosion of Niobium-1% Zirconium Alloy and Yttria
      by Lithium at High Flow Velocities, NASA-JPL Technical Report
      32-1233, December 1, 1967.

66.   Sessions, C. E., "Corrosion of Advanced Refractory Alloys in
      Lithium," AEC-NASA Liquid Metals Information Meeting, USAEC
      Report CONF-650411, 1966, pp. 143-148.

67. Metals and Ceramics Division Annual Progress Report for Period
     Ending June 30, 1967, USAEC Report ORNL-4170, November 1967, p. 95.

68. DeMastry, J. A., "Corrosion Studies of Tungsten, Molybdenum, and
     Rhenium in Lithium," Nuclear Applications, Vol. 3, February 1967,
     p. 127.

69. Harrison, R. W. and Holowach, J., Advanced Refractory Alloy
     Corrosion Loop Program, Quarterly Progress Report No. 12 for
     Quarter Ending April 15, 1968, NASA Contract NAS 3-6474 (to be
     published).

70. Weeks, J. R., "Liquidus Curves and Corrosion of Fe, Cr, Ni, Co, V,
     Cb, Ta, Ti, Zr in 500-750°C Mercury," Corrosion, Vol. 23, No. 4,
     April 1967, pp. 98-106.

71. DiStefano, J. R. and Litman, A. P., "Effects of Impurities in Some
     Refractory Metal-Alkali Metal Systems," Corrosion, December 1964,
     p. 392t.

72. McKisson, R. L., Eichelberger, R. L., Dahleen, R. C., Scarborough,
     J. M., and Argue, G. R., Solubility Studies of Ultra Pure Alkali
     Metals, NASA CR-610, October 1966.

73. Unpublished Data NASA Contract 3-8507, Solubility of Refractory
     Metals - Alkali Metals, Atomics International Division of North
     American Rockwell Corporation.

74. SNAP-8 Refractory Boiler Development Monthly Status Report for
     Period Ending February 10, 1968, NASA Contract NAS 3-10610.

75. Personal Communication, C. Klamut, Brookhaven National Laboratory.

76. Hanchett, James, "Whither Now SNAP-8," Nucleonics, May 1967.

77. Nevzorov, B. A., et al., "Corrosion Resistance of Constructional
     Materials in Alkali Metals," Paper A/Conf. 28/F/343, Third U. N.
     International Conference on the Peaceful Uses of Atomic Energy,
     Geneva, Switzerland (1964).

REFRACTORY METALS IN SPACE ELECTRIC POWER
CONVERSION SYSTEMS

J. W. Semmel, Jr.

ABSTRACT

The status of refractory alloy applications to space electric power
conversion systems is reviewed with emphasis on the work related
to alkali metal Rankine cycle systems for future power requirements exceeding 100 kw. Columbium and tantalum alloys are considered for alkali metal containment for periods exceeding 10,000
hours at temperatures near 2000°F. In addition, molybdenum alloys are considered for alkali metal vapor turbine wheel and blade
applications where weldability is not essential. Results and conclusions from corrosion and erosion tests of refractory alloys in
alkali metals are presented. The fabrication of power plant development components from refractory alloys is discussed.

J. W. Semmel, Jr. is affiliated with the Space Power and Propulsion Section of the Missile and Space Division, General Electric
Company, Cincinnati, Ohio.

INTRODUCTION

Comprehensive investigations of the behavior of several classes of materials are being conducted for their utilization in future space electric power plants having relatively high capabilities exceeding 100 kw for operating periods extending beyond 10,000 hours. Key materials for electrical devices, nuclear power sources, and high temperature structures are being evaluated, as are many other materials for special applications such as bearings, instrumentation, thermal control coatings, and working fluids. There has been considerable progress in these areas, and reviews of the overall material status have been presented recently [1-4]. The intent of this paper is to describe some of the current activities in the high temperature structural material category, which includes the refractory alloys—their strength, fabrication, and reactions with imposed environments. Summary descriptions of the reactions of refractory alloys with three of the pertinent environments, vacuum, inert gas, and alkali metals, are included in this symposium. Other papers on mechanical properties and alloy development also contain information on specific alloys that are significant to space electric power conversion systems. Consequently, the discussion of strengthening and environmental reactions will be curtailed somewhat in this paper and emphasis will be placed on some aspects of the fabrication of power conversion system development components from refractory alloys.

POWER CONVERSION SYSTEM CONDITIONS

Several types of space electric power conversion systems that would require refractory alloys have been described [5-7]. The potassium Rankine cycle power plant will be used here as a basis of discussion mainly because the development of structural refractory alloys has been most extensive for this type of system. Automatically, this includes structural refractory alloy considerations pertinent to other types of large power plants.

A description of the system and component design for a 1-Mw Rankine cycle power plant was presented recently by Zipkin, and the representative schematic diagram and component arrangement are shown in Figures 1 and 2 [8]. Liquid lithium, or possibly another alkali metal such as sodium or potassium, is heated to approximately 2100°F by a nuclear reactor. In a heat exchanger, the lithium is used to boil potassium or another turbine working fluid such as cesium or rubidium. As shown in Figures 1 and 2, potassium enters the turbine at 1950°F, and condensation of the 1300°F turbine discharge takes place in the tubes of a vapor fin, condensing radiator. An electromagnetic (EM) pump returns the 1250°F condensate to the pre-heater and boiler. A separate alkali metal circuit with a low temperature radiator provides a 600°F coolant to electrical components, bearings, and seals. Details concerning temperatures, working fluids, and components presented in Figures 1 and 2 are not considered final. Rather, they are a typical representation of conditions that would lead to a useful power plant, and they provide a framework for considering refractory alloys in this paper.

Three other types of large power conversion systems are based on the inert gas turbine Brayton cycle, the thermionic convertor, and the magnetohydrodynamic generator. Apart from important differences in operating temperatures and

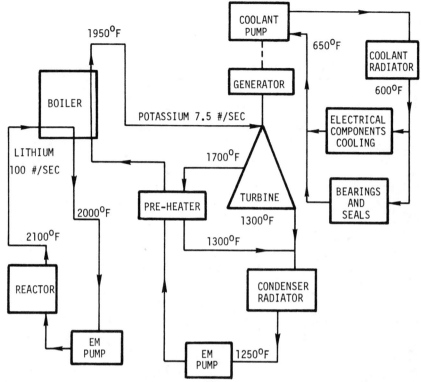

Fig. 1. Schematic diagram of a 1-Mw potassium Rankine cycle power plant.

material strength requirements that necessitate refractory alloys in all cases, the Brayton cycle power system obviates the alkali metal environment, and the latter two electrical devices avoid the rotating turbogenerator but retain the alkali metal environment. No comparison of the merits of these power conversion systems is intended here; however, it is pertinent to note that the refractory alloy considerations for the potassium Rankine cycle power system are quite broadly applicable to the other systems because they include both turbine and piping requirements.

ENVIRONMENTAL FACTORS

Alkali Metal Containment

It has been established that properly controlled refractory alloys have excellent resistance to corrosion by high-purity alkali metals under conditions simulating the potassium Rankine cycle system depicted in Figure 1. Reviews presented at recent symposia [9, 10] generally support this conclusion, and direct evidence was provided by a large-scale corrosion test with a Cb-1Zr* alloy loop

* Compositions of the refractory alloys mentioned herein are tabulated in the Appendix.

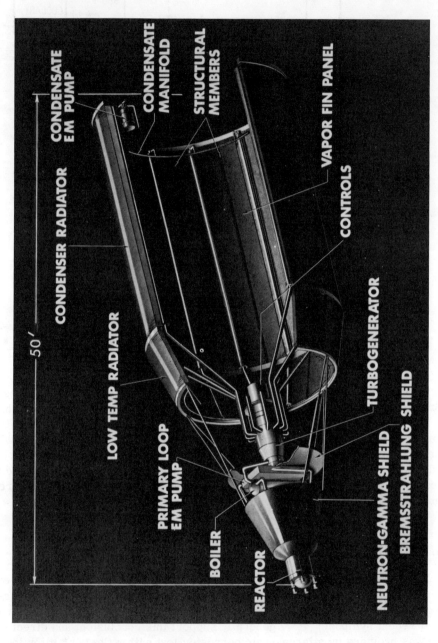

Fig. 2. General arrangement of components for a 1–Mw potassium Rankine cycle power plant.

specifically designed to simulate the power conversion system conditions. This corrosion experiment, which was conducted in a vacuum chamber for 5000 hours with potassium at 2000°F under NASA Contract No. NAS 3-2547 at the General Electric Company, Space Power and Propulsion Section (GE-SPPS), is illustrated in Figure 3 and the fabricated test loop is shown in Figure 4. Only minor amounts of corrosion were observed, and details are presented in the contract topical report [11] and Hoffman's paper at this symposium. The program is continuing at GE-SPPS under NASA Contract No. NAS 3-6474 in order to test a T-111 alloy loop of similar design for 10,000 hours with potassium at 2150°F.

To achieve a high degree of corrosion resistance, definite requirements must be placed on refractory alloy compositions and processing conditions. The presence of several hundred ppm of oxygen in solution in unalloyed columbium and tantalum can result in significant corrosion by the alkali metals. However, small additions of stable oxide-forming elements, such as zirconium or hafnium, to columbium- and tantalum-base alloys can combine with the oxygen in the alloy and thereby eliminate the rapid grain boundary and crystallographic attack of refractory metals which has been observed and studied by numerous investigators [9, 10].

This situation defines processing and fabrication requirements, inasmuch as it is necessary to assure that the oxygen content of the alloy does not exceed that amount which can be combined with the alloying addition. Furthermore, proper heat treatments must be used to precipitate the oxygen with the alloying element in a manner which does not embrittle the alloy significantly. These requirements can be satisfied in practice by (a) limiting the oxygen content in the ingots to a few hundred ppm, (b) limiting contamination during processing operations at high temperature or removing any contamination that occurs, (c) conducting heat treatments in vacuum furnaces at pressures below approximately 1×10^{-5} torr, (d) welding in an inert gas environment (or vacuum) with atmospheric contaminants below approximately 20 ppm, and (e) postweld annealing in a vacuum at temperatures somewhat above 2000°F, depending upon the particular alloy. It is also necessary to limit the oxygen content in the alkali metal, and long-time component tests are usually conducted in vacuum environments on the order of 1×10^{-8} torr. The technology for performing all of these operations is in hand.

There are, however, areas in which additional information on alkali metal containment is desired and these will be mentioned briefly. It is known that Mo-TZM has excellent resistance to corrosion by potassium near 2000°F, but the evaluation of molybdenum, tungsten, and rhenium alloys as a group has been less extensive than that conducted for columbium and tantalum alloys. The need for an alloying addition to form stable oxides has not been clearly established for these materials, which normally contain less oxygen than columbium and tantalum. The amount of testing of refractory alloys with lithium and potassium far exceeds that conducted with other alkali metals. Even in the case of lithium and potassium, however, some of the details describing the effects of various impurities in the refractory alloys and alkali metals are lacking. Finally, the upper temperature limits of the corrosion resistance of columbium and tantalum alloys are open to conjecture — 2200°F appears satisfactory while 2400°F is a moot question. As the temperature increases, mass transfer of the interstitial elements in columbium

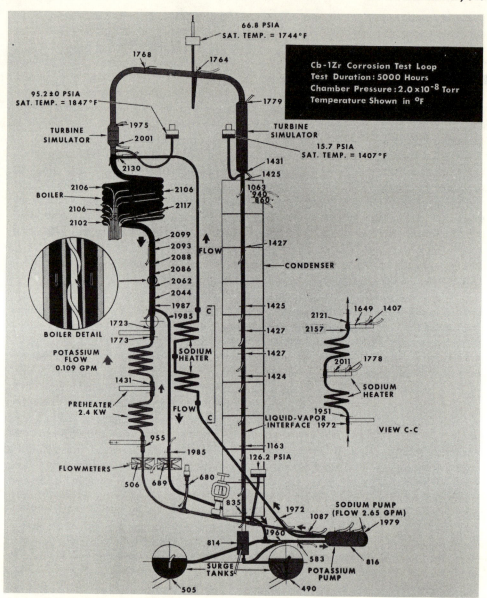

Fig. 3. Schematic diagram of a Cb–1Zr corrosion test loop designed and operated to simulate potassium Rankine cycle power conversion system conditions.

and tantalum alloys becomes much more prevalent, and alloys which utilize the interstitial elements for strengthening purposes are in jeopardy. Also at the higher temperatures, the corrosion inhibitors, zirconium and hafnium, tend to transfer; consequently, the interest expands to include the stronger tungsten and rhenium alloys which possibly may be more corrosion resistant but are certainly more difficult to fabricate.

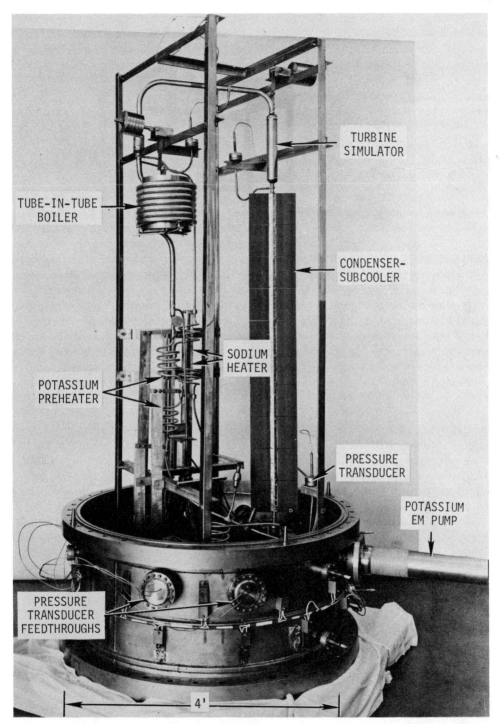

Fig. 4. Cb-1Zr alloy corrosion test loop installed in the vacuum test chamber spool piece.

There is work in progress in all of these areas which concern (a) the alternate choices in construction materials and working fluids, (b) the establishment of more definitive material specifications, and (c) the identification of a maximum temperature associated with corrosion resistance, which may or may not be the critical limitation on the system operating temperature. These items are important to future system optimization studies and second generation power plants; however, the major consideration from a designer's point of view now is that the demonstrated corrosion resistance makes the system illustrated in Figure 1 appear feasible with respect to containment of the working fluid.

Erosion by Alkali Metal

The possibility of impact erosion results from the condensation of alkali metal vapor within the turbine of the Rankine cycle system. Vapor qualities of 80 to 90% may be encountered in the latter turbine stages, and it is necessary to establish design and operating conditions which avoid excessive erosion of turbine components by the liquid alkali metal. Expectations are that most of the condensate passes harmlessly through the turbine as fine droplets entrained in the vapor; however, some of the condensate collects on the stator vanes and exits as relatively slow moving, large drops which are struck by the rotating turbine blades. Liquid accumulated on the turbine blades also can be ejected at a high velocity toward stationary components. Turbine blade tip speed, the possible inclusion of condensate extraction devices, and turbine materials selection are the major factors involved. Steam turbine practice is to limit turbine blade tip speeds to approximately 1000 fps and provide hard facing materials at locations subject to the more severe erosion conditions.

At the present state of development, space power system turbine materials are not being selected or modified specifically to enhance impact erosion resistance. Rather, candidate materials are being tested to determine whether they offer sufficient erosion resistance for practical turbine designs. Although it is difficult to achieve a good simulation of the turbine operating conditions, significant tests have been performed. Turbine simulators that contain nozzles and stationary blade shapes (Figure 5) are often included in the type of boiling and condensing test loop illustrated in Figures 3 and 4. Appropriate vapor velocities and average vapor contents can be obtained, but the liquid drop size and velocity remain unknown. Although the impact conditions may not be sufficiently severe in this type of test, it is encouraging that a material such as Mo-TZM was essentially unattacked in the 5000-hour system simulation test with potassium that was mentioned previously [11]. The continuing test program with the T-111 alloy loop will include Mo-TZC and Cb-132M nozzles and blade shapes.

More realistic test conditions were achieved in a two-stage potassium test turbine which was operated for 5000 hours at GE-SPPS under NASA Contract No. NAS 3-1143. These two turbine stages simulated stages three and four of a conceptual five-stage Rankine system turbine. A description of some of the materials aspects of the test has been presented [12], and the size of the test is illustrated in Figures 6 and 7, which show the turbine and the test facility. The boiler, condenser, turbine casing and piping were type 316 stainless steel, and the turbine was constructed of various superalloys. Turbine blades of Udimet 700, Mo-TZM,

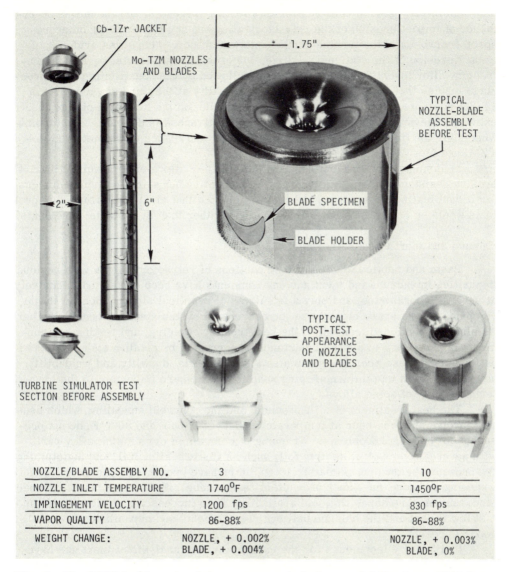

Fig. 5. Mo-TZM turbine simulator components before and following a 5000-hour exposure to potassium in a Cb-1Zr corrosion test loop.

and Mo-TZC were included in the 9.6-inch diameter, Udimet 700 second stage turbine wheel for erosion testing. Also, stationary erosion test inserts of Udimet 700, Mo-TZM, Mo-TZC, and AS-30 were located near the downstream side of the second stage blade tips. During successive periods of 2000 and 3000 hours, turbine was operated at approximately 18,500 rpm with a first stage inlet vapor temperature of 1500°F and a vapor quality of 99.5%. Inlet temperature at the second stage was 1390°F, the vapor quality was 96.7%, and the tip speed was 780 fps. The tur-

bine was in very good condition after this testing (Figure 8) and there was no evidence of impact erosion of the refractory alloys at any location. In other respects, corrosion of the turbine components was minor. There was evidence of
impact erosion of the stationary Udimet 700 erosion test inserts, but not the refractory alloy inserts. This test is being followed by the construction of a three-
stage turbine that is designed for 90% vapor quality at the third stage and for a
15,000-hour life. Provisions also have been made for the inclusion of condensate
extraction devices if an evaluation of their effectiveness is desired. The turbine
wheels for the first two stages will be Mo-TZM, and blades of Udimet 700, Mo-
TZM, and Mo-TZC will be included.

In summary, it appears that candidate turbine alloys offer adequate impact
erosion resistance when combined with conservative but suitable turbine designs.
Additional testing is required, and it is anticipated that any potential erosion problems would be solved by design modifications rather than by material development.

Vacuum and Inert Gas Environments

Basic and applied studies of the reactions of refractory alloys with residual
impurities in vacuum and inert gas environments have been conducted intensively
at several laboratories, and only a few items of practical significance for the development of the space electric power conversion systems will be considered here.
As already mentioned, control of the heat treating, welding, and testing environments is necessary to assure resistance to corrosion by alkali metals. Contamination during these operations can also affect material ductility and weldability
adversely, and it can influence aging reactions which are important to the long-
term stability of some alloys.

The heat treatment of mill products and the postweld annealing, which usually involve only one hour at temperatures between 2000° and 3000°F, do not present major technical problems. Wrapping of columbium or tantalum alloy parts
with several layers of protective foil, such as Cb-1Zr alloy foil, and maintaining
the pressure below approximately 1×10^{-5} torr have proven satisfactory. The foil
wrapping is more an added precaution than a necessity in most cases. Furnace
size, temperature capability, and availability have increased significantly in the
last few years, and the balance between requirements and capabilities is quite
good at present.

Satisfactory techniques for the purification and analysis of inert gas have
been developed to maintain adequate environments for welding columbium and
tantalum alloys. Methods used at two laboratories have been described [13, 14].
As an example, it is practical to make more than 150 tungsten-arc inert gas (TIG)
welds for the loop shown in Figure 4 to specifications which require (a) evacua-
tion of the welding chamber to 1×10^{-5} torr, (b) backfilling with helium having less
than 1 ppm oxygen and 1 ppm water vapor, and (c) never welding when contaminants
in the helium exceed either 5 ppm oxygen, 15 ppm nitrogen, or 20 ppm water vapor.
(Weld metal contamination was not detected by chemical analyses after welding in
helium of this purity [13].) This loop was welded in a chamber that was designed
especially for that purpose (Figures 9 and 10). Smaller, portable devices for the
construction of larger loops and for field welding are also in a continual state of
development and improvement for special applications.

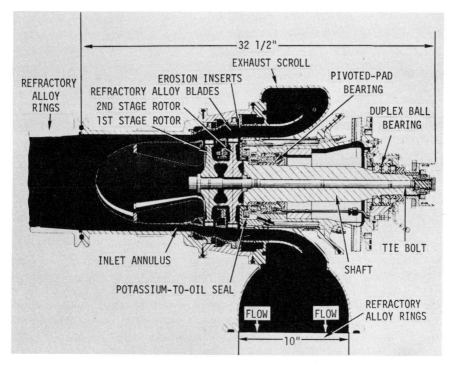

Fig. 6. Drawing of the two-stage potassium test turbine.

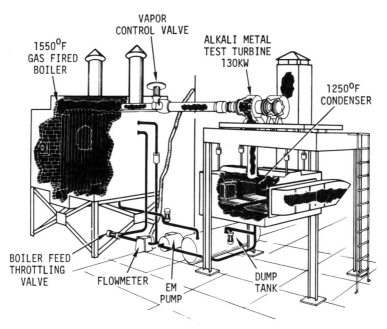

Fig. 7. Drawing of the two-stage potassium test turbine facility.

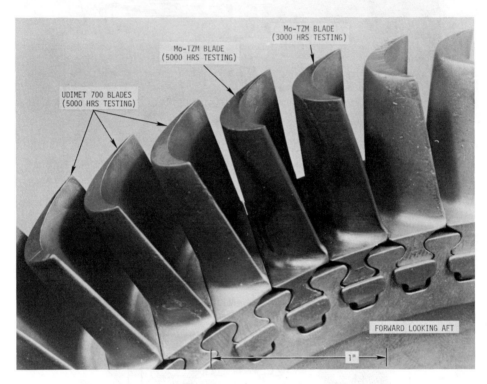

Fig. 8. Second stage turbine wheel assembly after 5000 hours of endurance testing with potassium, followed by vapor-blast cleaning.

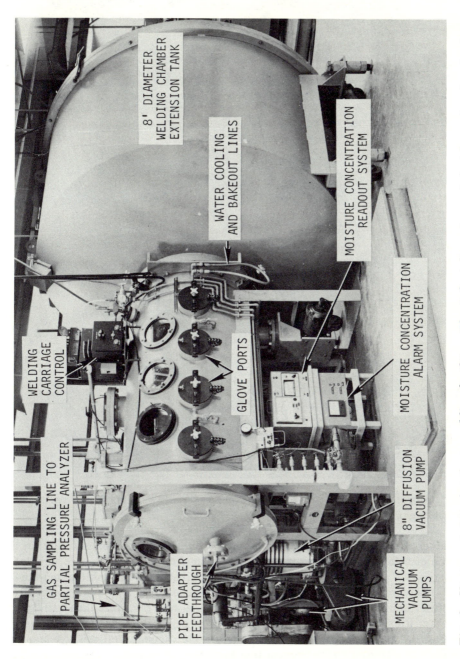

Fig. 9. Vacuum–purge, inert gas welding facility. The 3-foot diameter × 6-foot long chamber is shown with an 8-foot diameter × 6-foot long extension to the chamber in place.

Fig. 10. Corrosion test loop fixtured for final assembly welding in the vacuum-purge, inert gas welding facility.

The control of testing environments has been investigated extensively for the utilization of both vacuum and inert gas. Requirements have not been to simulate particular space conditions, but to restrict contamination of columbium and tantalum alloy components during tests of 1000 to 10,000 hours duration. Generally, much better results have been obtained with vacuum test facilities than with inert gas test facilities partly because fewer sources of contamination, such as ceramics for thermal insulation, have been included in the vacuum facilities. A significant technical advantage with vacuum test facilities, however, is that critical gaseous impurity levels can be measured easily with commercial gas analyzers, whereas comparable determinations cannot be made in an inert gas environment.

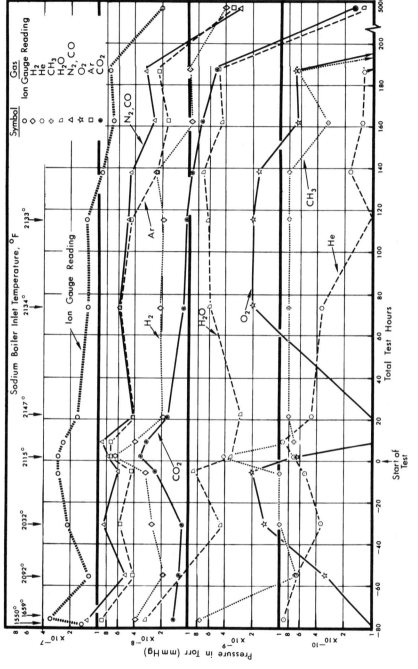

Fig. 11. Vacuum chamber pressure and residual gas composition during startup and operation of a Cb–1Zr corrosion test loop.

At GE-SPPS, the preference is for a vacuum test facility with provisions for automatically flooding the system with inert gas in the event of a vacuum failure. To
illustrate the state-of-the art, Figure 11 shows a part of the gas composition curve
for the 5000-hour test of the loop shown in Figure 4. The 4-foot diameter by 11-
foot high vacuum test facility included a 2400 l/sec getter-ion pump which could
reduce the pressure in the empty chamber to 2×10^{-10} torr and maintain the pressure at 2×10^{-8} torr with the loop at the test temperature. Contamination of 1.0-
inch OD $\times$ 0.10-inch wall Cb-1Zr tubing obtained from the loop after the 5000-hour
test amounted to approximately 100 ppm oxygen and 70 ppm carbon, and there was
no significant change in nitrogen or hydrogen concentration. This is considered
satisfactory.

Overall, the technology required for environmental control during heat treating, welding, and testing is very well established.

CONTAINMENT MATERIALS

Chronologically, the significant initial work on containment materials began
with detailed investigations of the Cb-1Zr alloy which proved to be very fabricable
and resistant to corrosion by alkali metals. Higher strength was desired, however, and this led to comparative evaluations of the alloys available in 1963. In
the same year, an alloy development program was sponsored by NASA at the Westinghouse Astronuclear Laboratory (WANL) to obtain a superior tantalum alloy.
The comparative evaluation led NASA to select T-111 for more detailed investigation, which is a major current activity, and the alloy development program led to
the identification of the stronger ASTAR group of alloys that are now in the stage
of evaluation and initial scale-up.

Creep strength data obtained from several sources [1-3, 15-18] are presented
in Figure 12 for some of the pertinent alloys. In general, the data were obtained
with appropriately heat treated material which was tested for periods of 1000 hours
or longer in a high vacuum environment that minimized the effects of contamination during the test. Although there are too few tests and heats of material involved to make the curves suitable for detailed design purposes, they can be used
to estimate the temperature capabilities of some representative alloys. Assuming, for example, a nominal requirement for an average 1% creep strength of
2000 psi during a 10,000-hour period of operation, it is evident that a strong, fabricable columbium alloy such as FS-85 appears serviceable up to approximately
2250°F. The T-111 tantalum alloy extends the temperature capabilities to approximately 2350°F, ASTAR 811C offers possibilities at 2650°F, and tungsten alloys
are required for temperatures approaching 2800°F. Other considerations involving weldability and corrosion resistance may limit the maximum temperature capabilities; however, it is clear that columbium and tantalum alloys offer adequate
strength for the conditions indicated by the Rankine cycle system outlined in Figure 1.

The weldability of many candidate alloys has been compared under a NASA-
sponsored program at WANL, and the data have been summarized [1-3, 19]. To an
extent dependent upon the interest in each alloy, the following topics were investigated: as-welded ductility in thick (0.375 inch) plate and thin (0.035 inch) sheet,
weld strength characteristics, the thermal stability of weldments, the effect of

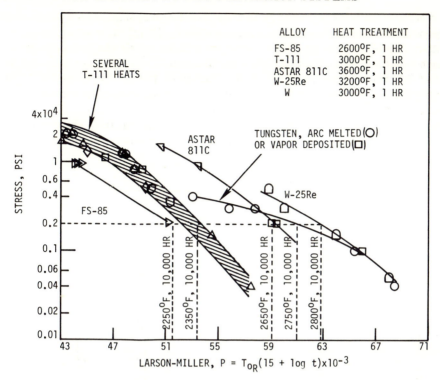

Fig. 12. Containment material 1% creep characteristics in high vacuum.

postweld annealing on the thermal stability, and the effect of oxygen contamination in the parent metal on weldability. Some of the results on as-welded ductility can be summarized conveniently in the form shown in Figure 13, in which the lengths of the bars represent the range of the ductile-to-brittle bend transition tempera-ture data obtained for several welding and test conditions. Significant conclusions are that Cb-1Zr and T-111 have excellent weldability, the more highly strengthened columbium and tantalum alloys (FS-85 and ASTAR 811C) will probably be accept-able on a weldability basis, and tungsten and the W-25Re alloy would present major problems in fabricating complex containment systems.

The more detailed evaluation of T-111 has included further studies of wel-dability and thermal stability in addition to corrosion investigations and the de-velopment of processing techniques for typical mill products, including large di-ameter tubing. In total, this work is generating a significant amount of useful in-formation concerning the alloy and, at the same time, it is uncovering the areas which require additional investigation.

To illustrate the range of T-111 mill products that have been produced suc-cessfully, Table I lists the materials already obtained for a T-111 corrosion test loop of the type illustrated in Figures 3 and 4. Construction of a larger loop for power plant boiler development at GE-SPPS under NASA Contract No. NAS 3-9426

requires significantly more T–111 (Table II), and the preparation of this material
is proceeding satisfactorily and is almost complete. Fabrication procedures for
larger diameter T–111 tubing (4.5-inch OD × 0.125-inch wall and 3.0-inch OD ×
0.080-inch wall) are being developed by WANL under NASA Contract No. NAS 3-
10602, and this work is reviewed in a recent paper by Davies and Stone [20]. Combined experience indicates that (a) high quality T–111 mill products can be made
in the variety of shapes and sizes indicated in Tables I and II, (b) each new producer
encounters new and specific problems which are solvable but frequently result in
deliveries that are late by several months, and (c) the material is more prone to
cracking than tantalum or Cb-1Zr.

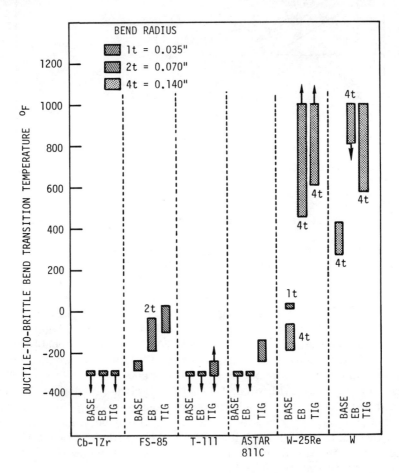

Fig. 13. Summary of as-welded bend ductility characteristics of 0.035-
inch thick containment material sheet.

TABLE I

T-111 ALLOY REQUIREMENTS FOR A CORROSION TEST LOOP

ITEM	WEIGHT, LBS	
1. ROD		
0.250-INCH DIA	1	
0.500-INCH DIA	11	
0.625-INCH DIA	5	
1.000-INCH DIA	40	
1.125-INCH DIA	10	
1.500-INCH DIA	13	
2.000-INCH DIA	85	
2.500-INCH DIA	93	
3.125-INCH DIA	74	
		332
2. BAR		
1.0-INCH X 1.0-INCH	30	
1.0-INCH X 2.0-INCH	115	
		145
3. WIRE		
0.062-INCH DIA	13	
0.094-INCH DIA	8	
0.125-INCH DIA	31	
		52
4. FOIL, SHEET, PLATE		
0.005-INCH X 3.5-INCH	2	
0.009-INCH X 3.5-INCH	1	
0.035-INCH X 1.0-INCH	1	
0.040-INCH X 12.0-INCH	29	
0.125-INCH X 6.0-INCH	5	
0.500-INCH X 6.125-INCH	41	
		79
5. TUBE		
0.375-INCH OD X 0.065-INCH WALL	66	
1.00-INCH OD X 0.065-INCH WALL	104	
2.25-INCH OD X 0.375-INCH WALL	40	
2.50-INCH OD X 0.450-INCH WALL	46	
3.00-INCH OD X 0.375-INCH WALL	50	
3.25-INCH OD X 0.250-INCH WALL	40	
3.25-INCH OD X 0.500-INCH WALL	73	
		419
TOTAL WEIGHT		1027

Cracking has been encountered in forging and extruding T-111 during ingot breakdown, in secondary working operations, in cutting finished products, and in weldments. To a considerable extent, cracking problems result from the fact that the T-111 fabrication technology is developing as an extension of the technology established for other refractory alloys, notably Cb-1Zr and unalloyed tantalum, and the T-111 alloy is probably not quite as resistant to thermal shock as the other materials. A specific example is evident in the difficulty encountered in cutting T-111 compared to Cb-1Zr. Use of an alumina abrasive cut-off wheel with T-111 readily introduces significant cracks which can be propagated (Figure 14), whereas Cb-1Zr is not similarly cracked. However, careful cutting of T-111 with a silicon

TABLE II

T-111 ALLOY REQUIREMENTS FOR A BOILER DEVELOPMENT FACILITY

ITEM	WEIGHT, LBS
1. ROD	
0.250-INCH DIA	2
0.690-INCH DIA	10
1.000-INCH DIA	38
1.375-INCH DIA	128
1.500-INCH DIA	34
2.250-INCH DIA	77
2.500-INCH DIA	236
3.063-INCH DIA	256
3.688-INCH DIA	121
4.438-INCH DIA	163
4.625-INCH DIA	176
	1241
2. WIRE	
0.062-INCH DIA	10
0.094-INCH DIA	15
0.125-INCH DIA	26
	51
3. FOIL, SHEET, PLATE	
0.005-INCH X 3.5-INCH	1
0.005-INCH X 5.5-INCH	1
0.040-INCH X 20.5-INCH	42
0.063-INCH X 6.0-INCH	3
0.100-INCH X 9.0-INCH	23
0.125-INCH X 22.0-INCH	51
0.250-INCH X 8.0-INCH	102
0.400-INCH X 6.0-INCH	44
	267
4. TUBE	
0.250-INCH OD X 0.050-INCH WALL	12
0.375-INCH OD X 0.065-INCH WALL	14
0.500-INCH OD X 0.075-INCH WALL	18
0.625-INCH OD X 0.008-INCH WALL	8
0.690-INCH OD X 0.045-INCH WALL	13
0.750-INCH OD X 0.040-INCH WALL	10
0.875-INCH OD X 0.100-INCH WALL	363
1.325-INCH OD X 0.100-INCH WALL	37
1.500-INCH OD X 0.100-INCH WALL	190
2.375-INCH OD X 0.220-INCH WALL	79
4.386-INCH OD X 1.343-INCH WALL	180
4.420-INCH OD X 0.537-INCH WALL	92
4.424-INCH OD X 1.362-INCH WALL	184
4.625-INCH OD X 0.225-INCH WALL	45
4.625-INCH OD X 0.275-INCH WALL	55
4.625-INCH OD X 0.409-INCH WALL	79
	1379
TOTAL WEIGHT	2938

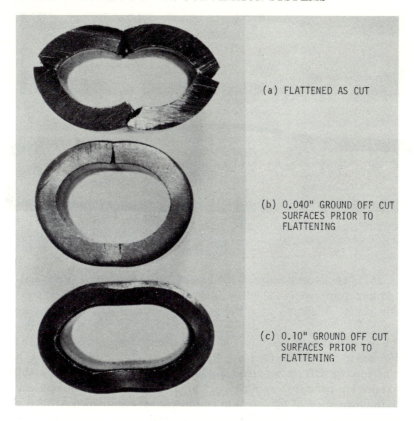

(a) FLATTENED AS CUT

(b) 0.040" GROUND OFF CUT SURFACES PRIOR TO FLATTENING

(c) 0.10" GROUND OFF CUT SURFACES PRIOR TO FLATTENING

Fig. 14. Samples of T-111 tubing (0.375-inch OD × 0.062-inch wall) showing the results of cracking induced by cutting with an alumina abrasive cut-off wheel using water cooling. Flattening propagated the cracks, and the cracked surface region could be removed by grinding.

carbide abrasive cut-off wheel essentially eliminates cracking, and, as an added precaution, the surface may be ground after cutting to remove any shallow cracks which may have formed. This is a typical case, and it is generally found that once the fabrication and joining details are established specifically for T-111, cracking can be avoided and the alloy appears highly fabricable for containment purposes.

The development of the ASTAR alloys is several years behind that of T-111. Alloys of major interest include:

ASTAR 811   Ta-8W-1Hf-1Re
ASTAR 811C   Ta-8W-0.7Hf-1Re-0.025C
ASTAR 811CN Ta-8W-1Hf-1Re-0.012C-0.012N

Scale-up is proceeding with the fabrication and evaluation of sheet and plate obtained from 4-inch diameter ingots at WANL under NASA Contract No. NAS 3-2542. The stability of the carbides and nitrides which strengthen some of the al-

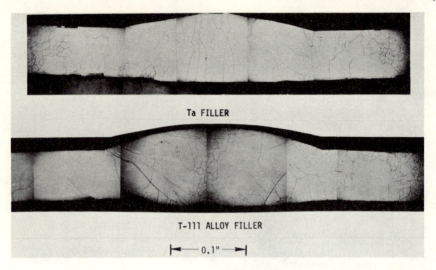

Fig. 15. Transverse microstructure of TIG welds made with unalloyed tantalum and T-111 alloy filler wire in 0.060-inch thick tantalum sheet. Etchant: $30gNH_4F+20mlH_2O+50mlHNO_3$.

loys will be examined in potassium and lithium environments by capsule testing
at GE-SPPS under NASA Contract No. NAS 3-6474. Assuming success, additional
scale-up and testing will be required, including an evaluation of the stability of
carbide and nitride particles with respect to irradiation from the nuclear reactor.

With the exception of the carbide and nitride strengthened ASTAR alloys,
the containment materials generally tend to be rather clean and free of particles
which often serve to reduce the as-cast grain size and impede grain growth in
other engineering materials. Even alloying with elements that remain in solid solution is quite limited in order to retain adequate ductility and weldability. Consequently, large grains, relative to component dimensions, can occur easily in
several circumstances. When the as-cast grain size is large, for example, and
the economics or equipment availability limits the ingot size, it can be difficult to
introduce sufficient cold work to obtain a relatively fine grain size in recrystallized material greater than a few inches in thickness or diameter. When these
large dimensions are required in stock used for machined components, such as
valve bodies, boiler header plates, electromagnetic pump ducts, turbine casings,
and pipe fittings, the grain size may be unusually large with respect to the thinner
sections of the machined component.

Welding can also result in rather large grains, particularly in unalloyed tantalum and Cb-1Zr, which are considered for the containment of mercury and lithium fluoride, respectively, in addition to their utilization with alkali metals. Figure 15 illustrates the large grain size in a tantalum weld and heat affected zone,
and it further illustrates that welding tantalum with a more highly alloyed filler
material such as T-111 does not offer a simple means of eliminating large grains.

A nonuniform grain structure and large grains can also result from critical
amounts of deformation and from strain gradients followed by annealing or thermal
exposure during application. A striking example is presented in Figure 16 for a Cb-1Zr

tube that was recrystallized at 2200°F, bent, and reannealed for 165 hours at 2200°F [21]. Somewhat similar behavior was observed in T-111 after equivalent bending and annealing for 1 hour at 3000°F, but not after bending and annealing for 1 hour at 2400°F.

Although it is evident that relatively large grain sizes can be encountered, it is not apparent that they have led to special difficulties or component failures. The situation is rather poorly documented, however. The effect of strain, temperature, and time on the type of grain growth illustrated in Figure 16 is not known, and it is expected that the behavior may be strongly influenced by the amount of impurities in the material at concentration levels within normally accepted commercial limits. Possible effects of an unusually large grain size on corrosion resistance and mechanical properties have not been examined in a systematic way. If significant problems arise in the future, they could require reduced operating stresses and involve solutions including (a) larger ingots which would permit additional working to achieve smaller grains in thicker material, (b) more conservative joint designs to accommodate larger grain sizes, and (c) bending material in a previously worked (instead of recrystallized) condition. All of these possibilities could influence fabrication procedures significantly; however, it does not appear that insurmountable problems are involved, inasmuch as refractory alloy containment materials have performed very well for thousands of hours in many test loops.

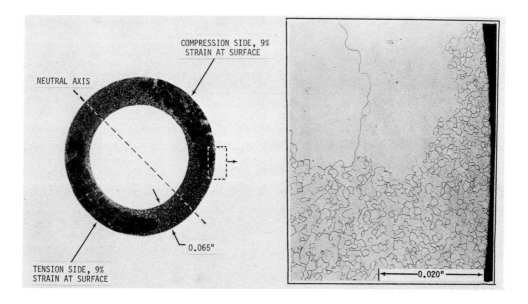

Fig. 16. Cross section of 0.375-inch OD × 0.065-inch wall Cb-1Zr tube. The recrystallized tube was bent around a 3.5-inch diameter die and subsequently annealed for 165 hours at 2200°F. Etchant: 60% Glycerine+20%HF+20%HNO$_3$.

TURBINE MATERIALS

Evaluation and development of the general metallurgical and processing aspects of candidate materials for the highly stressed turbine wheel and blade components have not been pursued as rapidly as for the containment materials. Instead, the resistance of turbine materials to impact erosion has been the subject of major interest to date, and the encouraging results were mentioned earlier. Creep and fatigue strength data have been obtained for several pertinent alloys, and considerably more design data must be obtained in conjunction with the development of fabrication procedures.

Creep strength data obtained at the TRW Equipment Laboratories (TRW) have been discussed in several papers [1-4, 17, 18], and results obtained for some of the candidate alloys are presented in Figure 17. Testing was conducted for periods longer than 1000 hours in high vacuum environments that minimized the effects of contamination. Although it is not essential, turbine designers prefer a 10,000-hour, 0.5% creep strength of at least 20,000 psi for rotating components made of molybdenum or columbium alloys with densities near 0.4 lb/in³. On this basis, the molybdenum alloys are attractive near 2000°F, and Cb-132M, the stron-

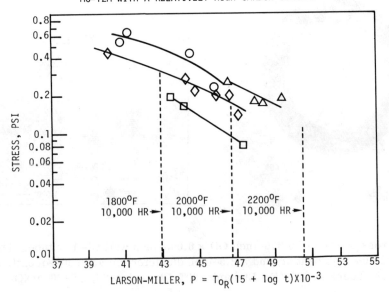

ALLOY	COMPOSITION	HEAT NUMBER	HEAT TREATMENT
O Mo-TZM*	Mo-0.6 Ti-0.1 Zr-0.035 C	KDTZM-1175	STRESS REL. 2300°F, 1 HR
◇ Mo-TZC	Mo-1.2 Ti-0.3 Zr-0.046 C	M-91	STRESS REL. 2300/2500°F, 1 HR
△ Mo-TZC	Mo-1.0 Ti-0.2 Zr-0.080 C	M-80	RECRYSTALLIZED 3092°F, 1 HR
□ Cb-132M	Cb-20Ta-15W 5Mo-2Zr-0.13 C	–	RECRYSTALLIZED 3092°F, 1 Hr

*Mo-TZM WITH A RELATIVELY HIGH CARBON CONTENT

Fig. 17. Turbine material 0.5% creep characteristics in high vacuum.

gest columbium alloy tested in the program at TRW, is promising at a temperature somewhat above 1800°F. ASTAR 811C has a comparable strength near 2000°F (Figure 12) in addition to useful weldability, but its high density of 0.6 lb/in^3 is a significant disadvantage in this application to rotating machinery where stress levels are proportional to the material density. Present indications are that the molybdenum alloys offer the best possibilities, and they are being investigated in additional programs.

Fatigue testing of Mo-TZC at TRW under NASA Contract No. NAS 3-6010 has led to interesting observations which could be pertinent in certain applications [22]. At a relatively high ratio of cyclic to static stresses (0.45), the fatigue life can become a more limiting design criterion than creep. Furthermore, this combination of cyclic and static stresses results in much more rapid creep during an initial period of approximately 30 minutes, and the steady-state creep rate is almost two orders of magnitude greater than that obtained with only a static load. Fortunately, the cyclic stresses in projected space power plant turbines are easily maintained below approximately 1000 psi in the hotter turbine stages because of the relatively low tangential vapor loading on the stubby blades. It appears, however, that additional data will be required for use in conjunction with turbine design studies.

The fabrication of turbine components from Mo-TZM and Mo-TZC has received only preliminary attention to date. Recent investigations at the AiResearch

Fig. 18. Mo-TZM turbine wheel prepared from a 12-inch diameter × 2.25-inch thick forging for application in a three-stage potassium test turbine. This first-stage wheel contains Udimet 700 blades with tip shrouds.

Manufacturing Company under USAF Contract No. AF33(615)-2289 and at GE-
SPPS under NASA Contract No. NAS 3-8520 have examined Mo-TZM turbine wheel
forgings with respect to properties obtained in various directions. The previously
mentioned creep testing program at TRW included the study of specially processed
Mo-TZM with a relatively high carbon content and two heats of Mo-TZC. As expected, substantial variations arise from differences in chemical composition, processing technique, and final heat treatment (Figure 17). The accumulation of definitive design data, therefore, must be obtained on specimens selected from processed shapes which simulate the intended application. A major objective of future
processing development work must be the achievement of suitable high temperature strength combined with sufficient ductility for various machining operations
and turbine start up at ambient or slightly elevated temperatures.

Attachment of the turbine wheels to each other in successive stages presents
an interesting joining problem and also influences the shape of the wheel forging
that is required. Although welding is desirable for design simplicity and light
weight, it is unlikely to succeed with molybdenum alloys. Interstage welding could
probably be accomplished by using a strong ASTAR tantalum alloy for the hotter
stages and changing to a weldable columbium alloy such as FS-85 in the cooler
stages, but this choice of alloys results in a relatively heavy rotor. Brazing, diffusion bonding, and friction welding are alternate possibilities for preparing metallurgical joints. The use of individual torque tube bolting, or a curvic coupling
arrangement combined with a central tie bolt, obviates metallurgical joints and
provides the most attractive compromise with molybdenum alloys at present. For
the previously mentioned three-stage potassium test turbine at GE-SPPS, Mo-
TZM wheels have been prepared with curvic couplings as shown in Figure 18. Individual torque tube bolting appears to provide a more attractive method of mechanical attachment for projected space power plant turbines; it is more consistent
with conventional jet engine design, and it also permits simpler wheel forgings
which are required in diameters ranging from approximately 5 to 10 inches. Current design studies are directed toward the formulation of development programs
to establish suitable joining procedures.

In summary, candidate turbine materials have demonstrated promising corrosion and erosion resistance to the alkali metal environment. Molybdenum alloys currently offer the most attractive strength characteristics for practical turbine designs. Further development of turbine wheel fabrication procedures and
interstage joining techniques is required in conjunction with design data procure-
ment. Provision of adequate ductility in all directions of relatively large sections
of the stronger molybdenum alloys should be a major goal of future work.

FABRICATED COMPONENTS

Many test loops and components have been fabricated from Cb-1Zr during
the past few years. More recently, however, loop facilities are being constructed
from the stronger refractory alloys, mainly T-111, and the test components are
becoming more complex in their material and joining requirements. For example,
a corrosion test loop similar to that shown in Figure 4 has already been fabricated
from the T-111 alloy, and the larger boiler development loop that was mentioned
previously also is being constructed with T-111. The more intricate loop compo-

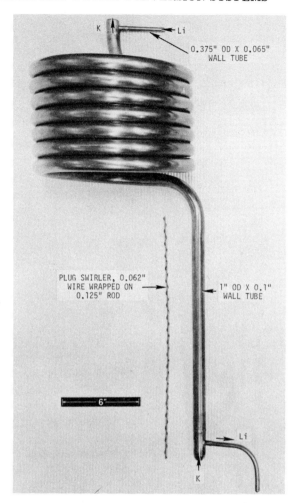

Fig. 19. Lithium heated, tube–in–tube potassium boiler fabricated for a T–111 corrosion
test loop.

nents include boilers, valves, electromagnetic pumps, and pressure transducers.
Many of the components now in use and their functions in test loops were described recently [23]. These components have been developed to the extent required
for reliable test loop operation to obtain basic data. Flight-type components are
following in the design or prototype stage [8]. For illustrative purposes here, a few
of the fabricated components will be described.

The T-111 boiler for the corrosion test loop mentioned above is shown in
Figure 19. A tube-in-tube construction was selected in a coiled configuration to
accommodate a 250-inch long boiler within the vacuum chamber. This boiler consists of a 0.375-inch OD × 0.065-inch wall center tube for boiling potassium and a
1.00-inch OD × 0.100-inch wall outer tube which provides an annulus for the lithium heating fluid. Concentricity is maintained by spacers located between the

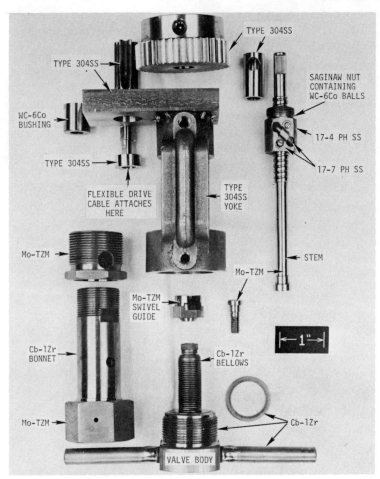

Fig. 20. Modified Hoke, Inc., refractory alloy valve and associated drive components fabricated for operation with potassium in a Cb–1Zr corrosion test loop.

tubes at 10-inch intervals. Forming of the 11-inch OD helical boiler was conducted with the center tube and annulus packed with sugar as an aid to maintaining concentricity. A boiler plug swirler is located in the straight section of the
center tube. All joining operations were performed by TIG welding in the facility
illustrated in Figure 9. It is anticipated that flight-type boilers will require a
compact, multitube-in-shell arrangement involving tube-to-header joints.

The valves used in the Cb–1Zr corrosion test loop of Figure 4 were designed
by Hoke, Incorporated, and modified to permit alkali metal contact only with refractory alloy components. Operation in an external vacuum environment was also
an important requirement. Components of the valve and the material selections
are illustrated in Figure 20, except that a Mo-TZM plug was contained within the
bellows region and mated with the seat in the Cb–1Zr valve body. Dissimilar materials were chosen for the plug and seat to minimize the possibility of diffusion

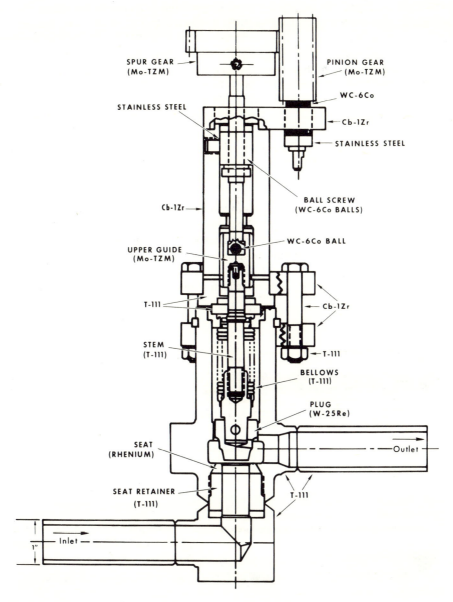

Fig. 21. Drawing of a modified Hoke, Inc., refractory alloy valve prepared for high temperature evaluation with alkali metal.

bonding in the closed position. The critical 0.35-inch ID × 0.52-inch OD Cb-1Zr
bellows were formed by the Standard-Thomson Corporation from 0.375-inch OD ×
0.0075-inch wall tube blanks, and EB welding of the bellows to the other Cb-1Zr
components was performed at GE-SPPS. Although this type of valve has performed
satisfactorily in loop tests at temperatures near 1000°F[11], it was not designed
specifically to provide a high temperature capability. More recently, a larger
valve was designed for higher temperature operation by Hoke, Incorporated, for
GE-SPPS under NASA Contract No. NAS 3-8514. This valve, which is illustrated
in Figure 21, provides a T-111 bellows for higher strength and a W-25Re alloy
plug versus an unalloyed rhenium seat combination for minimizing diffusion bonding. Forming of the 0.59-inch ID × 0.85-inch OD bellows was conducted by the
Mini-flex Corporation with 0.625-inch OD × 0.0085-inch wall tube blanks prepared
in 15-inch lengths by the Superior Tube Company. Fabrication of two valves is
complete, and testing will be conducted in alkali metal to determine their temperature capabilities.

The helical induction type of electromagnetic pump is frequently used in
high temperature alkali metal loops because a metallurgical joint for electrical
conduction to the refractory alloy pump duct is not required. It is also easy to
isolate the pump duct in a vacuum test chamber, away from the associated electrical equipment materials, which can be maintained in the external air environment[23]. The Cb-1Zr pump duct shown in Figure 22 is the largest refractory al-

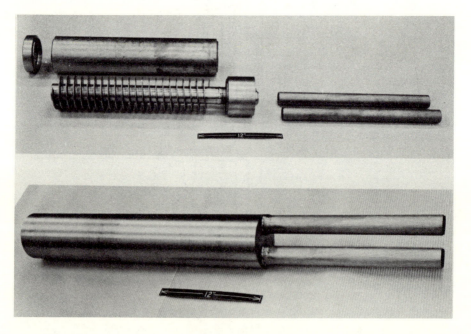

Fig. 22. Helical induction, electromagnetic pump duct fabricated from Cb-
1Zr for use with sodium. The duct components are shown in the top view
prior to shrink fitting and welding into the final configuration shown in the
bottom view.

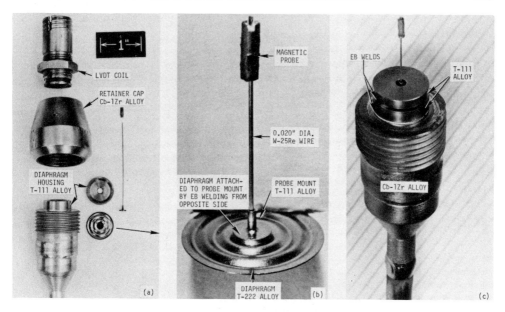

Fig. 23. Refractory alloy, stressed diaphragm pressure transducer at various stages of fabrication for use with potassium in a Cb-1Zr corrosion test loop.

loy duct that has been made, to the author's knowledge. This 5.5-inch OD × 32-inch long duct is approximately twice as large as similar Cb-1Zr and T-111 alloy ducts fabricated for the previously mentioned corrosion test loops. Shrink fitting of the helical duct into the outer tube shell to obtain a 0.001- to 0.005-inch diametrical interference is a critical operation that can be accomplished by cooling the helical duct in liquid nitrogen before inserting it into the tube shell. In current work at GE-SPPS under NASA Contract No. NAS 3-9422, a 4.1-inch OD × 28-inch long prototype pump duct is being fabricated from T-111 for the development of flight-type components.

Two kinds of refractory alloy pressure transducers were used in the Cb-1Zr corrosion test loop mentioned above. One was a slack diaphragm transducer of the type manufactured by the Taylor Instrument Companies, and the other was a stressed diaphragm transducer of the type manufactured by the Consolidated Controls Corporation. These transducers performed somewhat different functions in the test loop [11, 23, 24], and the stressed diaphragm transducer is shown here in Figure 23 because it involved the more intricate fabrication with refractory alloys. This transducer contains a 0.690-inch OD × 0.006-inch thick T-222 tantalum alloy corrugated diaphragm which is deflected by a change in the alkali metal pressure. Movement of the diaphragm is transmitted to a linear variable differential transformer by a stiff W-25Re alloy wire. The wire was EB welded to the T-111 probe mount which then was EB welded to the T-222 diaphragm, and, finally, the diaphragm was EB welded into the T-111 housing. Other dissimilar metal joints between the T-111 housing and the Cb-1Zr fitting to the loop were made by TIG welding with Cb-1Zr filler material at the housing ID and EB welding at the housing OD to make a double seal. All of these dissimilar metal joints performed well at a temperature near 700°F during the 5000 hours of loop operation.

Joining refractory alloys to other materials is an important, practical requirement in most loop test facilities, and it will probably be necessary for the construction of some flight systems in the future. As exemplified by the pressure transducer, the welding of different classes of refractory alloys to each other has met with some success, and this has been done in other cases where the service conditions are not severe. Knowledge of the bimetal weld properties, thermal stability, postweld heat treatment requirements, etc. is lacking in detail, especially when compared to the previously mentioned evaluation of containment material weldments made with individual alloys. In addition to joining different types of refractory alloys to each other, it is often necessary to join a refractory alloy to a stainless steel or a superalloy in the form of a tube–to–tube joint. Such joints have been used in a few bimetallic loops and more extensively for transferring al-

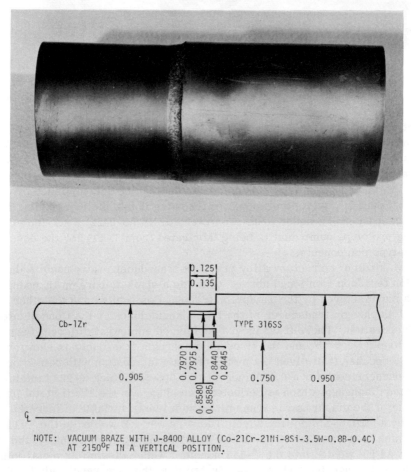

NOTE: VACUUM BRAZE WITH J-8400 ALLOY (Co-21Cr-21Ni-8Si-3.5W-0.8B-0.4C) AT 2150°F IN A VERTICAL POSITION.

Fig. 24. Tongue–in–groove design and joint between Cb–1Zr and Type 316SS tubing.

kali metals and inert gases from external stainless steel containers to refractory alloy loops located in vacuum systems. For example, the corrosion test loop shown in Figure 4 employed eight 0.375-inch OD Cb-1Zr tube joints made to stainless steel by brazing. This type of tongue-in-groove joint, which was developed at GE-SPPS[25], is illustrated in Figure 24 for a larger tube size. In designing these joints, it is important to assure that differential thermal expansion provides an adequate annular opening between the tongue and groove at the brazing temperature to achieve complete flow of the braze metal. Coextruded tube joints between Cb-1Zr and stainless steels have also been prepared successfully by Nuclear Metals, Incorporated. In making this type of tapered joint, it is important to use two alloys with reasonably well matched extrusion characteristics at a low enough temperature to avoid excessive metallurgical reactions between the two alloys during the extrusion process. Both types of joints have performed well in sizes up to approximately 3 inches OD, and they have fulfilled many useful functions in test facilities. In general terms, the joints are quite resistant to thermal cycling, but their true service capabilities usually must be determined for each specific application because of widely varying sizes and operating conditions.

Obviously, additional refractory alloy components could be mentioned; however, the above examples serve as an adequate illustration of the status. Considerably more details of component fabrication are continued in References 11, 21, 23-25, and an appreciation of future requirements can be obtained from References 6-8.

CONCLUDING REMARKS

As mentioned in the introduction and comments on system conditions, a major objective of this paper has been to review some of the factors affecting the fabrication of refractory alloy components, drawing upon the potassium Rankine cycle power conversion system development programs for typical examples. This supplements the cited review papers which elaborate on the interaction of refractory alloys with nuclear fuels and radiosotopes and also compare the strength, weldability, and corrosion resistance of many alloys for the purpose of material selections[1-4]. In addition to the work discussed here, it should be noted that, because of their unique corrosion resistance, moderate strength refractory metals and alloys are also candidates for the construction of some relatively low temperature components in space power plants considered for the 10 to 100 kw power range. As examples, unalloyed tantalum is a candidate for mercury containment in the boiler of the SNAP-8 mercury Rankine cycle power plant, and Cb-1Zr is a candidate for lithium fluoride containment in the solar heat receiver—storage unit of an inert gas Brayton cycle power plant[6]. Projected work on these systems undoubtedly will extend the fabrication technology for some of the relatively low strength materials, and this, in part, will be applicable to the more advanced refractory alloys. Furthermore, this work presents the possibility that the first applications of refractory metals and alloys in operational space electric power conversion systems may result from their corrosion resistance to special working fluids, independent of their high temperature strength capabilities.

Many examples in this paper have been chosen from NASA-sponsored programs to illustrate the state-of-the-art mainly because of that Agency's concern with power conversion systems and because of the author's participation in some of the NASA programs. It should be noted, however, that both the U.S. Air Force and AEC also have sponsored major programs on refractory alloy development and evaluation for application to space electric power plants, particularly in connection with nuclear materials and alkali metal containment. In a paper of this length, it would be impractical to reference and discuss all of the pertinent work conducted at the following laboratories: Aerojet-General Nucleonics, AiResearch Manufacturing Company, Allison Division-General Motors Corporation, Argonne National Laboratory, Atomics International, Battelle Memorial Institute, Brookhaven National Laboratory, General Electric Company-Nuclear Materials and Propulsion Operation and-Space Power and Propulsion Section, Jet Propulsion Laboratory, Lawrence Radiation Laboratory, Los Alamos Scientific Laboratory, NASA-Lewis Research Center, Oak Ridge National Laboratory, Pratt & Whitney Aircraft-CANEL, TRW Equipment Laboratories, United Nuclear Corporation, and Westinghouse Astronuclear Laboratory. Reports issued by these laboratories are recommended for detailed information, and References 9 and 10 include recent status reports on many of the AEC-sponsored programs. That work generally supports the comments made in this paper.

In conclusion, columbium and tantalum containment materials indicate a satisfactory combination of strength, weldability, and corrosion resistance for the containment of alkali metal working fluids for periods exceeding 10,000 hours at temperatures near 2000°F, and probably a few hundred degrees higher if necessary. The strength and erosion resistance of molybdenum alloys are also promising for rotating turbine components, and the fabrication of relatively large, complex shapes with improved ductility appears worthy of additional effort. A variety of quite intricate components have been fabricated successfully from Cb-1Zr and more advanced alloys. The associated technology required to limit contamination by oxygen and other impurities to acceptable levels during melting, forming, heat treating, welding, and testing appears highly satisfactory. Additional work is required to obtain more detailed design data and to establish upper limits on temperature capabilities for the purpose of system optimization. The fabrication of larger and more sophisticated components from both containment and turbine materials also will require continued development. Overall, it appears that the refractory alloys have an excellent opportunity to perform critical functions satisfactorily in a variety of projected space electric power conversion systems.

ACKNOWLEDGEMENTS

The author is grateful to many of his associates at GE-SPPS for their assistance in preparing this paper. Much of the work described was supported by NASA under the direction of T. A. Moss and R. L. Davies, and their contributions are gratefully acknowledged.

APPENDIX

Nominal Compositions of Refractory Alloys

Alloy	Cb	Ta	Mo	W	Re	Ti	Zr	Hf	C	N
				Composition in Weight Percent						
AS-30	bal			20			1		0.1	
Cb-1Zr	bal						1			
Cb-132M	bal	20	5	15			2		0.13	
FS-85	bal	28		10.5			0.9			
ASTAR 811		bal		8	1			1		
ASTAR 811C		bal		8	1			0.7	0.025	
ASTAR 811CN		bal		8	1			1	0.012	0.012
T-111		bal		8				2		
T-222		bal		9.6				2.4	0.01	
Mo-TZC			bal			1.25	0.15		0.12	
Mo-TZM			bal			0.5	0.08			
W-25Re				bal	25					

REFERENCES

1. Moss, T. A., Nuclear Applications, 1967, vol. 3, no. 2, pp. 71-81.
2. Moss, T. A., Davies, R. L., and Moorhead, P. E., Material Requirements for Dynamic Nuclear Power Space Systems, NASA TM X-52344, National Aeronautics and Space Administration, Washington, 1967.
3. Rosenblum, L., Englund, D. R., Hall, R. W., Moss, T. A., and Scheuermann, C., "Potassium Rankine System Materials Technology", in Space Power Systems Advanced Technology Conference, NASA SP-131, National Aeronautics and Space Administration, Washington, 1966, pp. 169-199.
4. English, R. E., Cummings, R. L., Davies, R. L., Moffitt, T. P., and von Glahn, U. H., "Potassium Rankine Systems Technology", in Space Power Systems Advanced Technology Conference, NASA SP-131, National Aeronautics and Space Administration, Washington, 1966, pp. 201-238.
5. Corliss, W. R., Propulsion Systems for Space Flight, McGraw-Hill Book Co., New York, 1960.
6. Space Power Systems Advanced Technology Conference, NASA SP-131, National Aeronautics and Space Administration, Washington, 1966.
7. Advances in Energy Conversion Engineering, Papers Presented at 1967 Intersociety Energy Conversion Engineering Conference, Am. Soc. Mech. Eng., New York, 1967.
8. Zipkin, M. A., J. Spacecraft, 1967, vol. 4, no. 7, pp. 852-858.
9. AEC-NASA Liquid Metals Information Meeting Held in Gatlinburg, Tennessee, April 21-23, 1965, Conference No. CONF-650411, Atomic Energy Commission Report No. TID-4500, Oak Ridge National Laboratory, Oak Ridge, Tenn., 1965.
10. Alkali Metal Coolants, International Atomic Energy Agency, Vienna, 1967.

11. Hoffman, E. E. and Holowach, J., Cb-1Zr Rankine System Corrosion Test Loop, Topical Report No. 7 on NASA Contract No. NAS 3-2547, General Electric Co. Report No. R67SD3016, General Electric Co., Cincinnati, 1968.

12. Zimmerman, W. F. and Rossbach, R. J., Metallurgical and Fluid Dynamic Results of a 2000-Hour Endurance Test on a Two-Stage, 200-Horsepower Turbine in Wet Potassium Vapor, Am. Soc. Mech. Eng. Publication No. 67-GT-9, Am. Soc. Mech. Eng., New York, 1967.

13. Lyon, T. F., Purification and Analysis of Helium for the Welding Chamber, Topical Report No. 1 on NASA Contract No. NAS 3-2547, National Aeronautics and Space Administration Report No. NASA-CR-54168, General Electric Co., Cincinnati, 1965.

14. Stoner, D. R. and Lessmann, G. G., Welding J., 1965, vol. 44, no. 8, Research Suppl., pp. 337-s-346-s.

15. Titran, R. H. and Hall, R. W., High-Temperature Creep Behavior of a Columbium Alloy, FS-85, NASA TN D-2885, National Aeronautics and Space Administration, Washington, 1965.

16. Titran, R. H. and Hall, R. W., "Ultrahigh-Vacuum Creep Behavior of Columbium and Tantalum Alloys at 2000° and 2200°F for Times Greater Than 1000 Hours", presented at the Fourth Symposium on Refractory Metals sponsored by the Refractory Metals Committee, AIME, at French Lick, Ind., 1965, to be published. Also available as NASA TM X-52130, National Aeronautics and Space Administration, Washington, 1965.

17. Sawyer, J. C. and Steigerwald, E. A., J. Materials, 1967, vol. 2, no. 2, pp. 341-361.

18. Sawyer, J. C. and Steigerwald, E. A., Generation of Long Time Creep Data on Refractory Alloys at Elevated Temperatures, Final Report on NASA Contract No. NAS 3-2545, TRW Equipment Laboratories Report No. ER-7203, TRW Equipment Laboratories, Cleveland, 1967.

19. Lessmann, G. G., Welding J., 1966, vol. 45, no. 12, Research Suppl., pp. 540-s-560-s.

20. Davies, R. L. and Stone, P. L., "Space Power Applications for Refractory Metal Tubing", presented at the Fifth Annual Western Metal and Tool Exposition and Conference sponsored jointly by the ASM and ASTME at Los Angeles, 1968.

21. Hoffman, E. E. and Holowach, J., Sodium Thermal Convection Loop, Topical Report No. 5 on NASA Contract No. NAS 3-2547, General Electric Co. Report No. R67SD3014, General Electric Co., Cincinnati, 1967.

22. Honeycutt, C. R., Martin, T. F., Sawyer, J. C., and Steigerwald, E. A., Trans. Am. Soc. Metals, 1967, vol. 60, no. 3, pp. 450-458.

23. Hoffman, E. E. and Holowach, J., "New Components for Refractory Metal-Alkali Metal Corrosion Test Systems", in AEC-NASA Liquid Metals Information Meeting Held in Gatlinburg, Tennessee, April 21-23, 1965, Conference No. CONF-650411, Atomic Energy Commission Report No. TID-4500, Oak Ridge National Laboratory, Oak Ridge, Tenn., 1965, pp. 149-214.

24. Hoffman, E. E. and Holowach, J., Cb-1Zr Pumped Sodium Loop, Topical Report No. 6 on NASA Contract No. NAS 3-2547, General Electric Co. Report No. R67SD3015, General Electric Co., Cincinnati, 1967.

25. Kearns, W. H., Young, W. R., and Redden, T. K., Welding J., 1966, vol. 45, no. 9, pp. 730-739.

RECENT ADVANCES IN COLUMBIUM ALLOYS

R.G. Frank

Abstract

The status of columbium alloys is reviewed with respect to major
applications, production, cost and selected properties. Alloys with
low to moderate strength and good ductility and that are readily
fabricable and weldable are commercially available. These alloys were
the first of the columbium base alloys to receive serious consideration for use in components on actual flight systems. Emphasis in
the development of new columbium alloys over the past several years
has been directed toward the achievement of greater creep strength
in the 2000°F to 2400°F temperature range and more recently, improved
oxidation resistance in combination with varying levels of strength
and ductility.

R.G. Frank is Manager, Refractory Metals Development, Nuclear
Systems Programs, General Electric Company, Cincinnati, Ohio (45215)

Introduction

Serious development of columbium as an engineering material
began approximately 13 years ago with the advent of the discovery of
large deposits of columbium containing ore in the Western Hemisphere,
namely Canada and Brazil. These discoveries nearly doubled the known
resources of columbium metal in the free world from approximately
4×10^6 tons to 7×10^6 tons. Although the majority of the then newly
discovered ore deposits were of the low grade (0.2 - 2% Cb_2O_5)
pyrochlore variety $[(Na,Ca)_2 Cb_2O_6F]$ which made the winning of pure
columbium metal difficult and costly, the pyrochlore ores have become
a major source of the world's ferro-columbium supply; thus freeing the
rarer columbite ores (55 - 67% Cb_2O_5), obtained primarily from Nigeria,
for the production of the pure metal.

The emphasis in columbium during the first 5 years of intensive
study was in compositional development directed toward high strength
and oxidation resistance. Alloys evolved in which oxidation resistance was improved by 2 orders of magnitude and strength by one order
of magnitude. In the same time span, the cost of columbium decreased
by approximately 1/2 an order ($150/lb to $30/lb). The following 5
years saw a change in emphasis to the development of more fabricable
alloys, their application in engineering systems and the initiation of
basic studies in strengthening mechanisms and thermal mechanical processing. There was a general de-emphasis in compositional development.
In the last 3 years, the emergence of a need for a high temperature,
short life turbine blade spurred new vigor in the compositional development of ductile and high strength alloys of columbium with improved
oxidation resistance.

The salient properties of columbium which prompted its rapid rise
as an engineering material have been reviewed many times [1-11] and it
isn't necessary nor is it the intent of this paper to present a comprehensive account of the properties and characteristics of columbium
and its alloys. Instead, the current status of the major alloy
systems of columbium and some aspects on the properties, stability
and constitution of the more advanced alloys will be reviewed.

Applications

It is difficult to present a review of columbium alloys without some mention of the important physical properties and characteristics of columbium in comparison to the other three principle refractory metals - tungsten, molybdenum and tantalum (Table 1). Columbium has a unique combination of physical properties that includes a high melting point, the lowest density and thermal neutron cross-section of the refractory metals, relatively high thermal expansion coefficients which is an important consideration with respect to protective coating systems, a high thermal conductivity which increases with temperature, a high transition temperature for superconductivity, good ductility at temperatures as low as -200°C, and general chemical inactivity and good resistance to liquid metal corrosion. Special consideration must be given to the relatively low elastic modulus of columbium. Although a low elastic modulus is beneficial from the standpoint of low thermal stresses induced by large thermal gradients, it poses problems in structural design with respect to buckling considerations. The major deterrent preventing greater use of columbium at elevated temperatures, where its use is of most interest, is its poor resistance to oxidation. Although the oxide of columbium is relatively stable (melting point of 1490°C/2714°F), at temperatures above 600°C(1112°F) oxidation of the metal is extremely rapid and occurs at a linear rate. Diffusion of oxygen into the substrate results in a loss in ductility.

With the exception of its poor high temperature oxidation characteristics, the excellent combination of properties have resulted in intensive investigations of columbium and its alloys over a wide range of applications, as indicated in Table 2. It is significant to note that columbium and alloys of columbium are currently in use in a number of major advanced engineering systems. The United Kingdom AEA was the first to use columbium in a fast reactor[12]. The good creep strength, excellent weldability and fabricability, good corrosion resistance to liquid sodium and compatibility with the fuel make columbium an excellent choice for this application. However, there are several reservations concerning the future use of columbium as a cladding or structural material in the liquid metal fast breeder reactor (LMFBR). The corrosion rate of columbium is excessive in sodium with a high oxygen content and it has not been clearly demonstrated that in large engineering systems the oxygen content in the sodium can be maintained at the necessary purity level, i.e., on the order of 10 ppm or less. The nuclear characteristics of columbium in the fast reactor are less attractive than the nuclear characteristics of stainless steel or vanadium. The relatively high neutron absorption cross-section of columbium in the fast reactor results in a higher critical mass and lower breeding ratio. Although this is not limiting in the small metal fueled reactor such as the United Kingdom's Dounreay reactor, it does become limiting in the larger oxide fueled reactors. Also, the effect of fast flux irradiation on the properties of columbium are not well established. Considerably more research is required to determine the effect of fast flux irradiation at high temperature in sodium on the properties of cladding materials, the type and level of impurities that can be tolerated in sodium to maintain acceptable corrosion resistance and compatibility with the fuels.

Table 1. Comparative Properties of the Principal
Refractory Metals

Property	Cb	Mo	Ta	W
Melting point, °C	2468	2610	2996	3410
°F	4474	4730	5425	6170
Density, g/cc	8.57	10.22	16.6	19.3
lb/cu in.	0.31	0.369	0.60	0.697
Thermal neutron cross section				
Barns/atom	1.1	2.4	21.3	19.2
Thermal expansion in./in./°F x 10^6				
RT - 400°F	4.0	3.0	3.7	2.25
RT - 2000°F	4.4	3.3	3.9	2.45
Thermal conductivity BTU/ft^2/hr/°F/ft				
Room temp.	26	80	33	80
2000°F	36	58	43	63
Elastic Modulus, psi x 10^6				
Room temp.	15	46	27	58
2000°F	12	28	24	52
Superconducting transition temp. °K	9.46	--	4.48	--
Melting point of stable oxide, °C	1490	795	1890	1473
°F	2714	1463	3434	2683

Table 2. Columbium Applications Status - 1968[a]

Nuclear Power Reactors	1
Radioisotope Capsules	2
Solar/Thermal Energy Storage	2
Re-entry Vehicles	2
Launch Vehicles	1
Space Craft	1
Ramjets	2
Open Cycle Gas Turbines (Air Breathing)	3
Closed Cycle Gas Turbomachinery	2
Closed Cycle Metal Vapor Turbomachinery	2
Direct Energy Conversion Systems	2
Superconducting Devices	1
Corrosion Resistant Apparatus	2
Capacitors	3

(a) Code: 1 - Currently in Use; 2 - Good Chance; 3 - Some Chance

Columbium alloys Cb-1Zr, SCb-291, and C-103 currently are in use in attitude control engines on launch vehicles and alloy C-103 is used on the transstage nozzle extension of the Titan C-III. One of the largest structures fabricated from a columbium alloy (C-103) is the nozzle extension for the Apollo Spacecraft engine. The nozzle extension is a welded construction fabricated from C-103 alloy sheet and tube and strengthened with ribs fabricated from C-129Y alloy. The nozzle extension for the Lunar Excursion Module (LEM) and the liquid rocket nozzle for the Lunar Hopper also are fabricated from columbium alloy C-103. Columbium is in use in the form of columbium-zirconium alloy or Cb_3Sn intermetallic compound, superconducting wires or tape for use in experimental high field magnets.

Applications in components for space systems requiring good strength at elevated temperatures for extended periods of time in high vacuum or in special environments show excellent potential for columbium alloys. These applications include the pressure vessel for isotope fuel capsules - the generation of helium gas over long periods of time requires that the pressure vessel have good creep strength; containment and structural components for solar thermal energy storage systems - columbium alloys appear to have excellent compatability with thermal energy storage materials such as lithium fluoride; containment and highly stressed turbine components in closed cycle, inert gas (helium-xenon) and metal vapor (potassium) turbomachinery; and liquid metal containment components for thermonic direct energy conversion systems. Columbium alloys also have shown good potential for short time use at temperatures above $2000^\circ F$ in earth's atmospheric environment. These applications include thermal radiative panels and structural components for re-entry vehicles and structural components for ramjets. The fact that columbium is unattacked by nitric acid in all concentrations, dilute hydrochloric acid and dilute sulfuric acid and the fact that the cost of columbium sheet products is significantly lower than the cost of tantalum sheet should eventually result in increased usage of columbium in the chemical industry.

Other applications which show some promise for the use of columbium alloys are turbine components for open-cycle (air breathing) gas turbines. This application is directed toward the high temperature, short life turbine. There is little optimism toward the use of columbium alloys for long life turbojet engines because of the coating reliability problem. Another possible application for the use of columbium is in the capacitor field. However, low leakage, electrically insulative anodic films on columbium have been extensively studied and, to date, the properties have not been developed sufficiently well to compete with anodic films on tantalum.

Overall, it is estimated that an excess of 90% of the columbium metal production is directed towards aerospace applications. This in contrast to tantalum products which are more evenly distributed between the aerospace industry and the electronics industry, namely capacitor grade powder, wire and foil.

Availability of Columbium Containing Ores

The abundance of columbium metal in the earth's crust is estimated to be 24 parts per million[13]. In contrast, tungsten is estimated to be present in the quantities of 69 parts per million, molybdenum 15 parts per million, and tantalum 2.1 parts per million. Of the more abundant metals, aluminum is present in the quantity of 81,300 parts per million and iron 50,000 parts per million and of the more rare metals, iridium, ruthenium, osmium and rhenium are present in the quantities of approximately 0.001 part per million. On a more practical basis it is estimated that the free world's primary ore resources for columbium exceeds 9,000,000 tons of contained Cb_2O_5 in comparison to over 72,000 tons of contained Ta_2O_5 [14].

The principal source of columbium metal since War World II has been from the granitic cassiterite gravels of Nigeria [15]. The principal mineral is columbite $(Fe, Mn)(Cb, Ta)_2O_6$ and contains 47 to 78% Cb_2O_5. Other major sources of columbium metal is from Cb_2O_5 recovered as a by-product from the production of tantalum (tantalite). Columbium oxide also is recovered from tin slags when the tin concentrates are smelted directly. Principal sources of tin slags are Nigeria and Malaysia where the Cb_2O_5 concentration is 14% and 2.8 to 3.7% respectively.

The major resource for columbium metal in the free world is found in the carbonatite ores, and alkalic igneous rock containing 0.2 to 2% Cb_2O_5 in the form of the mineral pyrochlore $(Na, Ca)_2\ Cb_2O_6F$. The pyrochlore ore contains 26 to 73% Cb_2O_5. The principle deposits of this mineral are found in Brazil and in Ontario, Canada. The imports of columbium concentrates to the United States in 1967 in comparison to imports for tantalum concentrates are shown in Table 3.

Extraction and separation of Cb_2O_5 from the concentrates is accomplished primarily by the liquid/liquid process from a fluoride solution. Commercial reduction methods for the production of the pure metal are the carbon reduction of Cb_2O_5, sodium reduction of K_2CbF_7, and the sodium or hydrogen reduction of $CbCl_5$.

Production of Columbium Metal and Alloy

The growth of columbium and columbium alloy investigations and their applications is evident from the continued increase in the production of the pure metal (Figure 1). The production of columbium metal in the USA has increased from 10 tons per year in 1957 to approximately 200 tons per year in 1967, representing approximately a 2000% increase in the 10 year period [16,17].

Cost of Columbium Alloy Mill Products

The increase in production in columbium has resulted in a significant reduction in the price of columbium powder, ingots, and metal products. The years 1955 to 1958 saw a precipitous reduction in price in columbium products; afterwards there has been a steady decline in prices through 1968 (Figure 2). For example, 0.010-inch thick sheet of Cb-1Zr sold for $250 a pound in 1955, $110 a pound in 1960, and $60 a pound in 1968. Similarly, 0.250-inch thick plate of Cb-1Zr sold for approximately $250 a pound in 1955, $75 a pound in 1960, and $32 a pound in 1968. Although somewhat higher, the price of the low to the moderate strength alloys of columbium have followed the same trend as Cb-1Zr alloy has over the past four years. The 1968 price of 0.010-inch thick sheet of FS-85 and SCb-291 alloys is approximately $72 a pound and the price of 0.010-inch thick sheet of Cb-752 alloy is approximately $90 a pound.

Table 3.   Major Sources of Columbium and Tantalum Ore Concentrates
USA Imports - 1967[a]

Columbium Conc.		Tantalum Conc.	
Country	lbs x 10^3	Country	lbs x 10^3
Brazil	3,535	Congo	589
Nigeria	2,519	Brazil	356
Canada	890	Mozambique	241
Malaysia	201	Thailand	138
West Germany	80	Nigeria	135
Congo	66	Portugal	99
Other	140	Other	393
Total	7,431	Total	1,950

(a)   U. S. Bureau of the Census.

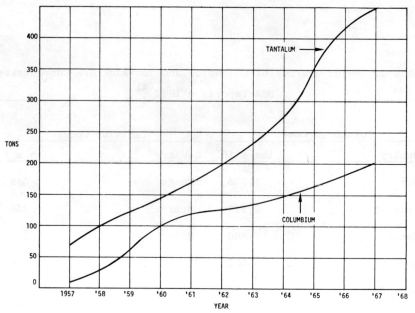

Fig. 1. Trend in USA production of columbium and tantalum metal (Ref. 16, 17).

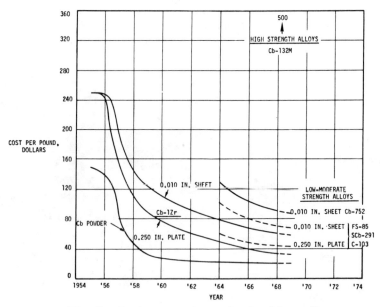

Fig. 2. Cost comparison of columbium alloys.

Price information on the experimental high strength alloys of columbium
is sparse and depends upon the users' requirements; however, prices can
be expected to be around $500 a pound and higher.

The electron beam melted ingots of columbium and Cb-1Zr alloy can
be purchased for approximately $12 a pound with mill products ranging
from 3 to 5 times higher. Since significant reductions in cost have
been estimated for producing columbium ingots in tonnage quantities,
it should be possible to procure low to moderate strength columbium
alloy mill shapes at prices of $12 to $20 a pound should the need warrant
production in those quantities. The cost for other basic metals ranges
from 3¢ a pound for iron, 25¢ a pound for aluminum, $3.75 a pound for
molybdenum, $600 a pound for rhenium, $1100 a pound for iridium, to
$2400 a pound for osmium[13].

Alloys of Columbium

Alloys of columbium that have received considerable attention or
that have been developed in the past five years, together with their
nominal chemical composition, are listed in Table 4. The alloys were
considered commercial if relatively large quantities of the various
mill products have been produced and if the mill products of the alloys
can be readily obtained from the major metal producers to meaningful
specifications. The commercial alloys listed in Table 4 either are in
use in current aerospace applications (Cb-1Zr, SCb-291, Cb-752, C-103,
C-129Y), had previously been used or produced for recent aerospace
programs (D-36, D-14) or are being considered for specific applications
in space hardware where moderate strength and fabricability are required
(FS-85, D-43). There currently is less interest in the remaining three
alloys listed namely, B-33, B-66, and Cb-753. The probable reason for
the lower level of interest in these alloys is that the vanadium addition present in all three alloys has little effect on the diffusivity
of columbium; as a result, there is little or no improvement in the
creep properties at elevated temperatures. In addition, the ductile-
to-brittle bend transition temperature of weldments in B-66 alloy is
relatively high (0° to +75°F) in comparison to the ductile-to-brittle
bend transition temperature of weldments in alloys with superior creep
properties, i.e., FS-85 alloy (-175°F), also resulting in a general lack
of interest in the alloy.

In reviewing the nominal chemical composition of all the commercial
alloys, it is observed that the alloys have moderate amounts of substitutional solute elements (with the exception of the tantalum in FS-85
alloy), additions of reactive elements and, with the exception of D-43
alloy, no intentional additions of interstitial elements.

Recent alloy development programs that were conducted in the USA
have been directed toward high strength for gas turbine applications.
Alloys that were identified and that have shown a promise for use in
high temperature turbines are Cb-1 and Cb-2 developed by General
Electric[18], B-88 and VAM-79 developed by Westinghouse Astronuclear
Laboratory[19,20], Cb-132[21] developed by Battelle Memorial Institute
and Pratt & Whitney Aircraft Division and Cb-132M[22,23]. The alloys
are similar in composition in that they all have a high level of
tungsten or a combination of tungsten and molybdenum for solid solution
strengthening and, with the exception of Cb-132 alloy, a similar reactive

Table 4. Nominal Chemical Composition of Columbium Alloys

Alloy Designation	Original Investigator	Nominal Composition w/o									ppm		Status (a)
		Cb	W	Mo	Ta	V	Hf	Zr	Ti	C	O	N	
USA													
Cb-1Zr	-	Bal	-	-	-	-	-	1	-	0.005	100	50	C
SCb-291	Fansteel	Bal	10	-	10	-	-	-	-	0.006	100	100	C
FS-85	Fansteel	Bal	10	-	28	-	-	1	-	0.004	60	50	C
Cb-752	Union Carbide	Bal	10	-	-	-	-	2.5	-	0.004	60	80	C
Cb-753	Union Carbide	Bal	-	-	-	5	-	1.25	-	0.005	100	75	C
C-103	Wah Chang/Boeing	Bal	-	-	-	-	10	0.7	1	0.015	225	150	C
C-129Y	Wah Chang/Boeing	Bal	10	-	-	-	10	-	(0.2Y)	0.015	225	150	C
B-33	Westinghouse	Bal	-	-	-	5	-	-	-	0.006	120	60	C
B-66	Westinghouse	Bal	-	5	-	5	-	1	-	0.006	120	60	C
D-14	DuPont	Bal	-	-	-	-	-	5	-	0.006	100	40	C
D-36	DuPont	Bal	-	-	-	-	-	5	10	0.006	100	40	C
D-43	DuPont	Bal	10	-	-	-	-	1	-	0.1	100	40	C
F-48	General Electric	Bal	15	5	-	-	-	1	-	0.05/0.1	300	100	PP
PWC-11	Pratt & Whitney	Bal	-	-	-	-	-	1	-	0.1	150	100	PP
AS-30	General Electric	Bal	20	-	-	-	-	1	-	0.1	100	100	D
AS-55	General Electric	Bal	5	-	-	-	-	1	(0.2Y)	0.06	200	250	D
Cb-1	General Electric	Bal	30	-	-	-	-	1	-	0.06	100	300	D
Cb-2	General Electric	Bal	30	-	-	-	-	1	5	0.06	100	300	D
B-88	Westinghouse	Bal	28	-	-	-	2	-	-	0.067	30	40	D
VAM-79	Westinghouse	Bal	22	-	-	-	2	-	-	0.067	50	50	D
Cb-132	Pratt & Whitney	Bal	15	5	20	-	-	-	-	0.004	100	40	D
Cb-132M	Pratt & Whitney	Bal	15	5	20	-	-	2.5	-	0.13	60	40	D
UK													
SU-16	Imperial Metal Ind.	Bal	11	3	-	-	2.0	-	-	0.08	-	-	PP
SU-31	Imperial Metal Ind.	Bal	17	-	-	-	3.5	-	-	0.1	-	-	D
USSR													
RN-5	-	Bal	10	5	-	-	-	1.2	-	0.034	250	-	-
RN-6	-	Bal	5	5	-	-	-	1.2	-	0.026	280	-	-
-	-	Bal	16.5	-	-	-	-	0.5	-	0.1	-	-	-

(a) C indicates commercial; PP, pilot production; D, development.

metal/carbon atom ratio. The objective of a current program at the
Westinghouse Astronuclear Laboratory is to study the creep behavior of
Cb-1 and B-88 alloys [24]; in addition, the thermal mechanical processing
schedule is being optimized to develop the best combination of creep
properties and low temperature ductility. Alloys VAM-79 and Cb-132M
are being utilized by TRW, Inc.[25] as the core materials in the development
of clad-forged turbine blades.

Alloys under development in the United Kingdom and the USSR are
very similar to those developed in the USA, i.e., additions of tungsten
or combinations of tungsten and molybdenum are employed for solid solution strengthening and controlled additions of zirconium or hafnium and
carbon are added to achieve additional strength through the formation of
dispersed carbide phases. Alloys SU-16 and SU-31 developed by Imperial[11,26]
Metal Industries, Ltd. of the UK employ similar hafnium/carbon
atom ratios as used in the USA alloys Cb-1 and B-88. However, the level
of tungsten and molybdenum additions for substitutional solid solution
strengthening is considerably less than the USA alloys in an attempt to
achieve an acceptable level of strength and improved lower temperature
ductility with some degree of weldability. The USSR alloys, RN-5 and RN-6,
employ a considerably higher zirconium/carbon atom ratio than used in
the USA which should result in achieving good low temperature ductility
at the sacrifice of high temperature creep strength[27]. The USSR alloy
Cb-16.5 W-0.5 Zr-0.1 C[28] employs a very low zirconium to carbon atom
ratio and it would be expected that the low temperature ductility of this
alloy would be very poor.

It is significant to note that advances in the melting and processing of columbium alloys in the past three to five years have resulted in
excellent control of the residual interstitial element contents in the
alloys. It is possible with commercial processes to maintain the carbon
level at less than 60 ppm, the oxygen and nitrogen levels between 50 to
100 ppm each and the hydrogen content less than 10 ppm.

Alloy Design

It generally is recognized that columbium can be alloyed to acquire
varying degrees of strength, ductility and oxidation resistance. Alloying to maximize any one property usually results in a sacrifice in one or
both of the other two properties. All of the columbium alloys of current
interest whether they be commercial, pilot-production or development can
be characterized in the following three classifications: 1) high strength,
2) moderate strength and ductility and 3) low strength and high ductility
(Table 5). The alloys characterized in Group 1 - high strength, all have
high levels of tungsten or a combination of tungsten and molybdenum for
solid solution strengthening and intentional additions of carbon and the
carbide forming elements hafnium or zirconium. Careful control of the
thermal mechanical processing history of these alloys is required in
order to assure an optimum combination of elevated temperature strength
and low temperature ductility. The primary difference in the Group 2
alloys - moderate strength and ductility, is the decreased level of solid
solution strengthening elements and, in the majority of the alloys, no
intentional additions of carbon. In the Group 3 alloys - low strength
and high ductility, there are no additions of effective solid solution
strengthening elements nor intentional additions of carbon.

Table 5. Columbium Alloy Design

1. **High Strength**

Cb-1	0.372 lbs/in.^{3}
B-88	0.373
VAM-79	0.356
Cb-132M	0.385
AS-30	0.347
F-48	0.340
SU-31	0.345

2. **Moderate Strength and Ductility**

SU-16	0.335
FS-85	0.383
D-43	0.326
Cb-752	0.326
SCb-291	0.347
C-129Y	0.343
B-66	0.305
AS-55	0.317
PWC-11	0.31

3. **Low Strength and High Ductility**

Cb-753	0.303
C-103	0.320
B-33	0.306
D-14	0.310
D-36	0.286
Cb-1Zr	0.31

4. **Strength and Oxidation Resistance**

Under Development

5. **Ductility and Oxidation Resistance**

Under Development

Although a few of the alloys listed in the first three groups of Table 5 possess somewhat superior oxidation resistance than others, i.e. C-129Y, C-103, and D-36, none of the first generation alloys that were developed for improved oxidation resistance have received attention in recent years. Significant improvements in oxidation resistance was achieved by alloys with additions of chromium and aluminum in combination with relatively high titanium (C121: Cb-20Ta-10Ti-5Cr-2Al) or combinations of high titanium and tungsten (Cb-16: Cb-20W-10Ti-3V) however, the improved oxidation resistance generally was not worth the sacrifice in strength or fabricability.

Development programs have been initiated by the US Air Force Materials Laboratory in the past twelve to eighteen months with the goal of achieving columbium alloys with improved oxidation resistance for use in high temperature, short life, aircraft gas turbine blading. The program at Union Carbide Corp. is designed to develop alloys with combined strength and oxidation resistance (Group 4, Table 5). Two base alloys have been identified and consist of Cb-11W-35Hf and Cb-15W-35Hf[29]. Programs underway at Westinghouse Astronuclear Laboratory and TRW Inc. are directed toward the development of alloys with good ductility and oxidation resistance to be used as claddings for high strength, oxidation resistant turbine blades[25,30]. The base alloy that has been identified for further studies at Westinghouse is Cb-15Ti-10Ta-10W-2Hf-3Al (Group 5 of Table 5).

In a continuing program being conducted at the Bureau of Mines[31,32], alloys of similar composition to those being developed at Union Carbide Corp. for combined strength and oxidation resistance have been identified. Of the alloys studied, the compositions showing the most attractive combination of properties were: Cb-32.2Hf-7.5W-4.1Ti and Cb-24.4Hf-8.4W-0.8Zr-1.0Al.

Although it is anticipated that useful alloys with improved oxidation resistance will be identified from the results of the programs listed in the previous paragraphs, in no case is it expected that alloys will be developed that can be used in turbine components without protective coatings. The intended purpose of achieving greater oxidation resistance in the cladding alloy or base alloy is to achieve a greater reliability of the overall material system.

Properties

Strengthening Mechanisms

Before proceeding with a discussion of select properties of the advanced alloys, it is appropriate to review recent work by McAdam on strengthening mechanisms of columbium[33,34]. In his work, McAdam attempted to establish a quantitative relationship between the substitutional solid solution alloying elements and the carbide and boride dispersoids on the creep properties of columbium. It was concluded that solid solution strengthening of columbium at low temperatures, specifically at room temperature, can be explained in elastic misfit, i.e., the difference between the shear modulus of columbium and the solute element. In this respect, tungsten, osmium, rhenium, iridium and molybdenum were the most effective strengtheners and hafnium, titanium and vanadium the least effective strengtheners. At temperatures on the order of one half the absolute melting point, the most important factor affecting the creep properties was the melting point of the

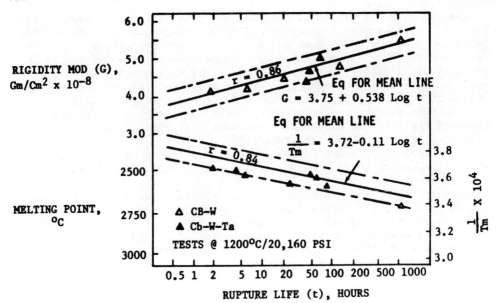

Fig. 3. Correlation of melting point and rigidity modulus with rupture life for Cb–W and Cb–W–Ta alloys (Ref. 33).

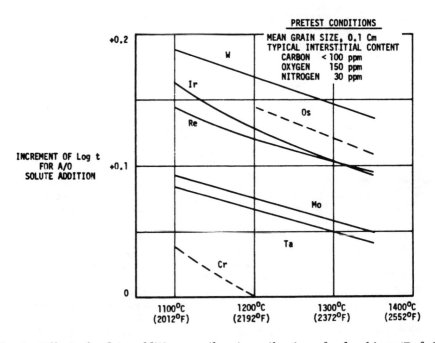

Fig. 4. Effect of solute additions on the strengthening of columbium (Ref. 33).

solute element; the elastic misfit of the solute atom was considered
less important. The effect of the atomic size mismatch was considered
of secondary importance at both low and elevated temperatures.

At temperatures on the order of 1200°C (2192°F), tungsten, osmium
and iridium were the most effective strengtheners because of both their
high melting point and high shear modulus. Tantalum and ruthenium have
intermediate strengthening effects: the strengthening effect of tantalum
due to its high melting point is moderated by its lower shear modulus;
the strengthening effect as a result of the high shear modulus of
ruthenium is moderated by the fact it has a lower melting point than
columbium. Elements with both low melting points and shear moduli such
as titanium, vanadium and zirconium result in reduced creep strengths of
columbium. In the case of hafnium, where the atomic size mismatch with
columbium is relatively large, low strengths in the hafnium-columbium
alloys are observed and are attributed to the low shear modulus of
hafnium.

The correlation of melting point and shear modulus with the rupture
life for columbium-tungsten and columbium-tungsten-tantalum alloys at
1200°C (2192°F) are shown in Figure 3. The overall effect of solute
additions on the strengthening of columbium is shown in Figure 4. Based
on this work an empirical equation for the creep rate of columbium
alloys was derived:

$$\dot{e} = 9.05 \times 10^{-35}\ S^{4.7}\ \exp. \left(\frac{-102,000}{RT}\right)\ \sec^{-1} \qquad (1)$$

$\dot{e}$ = Creep rate, $\sec^{-1}$

S = Stress, $Dyne/cm^2$

R = cal/degree/mole

T = °K

More recently, McAdam reported results of his work on the effect
of carbide and boride dispersions in a Cb-20 W base. It was established
that the reactive metal/carbon or boron atom ratio had a pronounced
effect on the rupture life and creep rate of the base alloy. A reactive
metal/carbon or boron ratio of 1.3/1 was found to be optimum for the
maximum rupture strength and minimum creep rate. It also was established
that a 2 mole percent of carbide was the most effective level of carbide
dispersion at a reactive metal/carbon atom ratio of 1.3 to 1 for maximum
rupture strength and minimum creep rates. These results correlate
reasonably well with those reported by Clark[35] for a Cb-20 W base
alloy. Clark found an optimum reactive metal/carbon atom ratio of 1
for maximum creep rupture properties. Both investigators found that
low reactive metal/carbon atom ratios, i.e., less than 1, resulted in
alloys with poor low temperature ductility. Another important conclusion from McAdams' work was that alloys containing both tantalum and
tungsten had superior creep rupture properties to those alloys containing additions of tungsten alone.

A summary of McAdams' work on the effect of substitutional solid
solution strengthening elements and carbide and boride dispersions on
the creep rupture properties of columbium is shown in Figure 6. It can
be observed that the carbide and boride compounds align themselves in
decreasing order of effectiveness with respect to the absolute melting

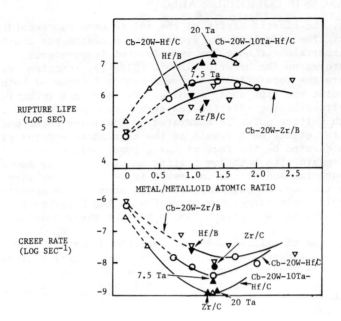

Fig. 5. Effect of metal/non-metal interstitial ratio (~1 mol. % carbide or boride) on creep-rupture properties of columbium alloys at 20,160 PSI and 1200°C (2192°F) (Ref. 34). Note: Pretest condition, 1800°C (3272°F) anneal; oxygen, 100-250 PPM; nitrogen 50 PPM.

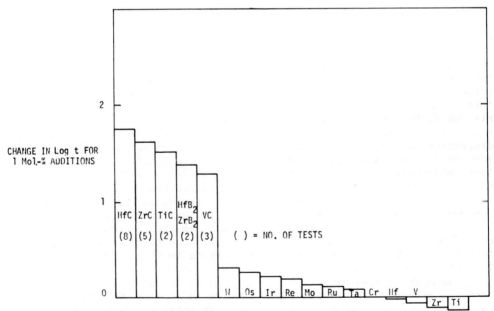

Fig. 6. Relative strengthening effect of compounds and elements in columbium at 1200°C (2192°F) (Ref. 34).

points in a similar manner as for the solute elements. It also is significant to note that with respect to rupture properties at 1200°C (2192°F), hafnium carbide is 20 times more effective than tungsten, the best substitutional solid solution strengthening element. The negative strengthening effect of titanium, zirconium, vanadium and hafnium is graphically illustrated in Figure 6 and, as previously stated, is attributed to the low shear modulus of these elements and their effect on increasing the diffusivity of columbium.

From an analysis of the data, it appears that the creep rupture properties of the dispersion strengthened Cb-20 W alloy follow a hyperbolic sine function of stress. The following general equation for the 100-hour rupture stress was developed:

$$\text{Log sinh } 0.121 \ \sigma_{100} = \text{Log sinh } 0.121 \ \sigma + 0.16 \ \text{Log } t - 0.32 \qquad (2)$$

$$t = \text{Rupture life, hours}$$

$$\sigma = \text{Rupture stress, tons/sq in.}$$

$$\sigma_{100} = \text{100-hour rupture stress, ton/sq in.}$$

Creep Rupture Properties

Because of the importance of the test environment and the structural condition of the alloy on the creep rupture properties, care was exercised in compiling the data in that only those data which were obtained at pressures of 10^{-6} torr or lower and with material whose processing history was reasonably well established were recorded - exceptions are noted. However, in regard to the latter, only final heat treatments were reported here as a compilation of the complete thermal-mechanical processing history of each alloy was not considered practical within the scope of this paper.

The stress rupture properties of the high strength columbium alloys are presented in the Figure 7, on the basis of stress as a function of the Larson-Miller parameter. The highest strength alloys reported in the literature are Cb-1 and B-88[18,19,20]. The best combination of elevated temperature creep strength and room temperature tensile ductility reported for Cb-1 were developed by a high temperature primary extrusion operation at 3750°F (2065°C) to assure complete solution of carbide phases, followed by a two step lower temperature swaging operation of 2750°F/2500°F (1510°C/1371°C) during which time a fine dispersion of face centered cubic, zirconium rich (Zr, Cb)C was achieved. The final treatment was a stress relief anneal of 2500°F (1371°C) for 1 hour. The 2400°F (1316°C)/100-hour stress rupture strength of the material in this condition was 29,000 psi (78,000 inches on a density corrected basis). Higher strengths can be achieved in this alloy in the solution annealed condition (3150°F/1732°C for 1 hour); the higher strength is attributed to strain induced precipitation of the face centered cubic, zirconium rich (Zr, Cb)C during the creep test. The 2400°F (1316°C)/100-hour stress rupture strength reported for Cb-1 in the solution annealed condition is 34,000 psi (91,000 inches on a density corrected basis).

The best combination of elevated temperature creep strength and room temperature tensile ductility of B-88 alloy were reported to have been developed by a high temperature extrusion operation at 3500°F

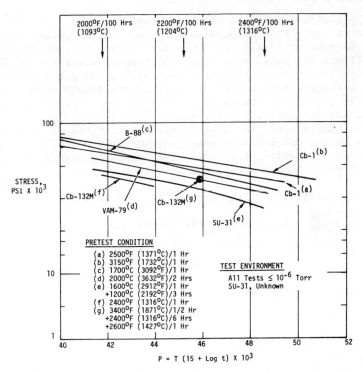

Fig. 7. Stress rupture properties of high strength columbium alloys.

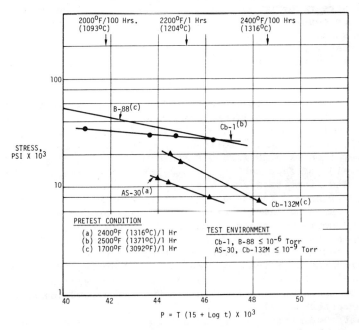

Fig. 8. Creep properties (1% strain) of high strength columbium alloys.

(1927°C) followed by lower temperature swaging operations at 2372°F
(1300°C) and a high temperature final anneal at 3092°F (1700°C) for
1 hour. The 3092°F anneal produces a recrystallized fine grain structure
which is believed responsible for the room temperature ductility of the
alloy. The elevated temperature creep strength is attributed to stabilization of dislocation networks by the face centered cubic monocarbide
precipitation at the sub-boundaries during primary creep. The 2400°F
(1316°C)/100-hour stress rupture strength of B-88 alloy in the above
stated condition is 25,500 psi (68,000 inches on a density corrected
basis). Because of the importance in establishing thermal-mechanical
processing schedules, the Westinghouse investigators determined the
carbide solvus for B-88 alloy. Metallographic examination of samples
annealed for 2 hours at 2000°C plus 1 hour at temperatures of 1400° to
1800°C (2552° to 3272°F) followed by a rapid quench in a brine solution
revealed the carbide solvus to be 1675°C ± 25°C (3047°F ± 45°F).

Although the stress-rupture properties for Cb-1 are reported to be
slightly higher than the stress-rupture properties for B-88, it is
believed that both alloys should have similar strengths if processed in
the same manner. There is only a slight difference in the tungsten
content of the alloys, 30% for Cb-1 vs. 28% for B-88 and the reactive
element/carbon atom ratio of both alloys is similar, approximately 2/1.
The zirconium and hafnium contents of these alloys is in excess of the
optimum reactive element/carbon atom ratio of 1-1.3/1 for maximum creep
strength in order to assure measurable room temperature ductility. It
should be noted that the nitrogen content of Cb-1 alloy is significantly
higher than the nitrogen content of B-88 alloy and may contribute to the
creep rupture properties reported for Cb-1 alloy, especially at the lower
temperatures. However, it is quite apparent that the thermal-mechanical
processing history of these alloys has considerably more effect on the
creep rupture properties than the composition per se. Further studies
on the creep behavior on these alloys are in progress[24].

The 2400°F (1316°C)/100-hour stress-rupture properties of Cb-1 and
B-88 alloys compare favorably with the stress-rupture properties of
Mo-TZC alloy on a strength-to-density basis, i.e., 68,000 to 78,000
inches for B-88 and Cb-1 alloys and 55,000 to 103,000 inches for Mo-TZC
alloy, depending upon the process history.

The stress-rupture properties of VAM-79 with a lower level of solid
solutioning strengthening addition, i.e., 22% tungsten vs. 28% tungsten
for B-88, are lower than the stress-rupture properties of Cb-1 and B-88
alloys and appear to be similar to the stress-rupture properties of
Cb-132M in the high temperature annealed condition. The stress-rupture
properties of Cb-132M in the stress relieved condition are considerably
lower than the stress-rupture properties obtained with material annealed
at temperatures above the carbide solvus temperature. Primary extrusion
of the Cb-132M material tested in the stress relieved condition was
carried out at temperatures of approximately 3100°F (1705°C) followed
by final working at temperatures of 2400°F (1316°C). It is apparent
that Cb-132M alloy also is susceptible to strain induced precipitation
during creep-rupture testing.

Creep properties (1% strain) of the high strength columbium alloys
are presented in Figure 8[18,19,20,36]. The data show similar trends
as indicated by the stress-rupture properties.

The stress-rupture properties of the low to moderate strength

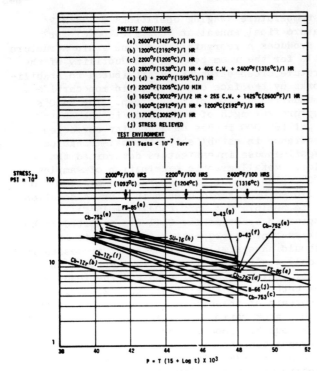

Fig. 9. Stress rupture properties of moderate and low strength columbium alloys.

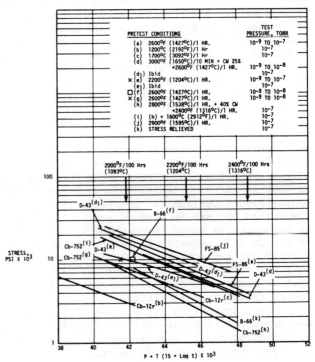

Fig. 10. Creep properties (1% strain) of moderate and low strength columbium alloys.

columbium alloys are presented in Figure 9. The spread in the 2200°F (1204°C)/100-hour stress-rupture strengths in these alloys ranges from 4000 psi for Cb-1Zr to 16,000 psi for SU-16 - a four fold spread in strength level[26,37-44]. On the basis of stress-rupture properties, the alloys would rank themselves from strongest to weakest in descending order: SU-16, FS-85, D-43, Cb-752, Cb-753, and Cb-1Zr. Data are reported for all alloys in the condition in which they are recommended for service by the producers or original investigators. The recommended final treatment for D-43 and Cb-752 alloys is a duplex anneal with an intermediate cold working operation. Data reported by ORNL[43] have shown that the stress-rupture properties of most of these alloys can be improved by high temperature annealing treatments and, significantly, without loss in room temperature ductility. For example at 1200°C (2192°F), the rupture life of Cb-1Zr can be increased 6 times by a high temperature anneal of 1700°C (3092°F). Similarly, the time to 1% strain can be increased 100 times[37]. The increase in rupture and creep strength is attributed to solution and reprecipitation of a second phase, presumably zirconium oxide, and the development of a substructure from the combination of the cold working operation and the high temperature anneal.

Creep properties (1% strain) of the low to moderate strength columbium alloys are presented in Figure 10.

The stress-rupture properties of columbium alloys developed in the UK and the USSR are presented in Figure 11. The data for SU-31 and SU-16 alloys, as represented by the broad bands, were reported by the Imperial Metal Industries Limited through Kawecki Chemical[26] and were obtained on material in the recommended condition, i.e., heat treated at 1600°C (2912°F) for 1 hour plus 1200°C (2192°F) for 3 hours. The data obtained by ORNL on SU-16 alloy[43] in the recommended condition are slightly below those reported by IMI. However, the stress-rupture properties of SU-16 as obtained by ORNL show the alloy to be superior in strength to all the moderate strength alloys of interest, as indicated in Figure 9. Although SU-31 alloy has lower stress-rupture properties than the other high strength columbium alloys, as indicated in Figure 7, the alloy has good strength and is reported to have a reasonable level of weldability. Although not well documented, it is reported that electron beam welded 0.040-inch thick sheet in the as-welded condition can be bent around a 6t radius without cracking[26]. In addition, 0.0015-inch thick foil has been produced from SU-31.

The stress-rupture properties reported for the USSR alloy RN-6[27] are similar to those of· SU-16 in the stress relieved condition (1200°C/ 2192°F for one hour). Although the recommended heat treatment for RN-6 alloy for a good combination of strength and room temperature ductility was described as an anneal between the 1300° to 1500°C (2372° to 2732°F) after 80% reduction, it is believed that the data reported in the cited reference was obtained on the material in the as-worked condition.

A study conducted by General Electric attempted to correlate creep strain as determined by uniaxial creep tests with creep strains bi-axially induced in thin wall tubes pressurized by potassium vapor[45]. D-43 alloy reflux capsules filled with high purity potassium were tested under conditions which resulted in from 5 to 10% strain during the 500-to 2000-hour exposure in the temperature range of 2000° to 2200°F (1093° to 1204°C). Reduced gauge sections were machined in the

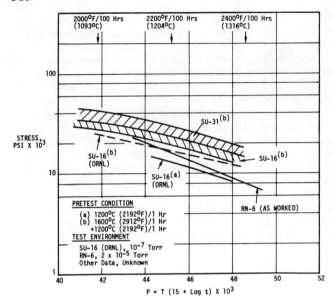

Fig. 11. Stress rupture properties of UK and USSR Columbium alloys.

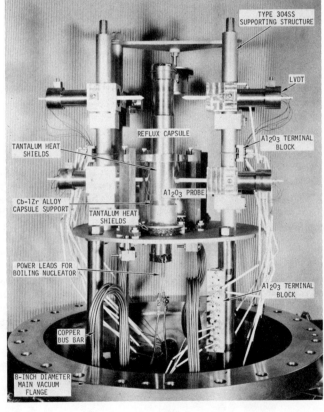

Fig. 12. Test facility to measure biaxially induced creep in D-43 alloy reflux corrosion capsules.

liquid and condensing zones with stress attenuating fillets based on designs developed by Grodezinski[46]. The test facility for the biaxial creep tests of D-43 alloy is shown in Figure 12. The changes in the diameter of the reduced gauge sections were measured by means of tungsten tipped, high purity Al_2O_3 probes in conjunction with a linear-variable-differential-transformer/recorder system. Assuming that the ratio of the principle stresses and their directions remain constant with time and that the test is performed isothermally, the equation for the change in diameter is:

$$\Delta d = \frac{2pr^2}{t} \left[\frac{1-\mu/2}{E} + \frac{3}{4} \frac{\epsilon e^c}{\sigma_e} \right] \qquad (3)$$

where:

$$\sigma_e = \frac{3}{2} \frac{pr}{t} \qquad (4)$$

Δd = diameter change

ϵe^c = uniaxial creep strain at any instant of time corresponding to an effective stress σ_e

r = mean tube radius

μ = Poissons ratio

d = tube diameter at neutral axis

p = internal pressure

E = Modulus of elasticity

t = tube wall thickness

The results of the biaxial creep tests are shown in Figure 13 along with the results of the uniaxial creep tests and results obtained from the literature. From these data it would appear that the 1% uniaxial creep strengths calculated from the biaxial test results are considerably lower than the strengths determined by conventional uniaxial tests. This program is continuing at General Electric in order to verify these results; however, the material to be tested will be a tantalum alloy, T-111 (Ta-8W-2Hf).

In creep tests conducted at 1200°C (2192°F) investigators at ORNL[47] found a correlation between constant and varying stresses on the creep rupture life of Cb-1Zr and D-43 alloys. This correlation is useful in predicting the rupture life of materials in applications such as unvented isotope fuel capsules. The continued generation of helium gas results in an increasing stress in the containment material over the life of the capsule. The realtionship is shown in equation (5):

$$A^n = \sigma^n t_r^{n+1}/(n + 1) \qquad (5)$$

where A, n = constants determined from conventional creep tests

$$t_r = (\frac{A}{\sigma})^n \qquad (6)$$

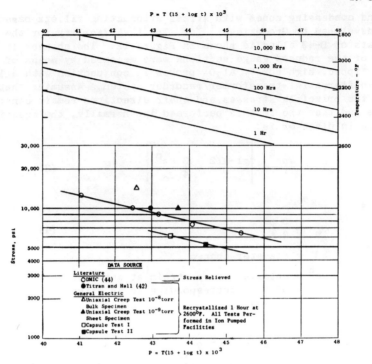

Fig. 13. Larson–Miller parameter curves for 1% uniaxial creep in D-43 alloy
(Ref. 45).

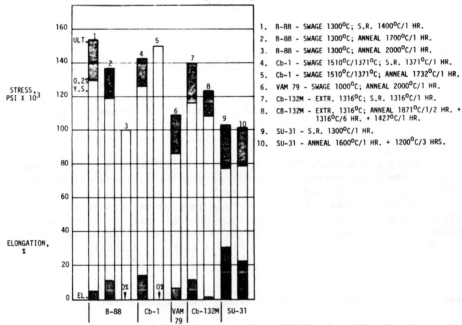

Fig. 14. Room temperature tensile properties of high strength columbium alloys.

σ = stress, psi

$\dot{\sigma}$ = stress rate, psi hour^{-1}

t_r= rupture life, hours

This relationship was found to hold from 1% strain to rupture. It was stated by the investigators that this relationship did not hold for FS-85 alloy.

Tensile Ductility

It is obvious that the creep strength of the alloys cannot be maximized without considering the effect on low temperature tensile ductility. The effect of structure on the tensile properties of the high strength columbium alloys, which are most affected by the thermal mechanical history, is shown in Figure 14. The optimum combination of elevated temperature creep properties and room temperature tensile elongation in B-88 alloy was achieved in the recrystallized condition (1700°C/3092°F for 1 hour) in which a well developed substructure was observed[19,20]. In this condition, the tensile elongation measured at room temperature was 11%. High temperature solution anneals at 2000°C resulted in a complete loss of ductility with no observable plastic deformation.

The optimum elevated temperature creep properties and room temperature tensile elongation for Cb-1 alloy were reported for material in the stress relieved condition (2500°F/1371°C for 1 hour); tensile elongation of 14% was measured[18]. High temperature solution anneals, i.e., 3150°F (1732°C)/1 hour, also resulted in complete loss of ductility with no observable plastic deformation at fracture. Similar effects have been observed with Cb-132M alloy[48,49].

It is interesting to note that measurable tensile elongation (6%) was observed in VAM 79 alloy after a high temperature solution annealing treatment of 2000°C (3632°F) indicating the effect of lower levels of substitutional solid solution strengthening elements[19]. The room temperature tensile elongation of SU-31 alloy in the stress relieved condition (1300°C/2372°F for 1 hour) is reported to be as high as 30%[26]. However, the recommended duplex final heat treatment of 1600°C (2912°F)/1 hour + 1200°C (2192°F)/3 hours results in a slightly reduced tensile elongation of 22%.

Similar data have been compiled for the moderate strength and low strength alloys and are reported in Figures 15 and 16. Good room temperature ductility, i.e., near 20% elongation, was reported for all moderate strength alloys in their recommended final condition[26,27,50-55]. Significant losses in ductility were observed in the carbon containing D-43 alloy solution annealed at 1650°C/1 hour and the USSR alloy RN-6 in the as-worked condition. The tensile elongation of the low strength columbium alloys approached or exceeded 30% elongation[10,38,50].

Ductile-to-Brittle Transition Temperature

The ductile-to-brittle transition temperature is of interest in structural applications of columbium alloys, particularly large structures containing weldments. Westinghouse Astronuclear Laboratories have reported data for the bend transition temperatures for a series of moderate strength columbium alloys[56]. Data were reported for the base metal and tungsten-inert-gas (TIG) and electron beam weldments which

Table 6. Ductile to Brittle Transition Temperature for Moderate
Strength Columbium Alloys

| Alloy | Base Condition (0.035 in. Sheet) | | 1 Hr Post Weld Anneal Temp, °F | | DBTT, (a) °F | | | | |
	Final Heat Treatment	ASTM Grain Size	TIG	EB	Base Material Long/Trans	TIG Weld (b) Long	TIG Weld (b) Trans	EB Weld (b) Long	EB Weld (b) Trans
B-66	Rx 2100°F/1 Hr	10	None	1900	>-300 <-250	0	+75	-225	-175
C-129Y	Rx 2400°F/1 Hr	10	2400	2200	<-320	-200	-225	-250	-250
Cb-752	Rx 2200°F/1 Hr	8-9	2200	2400	-275	-75	0	-200	-200
D-43	Rx 2600°F/1 Hr	5	2400	2400	>-150 <+75	+100	0	-225	<-125
D-43Y	Rx 2400°F/2 Hrs	8	2400	2400	>-275 <-175	-175	-250	-250	<-300
FS-85	Rx 2375°F/1 Hr	8	2400	2200	-250	-175	-175	-200	-200
SCb-291	Rx 2100°F/1 Hr	6	2200	None	-275	-275	-275	<-320	-250
SU-16	Unknown (c)	-	None/2192	None	+10(d)	+446(d)		+194(d)	

(a) 90° Bend Transition Temperature at 1T Bend Radius
    (FS-85, 2T Radius); Ram Speed, 1 inch/min

(b) Optimum Weld Conditions

(c) 0.060 In. Sheet

(d) 4T Bend Radius

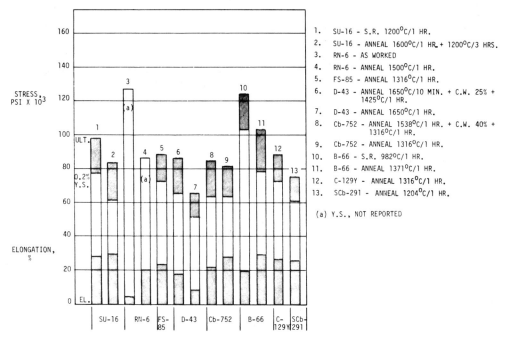

Fig. 15. Room temperature tensile properties of moderate strength columbium alloys.

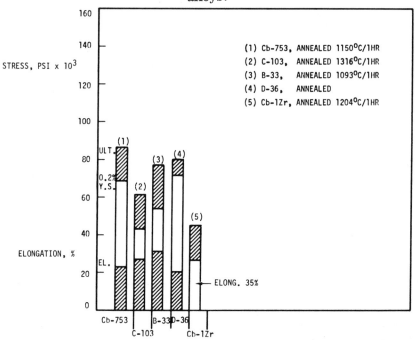

Fig. 16. Room temperature tensile properties of low strength columbium alloys.

were made to optimum conditions including any required post weld heat
treatments. With the exception of D-43 alloy, all the alloys exhibited
good ductility at cryogenic temperatures; the transition temperature
for D-43 alloy ranged between -150°F to +75°F (-99° to +24°C). The
ductile-to-brittle bend transition temperature for weldments generally
exhibited an increase over the base material with the highest ductile-
to-brittle transition temperature being measured for the weldments made
by the tungsten-inert-gas method. The investigators found that the
welding conditions which led to increases in the heat-affected-zone and
the grain size in the heat-affected-zone resulted in an increase in the
ductile-to-brittle transition temperature of the weldment. Therefore,
difference in alloying response, as measured by weld ductility, can be
related to the metallurgical characteristics of the alloy affecting
grain stability and growth phenomena. It was found that all columbium
alloys generally require post weld annealing to improve weld ductility.
The studies conducted at ORNL[57] show the ductile-to-brittle bend
transition temperature of SU-16 alloy, the highest strength alloy of
Group II - alloys of moderate strength and ductility, to be significantly
higher than the other alloys of Group II which were investigated at
Westinghouse.

Thermal Stability
      The thermal stability and the influence of changes in the structure
on the properties of columbium alloys are extremely important in the
application of alloys where long time service, on the order of 10,000-
50,000 hours, is a requirement. One measure of thermal stability is to
evaluate the effect of long time aging treatments at various temperatures
on the room temperature tensile properties. Data of this type for seven
moderate strength alloys are shown in Figure 17. The data were obtained
on specimens from the base material and weldments produced by the TIG
process and which were aged for 1000 hours at temperatures of 1500° to
2400°F (816° to 1316°C)[58]. Aging treatments were conducted under
close environmental control at pressures on the order of 10^{-8} torr. The
investigators reported that three of the seven alloys investigated
(FS-85, C-129Y and SCb-291) exhibited excellent stability within the
times and temperatures investigated; three alloys showed small changes
in tensile properties (B-66, Cb-752 and D-43Y). The D-43 alloy appeared
to overage in a classic matter. However, the changes in room temperature tensile properties,and particularly the room temperature tensile
elongation of all the alloys,were not considered severe or damaging.
      The influence of long time thermal exposures at 2000°F (1093°C) on
the stress-rupture properties of D-43 and AS-55 alloys is shown in
Figure 18[59]. The test specimens were machined from the walls of
corrosion test capsules after 5000 and 10,000 hours exposure to refluxing potassium at the indicated temperature. The long time thermal
exposures were conducted in a high vacuum at pressures of 10^{-8} to 10^{-9}
torr and the rupture tests were conducted at pressures of 10^{-7} to 10^{-8}
torr. The data show a significant decrease in rupture life of AS-55
alloy for both the 5000 and 10,000-hour exposures. The data for D-43
alloy suggest reasonable stability to times on the order of 5000 hours
and a significant decrease in rupture life after the 10,000-hour exposure. The decrease in rupture life for both alloys is attributed to
overaging phenomena resulting in coalescence of the carbides. The change

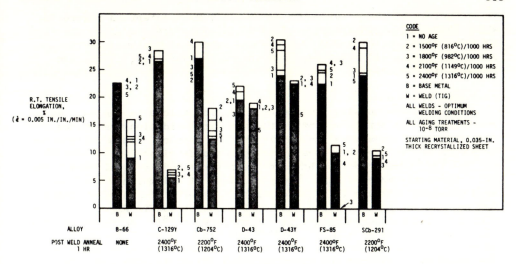

Fig. 17. Effect of 1000-hr aging treatment on room temperature ductility of columbium alloys.

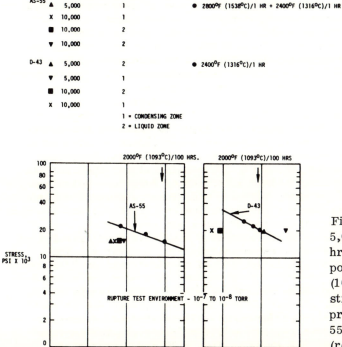

Fig. 18. Effect of 5,000 and 10,000-hr exposure to potassium at 2000°F (1093°C) on the stress rupture properties of AS-55 and D-43 alloys (ref. 59).

in morphology of the carbides in AS-55 alloy is apparent in Figure 19, which shows the microstructure of AS-55 alloy before and after the 10,000-hour exposure at 2000°F (1093°C). Chemical analysis of the specimens before and after exposure indicates that the total carbon content did not change.

Oxidation Behavior

As stated previously, studies of columbium alloys in the past five years have given little attention to improving oxidation resistance. Only recently have studies been initiated to develop alloys of improved oxidation resistance with usable strength and fabricability[25,29,30]; results from these programs have not been published in final form. The oxidation resistance of high strength alloys of columbium which have been developed for use in high temperature, short life turbine blades and represented by alloys Cb-1 and Cb-2 is presented in Figure 20 along with data for unalloyed columbium and molybdenum[18]. Exposure of Cb-1 alloy for 10 hours at 2400°F (1316°C) results in a metal recession of 20 mils per side and an oxygen contaminated zone of 80 mils per side. The addition of two percent titanium (Cb-2 alloy) reduces the metal recession under the same conditions to 7 mils per side and the oxygen contaminated zone to less than 40 mils per side. Not only do titanium additions result in decreased scaling and contamination rates but they apparently eliminate the inversion in scaling rates between 1800°F (982°C) and 2400°F (1316°C). The high oxidation rate of Cb-1 alloy at 1800°F is attributed to the transformation of the oxide and the resultant change in scale plasticity.

The current alloy development programs aimed at developing alloys of columbium with improved oxidation resistance are based on one or more of the following three approaches: 1) reducing the anion deficiency of the scale to reduce the diffusion of oxygen through the scale to the oxide-metal interface. This can be accomplished by the addition of elements of higher valence than columbium, i.e., molybdenum, tungsten, iridium, and vanadium. 2) minimize the ratio of the volume of the oxide to the volume of the metal (Pilling-Bedworth ratio) to inhibit spalling of the oxide. The ratio of the volume of Cb_2O_5 to columbium metal is 2.67 in comparison to Al_2O_3 to aluminum metal ratio of 1.28. Additions of elements of smaller ionic radius would decrease the oxide/metal volume ratio, i.e., vanadium, molybdenum, tungsten, silicon, and aluminum. 3) increasing the mechanical and chemical stability of the scale by large solute additions of tungsten, molybdenum, tantalum, titanium, or hafnium; alloying additions to promote optimum density of the scale to inhibit oxygen diffusion (lithium); and alloying additions to prevent transformation of the scale (silicon).

Oxidation data indicate that relatively high levels of titanium or hafnium are necessary to inhibit contamination of columbium. The improved contamination rates are attributed to the formation of a double oxide, i.e. $Cb_2O_5 \cdot TiO_2$. However, as discussed in a previous section of this paper under strengthening mechanisms, page 7, high levels of titanium or hafnium have a negative strengthening effect due to their low shear modulus and their effect on the diffusivity of columbium. Oxidation data on the first generation alloys developed for improved oxidation resistance also are shown in Figure 20 for purposes of comparison. These data illustrate the influence of higher titanium levels and additions of chromium and aluminum.

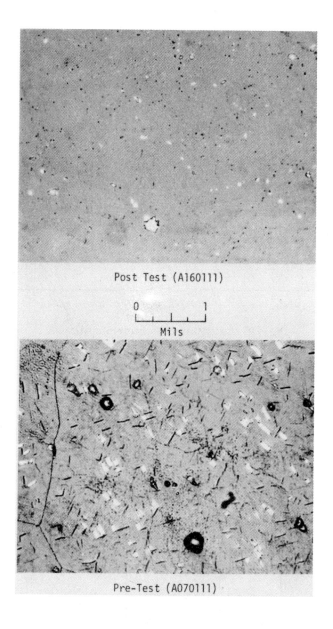

Post Test (A160111)

0   1
Mils

Pre-Test (A070111)

Fig. 19. Microstructure of AS-55 alloy before and after 10,000-hr exposure to refluxing potassium at 2000°F (1093°C) in a vacuum of 10^{-9} torr.

Etchant: Stain Etched   Mag.: 1000X
N.A.:   0.85

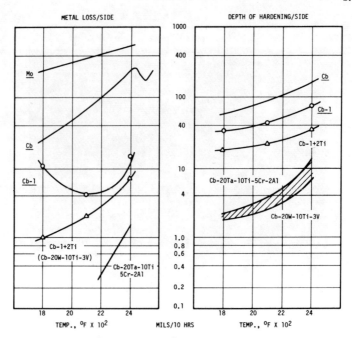

Fig. 20. Oxidation of columbium alloys.

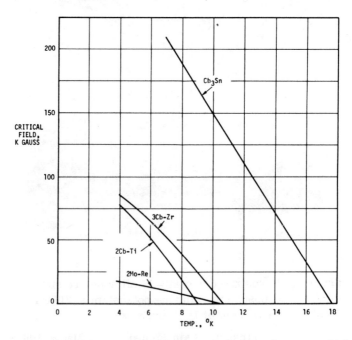

Fig. 21. Dependence of critical field on temperature for some high field
superconductors (ref. 60).

From information available at this time, the development of truly
oxidation resistant alloys of columbium for service at temperatures
above 2000°F and with usable engineering properties, i.e., strength
and ductility, appears very remote.

Superconductivity

The interest in superconducting materials is obvious when one
considers that magnetic coils operating at room temperature with fields
of 100 KO_e require megawatts of electrical power and elaborate facilities;
at 4.2°K, a high field superconducting magnet of 100 KO_e requires only
several cubic inches of space. The most widely used materials for
superconducting devices are the Cb_3Sn intermetallic compound, columbium-
zirconium alloys and columbium-titanium alloys. Of these materials,
the Cb_3Sn compound is the best superconducting material in that it
retains its superconducting properties at higher magnetic fields and
higher temperatures than any other materials. The dependence of
critical field on temperature for several high field superconductors
are shown in Figure 21[61]; the critical temperature for Cb_3Sn being
17.8°K. However, because of the brittle nature of Cb_3Sn and the
resultant difficulties in fabrication into wire or strip, the more
ductile columbium-zirconium and columbium-titanium alloys have been
used wherever possible. The Cb-25a/oZr alloy can be used for coils with
fields up to 70K Gauss and the Cb-44a/oTi alloy can be used with fields
approaching 100K Gauss as shown in Figure 22[11]. For magnetic fields
above 100K Gauss, it is necessary to employ use of the Cb_3Sn compound.
Although there have been many methods devised to produce wire or strip
of this material, one of the more unique and successful methods was
developed by Benz[61]. The superconducting tape developed by Benz
consists of an unalloyed columbium core over which a layer of Cb_3Sn
is formed by a diffusion coating process. Subsequently, the Cb_3Sn
coated columbium is laminated with copper, or copper and stainless
steel for added strength, by means of a lead-tin solder which brazes
at 453°K. The brazed tape, which is approximately 0.005-inch thick by
0.5-inch wide, can be flexed to very tight radii with no deleterious
effects. Figure 23 shows a schematic of the composite superconductor
developed by Benz. Since the cooling of the tape to temperatures of
4.2°K places the Cb_3Sn in compression due to the differences in the
thermal expansion properties of the materials, relatively large external tensile stresses can be applied to the superconducting
composite at 4.2°K without putting the Cb_3Sn in tension. Therefore,
the design stress for the Cb_3Sn composite tape is that stress which
places the Cb_3Sn in tension. The failure stresses for various Cb_3Sn
composites have been determined in a field of 50K Gauss such that the
magnetic field vector is transverse to the current direction. The
results are shown in Figure 23. The failure stress is defined as the
initiation of fracture of Cb_3Sn.

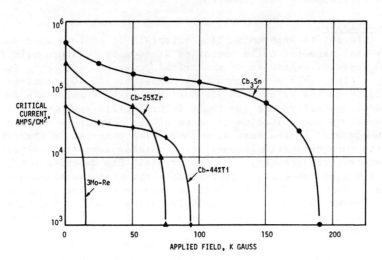

Fig. 22. Superconducting properties of Cb$_3$Sn, Cb–44%Ti and
Cb–25%Zr alloys at 4.2°K (ref. 11,60).

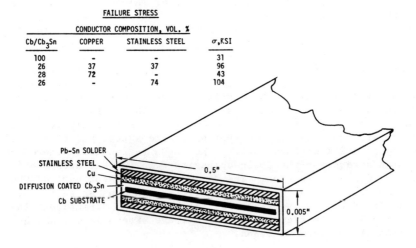

FAILURE STRESS

Cb/Cb$_3$Sn	CONDUCTOR COMPOSITION, VOL. %		σ,KSI
	COPPER	STAINLESS STEEL	
100	-	-	31
26	37	37	96
28	72	-	43
26	-	74	104

Fig. 23. Schematic view of composite superconductor (Ref. 61).

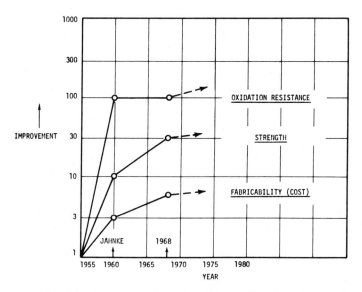

Fig. 24. Progress in columbium alloy development.

Concluding Remarks

The progress that has been made in alloying columbium with respect to improvements in strength, oxidation resistance and fabricability (represented by cost) is depicted in Figure 24 (after Jahnke) and was developed from factual data.   In 1960, Jahnke concluded that in the first five years of relatively intensive development, from 1955 to 1960, alloying improved the oxidation resistance by two orders of magnitude, strength was improved by one order and fabricability by less than 1/2 an order.   From studies conducted since 1960, it is concluded that essentially no further improvement was made in alloying for oxidation resistance while retaining usable engineering properties (strength and ductility); a three fold improvement in strength was achieved over the strongest alloy identified in 1960 (F-48) with the development of alloys Cb-1 and B-88 in combination with a greater understanding of the influence of microstructure on creep properties; and a two fold improvement in fabricability (cost) was brought about largely by advances in electron beam melting and the resultant reduction in interstitial impurity elements.   However, significant reductions in cost of columbium alloy mill products also can be attributed to reductions in the cost of the raw materials.

It can be expected that in the post-1968 years, advances in columbium alloys will depend largely on their acceptance in the wide range of potential applications and the funds available for their development.   Improvements in oxidation resistance of usable alloys should be forthcoming from development programs in progress.   Improvements in strength can be expected through process and compositional optimization of current alloys, employing maximum amounts of solute elements with high melting points and high shear moduli, or additions of tantalum to Cb-1/B-88 alloys.   Because of the fact that carbide dispersions become increasingly less effective with respect to creep properties at temperatures above $2200^\circ F$, it will be necessary to consider other means of achieving good strengths at the higher temperatures.   This should be possible through a combination of additions of the effective solid solution strengtheners (tungsten, osmium) and dispersion of inert oxides through powder metallurgy techniques.   Employment of powder metallurgy processes may also provide significant improvements in oxidation behavior through development of graded compositions and the ability to process brittle surface layers into usable shapes; in addition, reduced costs may result.

REFERENCES

1. A. G. Quarrell (Ed), Niobium, Tantalum, Molybdenum and Tungsten, Elsevier, 1961.

2. D. L. Douglass and F. W. Kunz (Eds), Columbium Metallurgy, Interscience, 1961.

3. M. Semchyshen and J. J. Harwood (Eds), Refractory Metals and Alloys, Interscience, 1961.

4. G. M. Ault, W. F. Barclay and H. P. Munger (Eds), High Temperature Materials II, Interscience, 1963.

5. M. Semchyshen and I. Perlmutter (Eds), Refractory Metals and Alloys II, Interscience, 1963.

6. R. I. Jaffee (Ed), Refractory Metals and Alloys III: Applied Aspects, Gordon & Breach, 1966.

7. N. E. Promisel (Ed), The Science and Technology of Tungsten, Tantalum, Molybdenum, Niobium and Their Alloys, Pergamon Press, 1964.

8. R. W. Fountain, J. Maltz and L. S. Richardson, High Temperature Refractory Metals - Part 2, Gordon & Breach, 1966.

9. F. T. Sisco and E. Epremian (Eds), Columbium and Tantalum, John Wiley and Sons, 1963.

10. T. E. Tietz and J. W. Wilson, Behavior and Properties of Refractory Metals, Stanford University Press, 1965.

11. T. E. Green and C. B. P. Minton, "Titanium, Zirconium and Niobium," Journal of the Institute of Metals, 1965.

12. Liquid Sodium Metal Cooled Fast Breeder Reactor (LMFBR) Program-Fuels and Materials Summary Report, Fuels and Materials Branch, USAEC Report Wash.-1065, January, 1966.

13. J. H. Keeler, "Commerical Applications of Refractory Metals," ASME Design Engineering Conference, New York, May 17 to 20, 1965.

14. F. H. Buttner and R. W. Hale, The Refractory Metals - An Evaluation of Availability, DMIC Memorandum 235, March, 1968.

15. F. T. Sisco and E. Epremian (Eds), Columbium and Tantalum, John Wiley and Sons, 1963, p. 5.

16. Ibid., p. vii.

17. U. S. Bureau of Mines, 1967.

18. W. H. Chang, Influence of Heat Treatment on Microstructure and
     Properties of Columbium Base and Chromium Base Alloys, ASD TDR-62-
     211, Part IV, March, 1966.

19. R. T. Begley, J. L. Godshall and D. Harrod, Development of Columbium
     Base Alloys, AFML TR-65-385, January, 1966.

20. R. T. Begley, J. A. Cornie and R. C. Goodspeed, Development of
     Columbium Base Alloys, AFML TR-67-116, November, 1967.

21. F. P. Talboom, A. D. Joseph and E. F. Bradley, "Columbium Systems
     for Turbine Blade Applications," SAE-ASME Air Transport and Space
     Meeting, New York, April 27 to 30, 1964.

22. A. L. Hoffmanner, "Thermal-Mechanical Processing of Precipitation
     Hardenable-Dispersion Hardened Columbium Alloy," AIME Refractory
     Metals Symposium, French Lick, Indiana, October, 1965.

23. C. R. Smeal, An Investigation of High Strength Refractory Metal
     Alloys - The Influence of Thermal-Mechanical Treatments on the
     Structure and Properties of High Carbon Cb-132M, TRW Inc. Summary
     Report ER 7081, January 31, 1967.

24. R. T. Begley and J. A. Cornie, Investigation of the Effects of
     Thermal-Mechanical Variables on the Creep Properties of High
     Strength Columbium Alloys, Westinghouse Electric Corp., Ctr.
     AF33615-67-C-1443, 1967.

25. S. T. Scheirer, Development of Columbium Alloy Combinations for
     Gas Turbine Blade Applications, TRW Inc., Contract AF33615-67-C-
     1688, 1967.

26. Personal Communication, Norman R. Gardner, Kawecki Chemical Company,
     March 8, 1968.

27. E. M. Savitskii, V. V. Baron and K. N. Ivanova, "Niobium Base
     Alloys and Their Properties," Sampe Journal, Vol. 4, No. 2,
     February/March 1968. Translated from Atomnaya Energia, Vol. 23,
     Academy of Science, USSR, 1967.

28. I. S. Malashenko, "Effect of Carbon on the Structure and Tensile
     Properties of Niobium and Its Alloys," IZV Akad Nauk SSSR Metally,
     No. 3, 1967, p. 159.

29. D. Geiselman, T. K. Roche and D. L. Graham, Development of Oxidation
     Resistant High Strength Columbium Base Alloys, Union Carbide Corp.,
     Ctr. AF33(615)-3856, 1966, 1967.

30. J. A. Cornie, Development of a Ductile Oxidation Resistant Columbium
     Alloy, Westinghouse Electric Corp., Ctr. AF33615-67-C-1689, 1967,
     1968.

REFERENCES (Cont)

31. H. R. Babitzke, R. E. Siemens and H. Kato, High Temperature Columbium and Tantalum Alloys, U. S. Dept. of the Interior, Bureau of
Mines Report 6777, 1966.

32. R. Yoda, H. R. Babitzke and H. Kato, Study of Columbium Base Alloys,
U. S. Dept. of the Interior, Bureau of Mines Report 6988, July, 1967.

33. G. D. McAdam, "Substitutional Niobium Alloys of High Creep
Strength," Journal of the Institute of Metals, December, 1965,
p. 559.

34. G. D. McAdam, "The Influence of Carbide and Boride Additions on the
Creep Strength of Niobium Alloys," The Journal of the Institute of
Metals, January, 1968, p. 13.

35. J. W. Clark, Recent Developments in Columbium Base Alloys, General
Electric Company, Aircraft Engine Group Brochure, February, 1962.

36. J. C. Sawyer and E. A. Steigerwald, "Creep Properties of Refractory
Metal Alloys in Ultra-High Vacuum," Journal of Materials, Vol. 2,
No. 2, June, 1967, p. 341.

37. H. E. McCoy, "Creep Properties of the Niobium-1% Zirconium Alloy,"
Journal of the Less Common Metals, Vol. 8, 1965, p. 20.

38. M. Schussler, Properties of Haynes Alloy Cb-753, Union Carbide
Corp., March 1, 1965.

39. R. L. Stephenson, Creep Rupture Properties of Cb-752 Alloy and the
Response to Heat Treatment, ORNL-TM-1577, August, 1966.

40. R. L. Stephenson, The Effect of Heat Treatment on the Creep Rupture
Properties of D-43 Alloy, ORNL-TM-1587, September, 1966.

41. R. H. Titran and R. W. Hall, High Temperature Creep Behavior of
Columbium Alloy FS-85, NASA TN D-2885, June, 1965.

42. R. H. Titran and R. W. Hall, Ultra-High Vacuum Creep Behavior
of Columbium and Tantalum Alloys at 2000° and 2200°F for Times
Greater than 1000 Hours, NASA TM X-52130, 1965.

43. R. L. Stephenson, "Creep Properties of SU-16," Metals and Ceramics
Division Annual Progress Report, ORNL-4170, November, 1967.

44. J. J. English and E. S. Bartlett, Recent Information on Long Time
Creep Data for Columbium Alloys, DMIC Memorandum 203, April 26, 1965.

45. R. W. Harrison, Compatibility of Biaxially Stressed D-43 Alloy with
Refluxing Potassium, NASA CR-807, June, 1967.

REFERENCES (Cont)

46. P. Grodzinski, "Investigation on Shaft Fillets," Engineering
     (London), Vol. 152, 1941, p. 321.

47. H. E. McCoy, "Correlation of Creep Properties for Constant and
     Varying Stresses," Metals and Ceramics Division Annual Progress
     Report, ORNL-3870, November, 1965, p. 122.

48. F. N. Lake and C. R. Smeal, Process Development for Precision
     Forging Columbium Base Alloys, AFML TR-67-94, April, 1967.

49. General Electric Data, February, 1967.

50. Columbium, Tantalum and Tungsten Alloys, Wah Chang Albany Technical
     Information, Vol. 3, January, 1968.

51. H. L. Kohn and R. M. Curcio, Columbium Alloy Sheet Rolling Program,
     Fansteel Metallurgical Corp. Final Report, Ctr. NOw 63-0231-C,
     October, 1964.

52. R. A. Nadler, Processing and Evaluation of Preproduction Quantities
     of Columbium Alloy Sheet, Westinghouse Electric Corp. Final Report,
     Ctr. N 600(19)-59546, January, 1964.

53. J. W. Davis and R. M. Curcio, Fabrication of Fansteel 85 Metal
     Sheet, Fansteel Metallurgical Corp. Final Report, Ctr. NOw 65-0498-f,
     March, 1966.

54. J. S. Clark, A. L. Mincher and G. N. Villee, The Development of
     Optimum Manufacturing Methods for Columbium Alloy Sheet, RTD TDR-
     63-4236, 1963.

55. J. G. Bewley, "Strengthening of Columbium Alloy Cb-752 by Duplex
     Annealing Process," AIME Conference Physical Metallurgy of Refrac-
     tory Metals, French Lick, Indiana, October 3 to 5, 1965.

56. G. G. Lessmann, "The Comparative Weldability of Refractory Metal
     Alloys," Welding Journal, Vol. 45, No. 12, December, 1966, p. 540-S.

57. R. G. Donnelly, "Welding of Advanced Refractory Alloys," Metals
     and Ceramics Division Annual Progress Report, ORNL-4170,
     November, 1967, p. 71.

58. G. G. Lessmann, The Effect of 1000-Hour Thermal Exposures on
     Tensile Properties of Refractory Metal Alloys, NASA-CR-72095,
     Topical Report No. 1, Ctr. NAS 3-2540.

59. L. B. Engel, Jr. and R. G. Frank, Evaluation of High Strength
     Columbium Alloys for Alkali Metal Containment, General Electric
     Company Final Report R66SD3015, Covering the Period of July 1962
     to December 1964, Ctr. NAS 3-2140.

REFERENCES (Cont)

60. V. L. Newhouse, <u>Applied Superconductivity</u>, John Wiley and Sons, Inc., New York, 1964.

61. M. G. Benz, <u>Mechanical and Electrical Properties of Diffusion Processed Nb_3Sn-Copper-Stainless Steel Composite Conductors</u>, General Electric Report No. 67-C-459, December, 1967.

BIBLIOGRAPHY

1. R. L. Stephenson and R. G. Donnelly, Effect of Aging on the Creep
    Rupture Properties of D-43 Welds, ORNL-TM-1708, January, 1967.

2. F. Ostermann and F. Bollenrath, Investigation of Precipitates in
    Two Carbon-Containing Columbium-Base Alloys, AFML TR-66-259,
    December, 1966.

3. T. K. Roche, "Aging Studies on the D-43 Alloy," Metals and Ceramics
    Division Annual Progress Report, ORNL-3870, November, 1965, p. 116.

4. B. Harris and D. E. Peacock, "Low Temperature Mechanical Properties
    of a Solution Hardened Niobium (Columbium) Alloy," AIME Transactions,
    Vol. 233, July, 1965, p. 1308.

5. B. Harris and D. E. Peacock, "Physical Properties of Some Niobium
    (Columbium) Alloys at Low Temperature," AIME Transactions, Vol. 236,
    April, 1966, p. 471.

6. B. Harris, "Structural Changes During Creep of a Solution Hardened
    Niobium Alloy," Journal of the Less-Common Metals, Vol. 9, 1965,
    p. 244.

7. E. J. Delgrosso, C. E. Carlson and J. J. Kaminsky, Development of
    Cb-Zr-C Alloys, PWAC-464, September, 1965.

8. C. E. Carlson and E. J. Delgrosso, Recovery and Recrystallization
    of Columbium and Several Columbium-Zirconium Alloys, PWAC-465,
    September, 1965.

9. J. G. Bewley and M. Schussler, Final Report on Process Improvement
    of Columbium (Cb-752) Alloy, AFML TR-65-63, March, 1965.

10. E. S. Bartlett and J. A. VanEcho, Creep of Columbium Alloys, DMIC
    Memorandum 170, June 24, 1963.

11. F. F. Schmidt and H. R. Ogden, The Engineering Properties of Colum-
    bium and Columbium Alloys, DMIC Report No. 188, September 6, 1963.

12. H. R. Babitzke, R. E. Siemens, G. Asai and H. Kato, Development of
    Columbium and Tantalum Alloys for Elevated Temperature Service,
    U. S. Department of the Interior, Bureau of Mines Report 6558, 1964.

13. H. R. Babitzke, M. D. Carver and H. Kato, Columbium and Tantalum
    Alloys Suitable for Use in High Temperatures, U. S. Department of
    the Interior, Bureau of Mines Report 6390, 1964.

14. D. J. Maykuth, Summary of Contractor Results in Support of the
    Refractory Metals Sheet Rolling Program, DMIC Report 231,
    December 1, 1966.

BIBLIOGRAPHY (Cont)

15. Materials Advisory Board, Final Report of the Refractory Metals
    Sheet Rolling Panel, National Academy of Sciences-National Research
    Council Publication, MAB-212-M, March, 1966.

16. S. R. Thompson, W. R. Young and W. H. Kearns, Investigation of the
    Structural Stability of Welds in Columbium Alloys, ML TDR-64-210,
    Part II, April, 1966.

17. J. M. Gerken, A Study of Welds in Columbium Alloy D-43, TRW, Inc.
    Report TM 3865-67, March 25, 1964.

18. J. D. W. Rawson and B. B. Argent, "The Effect of Oxygen and Carbon
    on the Creep Strength of Niobium," The Journal of the Institute of
    Metals, Vol. 95, July, 1967, p. 212.

19. N. S. Bornstein, E. C. Hirakis and L. A. Friedrich, Carburization
    of Cb-1Zr Alloy, Pratt and Whitney Aircraft Division Report TIM-927,
    August, 1965.

20. R. M. Bonesteel, J. L. Lytton, D. J. Rowcliffe and T. E. Tietz,
    Recovery and Internal Oxidation of Columbium and Columbium Alloys,
    AFML TR-66-253, November, 1966.

21. W. F. Scheely, "Mechanical Properties of Niobium-Oxygen Alloys,"
    Journal of the Less-Common Metals, Vol. 4, 1962, p. 487.

22. R. A. Rapp and G. M. Goldberg, "The Oxidation of Cb-Zr and Cb-Zr-Re
    Alloys in Oxygen at 1000°C," AIME Transactions, Vol. 236,
    November, 1966, p. 1619.

23. R. T. Bryant, "The Solubility of Oxygen in Transition Metal Alloys,"
    The Journal of the Less-Common Metals, Vol. 4, 1962, p. 62.

24. J. F. Hogan, E. A. Limoncelli and R. E. Cleary, Reaction Rate of
    Columbium -1 Zirconium Alloy with Oxygen at Low Pressures, Pratt
    and Whitney Aircraft Division Report TIM-901, September, 1965.

25. E. A. Limoncelli, High Temperature Refractory Alloy Testing in a
    Low Pressure Flowing Argon Environment, Pratt and Whitney Aircraft
    Division Report TIM-903, September, 1965.

26. J. Stoop and P. Shahinian, The Effect of Nitrogen on the Tensile
    and Creep Rupture Properties of Niobium, Naval Research Laboratory
    Report No. 6464, Ocotber 31, 1966.

27. R. J. Walter, J. A. Ytterhus, R. D. Lloyd and W. T. Chandler,
    Effect of Water Vapor/Hydrogen Environments on Columbium Alloys,
    AFML TR-66-322, December, 1966.

BIBLIOGRAPHY (Cont)

28. R. J. Walter and W. T. Chandler, "The Compatibility of Tantalum
     and Columbium Alloys with Hydrogen," AIAA Journal, Vol. 4, No. 2,
     February, 1966, p. 302.

29. J. R. Stephens and R. G. Garlick, Compatibility of Tantalum,
     Columbium, and Their Alloys with Hydrogen in the Presence of a
     Temperature Gradient, NASA TN D-3546, August, 1966.

30. R. W. Webb, Permeation of Hydrogen through Metals, Atomics Inter-
     national Division Report No. NAA-SR-10462, July 25, 1965.

31. C. C. Masser, Vapor Pressure Data Extrapolated to 1000 Atmospheres
     $(1.01 \times 10^8 N/m^2)$ for Thirteen Refractory Materials with Low
     Thermal Absorption Cross-Sections, NASA TN D-4147, October, 1967.

32. D. T. Bourgette, High Temperature Chemical Stability of Refractory
     Base Alloys in High Vacuum, ORNL-TM-1431, April, 1966.

33. C. E. Lundin and R. H. Cox, "The Age Hardening Characteristics of
     Niobium Base Zirconium Alloys," Journal of the Less-Common Metals,
     Vol. 13, No. 5 November, 1967, p. 501.

34. A. Taylor and N. J. Doyle, "The Solid Solubility of Carbon in Nb
     and Nb-Rich Nb-Hf, Nb-Mo and Nb-W Alloys," Journal of the Less-
     Common Metals, Vol. 13, No. 5, November, 1967, p. 511.

35. A. Taylor, Research for Solubility of Interstitials in Columbium,
     1- A Study of Columbium Rich Alloys in the Ternary System Cb-W-O,
     Cb-W-N, and Cb-W-C, AFML TR-65-48, March, 1965.

36. T. A. Roach and E. F. Gowan, Jr., Structural Fasteners for Extreme
     Elevated Temperatures, AFFDL TR-66-107, September, 1966.

37. T. Miyagishima, Investigation of High Strain Rate Behavior of
     Refractory Alloys and Coatings, The Marquardt Corp. Final Report
     No. PR-3009-F, Ctr. NAS 9-4905, February, 1966.

38. B. C. Allen and E. S. Bartlett, "Elevated Temperature Tensile
     Ductility Minimum in Silicide Coated Cb-10W and Cb-10W-2.5Zr,"
     ASM Transactions Quarterly, Vol. 60, No. 3, September, 1967, p. 295.

39. W. A. Gibeaut and E. S. Bartlett, Properties of Coated Refractory
     Metals, DMIC Report No. 195, January 10, 1964.

40. H. A. Hauser and J. F. Holloway, Jr., Evaluation and Improvement
     of Coatings for Columbium Alloy Gas Turbine Engine Components,
     AFML-TR-66-186, Part 1, July 1966.

BIBLIOGRAPHY (Cont)

41. R. B. Kaplan and F. A. Glaski, Research on Gas Plated Refractory Metal Coatings for Liquid Metal Compatibility Investigation, AFML-TR-66-72, April, 1966.

42. V. A. Kirillin, A. E. Sheindlin, and V. Yachekhovskoi and I. A. Zhukova, "The Thermodynamic Properties of Niobium in the Temperature Range from $0^\circ K$ to the Melting Point $2740^\circ K$," Teplofizika Vysokikh Temperatur , Vol. 3, No. 6, November to December, 1965, p. 860.

43. H. Conrad, Guiding Principles for Lowering the Ductile to Brittle Transition Temperature in the Body Centered Cubic Metals, Aerospace Corp. Report No. ATN-64(9236)-4, December 10, 1963.

44. C. W. Marschall, F. C. Holden, A. Gilbert, and B. L. Wilcox, Further Investigation of Notch-Sensitivity of Refractory Metals, AFML-TR-65-286, November, 1965.

45. A. Fourdeux, F. Rueda, E. Votava, and A. Wronski, Surface and Interfacial Effects in Relation to Brittleness in Refractory Metals, AFML-TR-65-226, April, 1966.

46. I. M. Nedokha, et al., Niobium-Metal of the Space Age, Efkiev, 1965, Joint Publications Research Service Translation No. 36718, July 28, 1966.

47. G. W. P. Rengstorff, High Purity Metals, DMIC Report No. 222, January 3, 1966.

48. A. N. Zelikman, O. E. Krein, and G. V. Samsonov, Metallurgy of Rare Metals, Izdatel'stvo Metallurgiya, Moskva, 1964, NASA TT-F-359, Israel Program for Scientific Translations, Jerusalem, 1966.

49. J. M. Williams, J. T. Stanley, and W. E. Brundage, "The Interaction of Radiation Produced Defects and Interstitial Impurity Atoms in Niobium," ORNL-4097, Radiation Metallurgy Section Solid State Division Progress Report, April, 1967, p. 30.

50. P. R. V. Evans, A. F. Weinberg, and R. J. VanThyne, "The Radiation Hardening in Columbium," Acta Metallurgica, Vol. 2, February, 1963, p. 143.

51. H. E. McCoy and J. R. Weir, "Effect of Irradiation on Bend Transition Temperature of Molybdenum and Niobium Base Alloys," Metals and Ceramics Division Annual Report, ORNL-3670, 1964, p. 87.

52. T. Kofstad, Studies of Oxidation of Niobium Alloys at Very High Temperatures, RTD ML-TR-67-40, April, 1967.

53. J. V. Peck, Resistance Welding of Refractory Metals, RTD ML-TR-67-130, June, 1967.

BIBLIOGRAPHY (Cont)

54. W. Batiuk, Fluidized Bed Techniques for Coating Refractory Metals,
     RTD ML-67-127, April, 1967.

55. A. Avguspinik, Reactions of Oxides and Carbides with Metals, UCRL
     Translations 10044, May, 1966.

56. J. A. DeMastry, Investigation of High Temperature Refractory Metals
     and Alloys for Thermionic Converters, RTD APL-TR-65-29, April, 1965.

57. M. Hoch, High Temperature Specific Heats of Refractory Metals and
     Alloys, RTD ML-TR-66-360, November, 1966.

58. A. P. Dannessa, "Characteristic Redistribution of Solute in Fusion
     Welding," Welding Journal, December 1966, p. 569.

59. J. Hernaez, et al., "Influence of Surface Condition on the Mechani-
     cal Properties of Polycrystalline Niobium," Rev. Met., Vol. 3,
     No. 2, March to April, 1967, p. 101 (Spanish).

60. J. A. Roberson, et al., "The Observation of Markers During the
     Oxidation of Columbium," AIME Transactions, Vol. 239, No. 9,
     September, 1967, p. 1327.

61. E. M. Savitskii, et al., "Effect of Vanadium Additions on the
     Structure and Superconducting Properties of Niobium-Zirconium
     Alloys," Russian Mets., No. 3, 1966, p. 100.

62. T. Doe, et al., "Peak Effect in Superconducting Nb-15 A/oZr-45
     A/o Ti Alloys," Journal of Applied Physics, Vol. 38, No. 10,
     September, 1967, p. 3811.

63. F. F. Schmidt, and E. S. Bartlett, "The Mechanical Behavior of
     Refractory Metal Alloys," Metal Treating, Vol. 18, No. 4,
     August to September, 1967, p. 12.

64. A. I. Evstyakhin, et al., "Study of Alloys in the Niobium-Zirconium
     System," High Purity Metals and Alloys, Fabrication, Properties
     and Testing, Plenum Press, 1967, p. 37.

65. A. C. Prior, "A Comparative Review of Materials for Construction
     of Superconducting Solenoids," Cryogenics, Vol. 7, No. 3,
     June, 1967, p. 131.

66. C. M. Yen, et al., "Superconducting H_C-J_C and T_C Measurements in
     the Niobium-Titanium-Nitrogen, Niobium-Hafnium-Nitrogen and
     Niobium-Vanadium-Nitrogen Ternary Systems," Journal of Applied
     Physics, Vol. 38, No. 5, April, 1967, p. 2268.

67. E. M. Savitskii, et al., "Dispersion Strengthening in Niobium-Base
     Alloys," IZV Akad Nauk SSSR Metally, No. 3, 1967, p. 152 (Russian).

BIBLIOGRAPHY (Cont)

68. Y. Sasaki, et al., "Mechanical Properties of Niobium-Hydrogen Alloys," Japanese Institute of Metals, Vol. 31, No. 4, April, 1967, p. 401 (Japanese).

69. K. Tachikawa, et al., "Fabrication of Multi-layered Nb_3Sn Super-conducting Wire," NRIM Transactions, Vol. 9, No. 1, 1967, p. 39.

70. C. P. Davis, et al., "Welding of Refractory Alloys," Machinery, Vol. 73. No. 9, May, 1967, p. 92.

71. R. A. Pasternak and B. Evans, "Adsorption, Absorption, and Degassing in the Oxygen-Niobium System at Very Low Pressures," Journal of the Electro-Chemical Society, Vol. 114, No. 5, May, 1967, p. 452.

72. H. W. Lavendel, et al., "Evaluation of Silicide Coatings on Columbium and Tantalum and a Means for Improving Their Oxidation Resistance," AIME Transactions, Vol. 239, No. 2, February, 1967, p. 143.

73. B. Harrison, "The Influence of Some Solutes on Young's Modulus of Niobium," Journal of the Less-Common Metals, Vol. 12, No. 3, March, 1967, p. 247.

74. E. J. DelGrosso, "Development of Nb-Zr-C Alloys," Journal of the Less-Common Metals, Vol. 12, No. 3, March, 1967, p. 113.

75. D. K. Bowen, et al., "Deformation Properties of Niobium Single Crystals," Canadian Journal of Physics, Vol. 45, No. 2, February, 1967, p. 903.

76. T. J. Sherwood, et al., "Plastic Anisotropy of Tantalum, Niobium, and Molybdenum," Canadian Journal of Physics, Vol. 45, No. 2, February, 1967, p. 1075.

77. D. A. Prokoshkin, et al., "Research into the Oxidiation of Niobium Alloyed with Vanadium, Titanium and Zirconium," Nauka, Moskow, 1966, p. 285 (Russian).

78. H. E. McCoy, "Carburizing of Niobium and Tantalum Base Alloys," Journal of the Less-Common Metals, Vol. 12, No. 2, February, 1967, p. 139.

79. D. C. Briggs, et al., Aging in Niobium Rich Niobium-Hafnium-Carbon Alloys, Report No. R 185 of Department of Energy, Mines and Resources, Mines Branch, Ottawa, Canada, September, 1966.

80. R. A. Meussner, et al., "Oxidation and Self Repair of the Zinc Based High Temperature Coating for Niobium," Corrosion Science, Vol. 7, No. 2, February, 1967, p. 103.

BIBLIOGRAPHY (Cont)

81. J. Chelius, Use of Refractory Metals in Corrosive Environment
     Service, ASM-ASTME Technical Report WES-7-56, 1967.

82. M. S. Duesdery, et al., "The Plasticity of the Pure Niobium Single
     Crystals," Journal of Physics, Vol. 27, No. 7-8, July to August,
     1966, p. 193.

83. A. McClure, "Metal Fasteners in Ultra High Temperatures,"
     Machinery, Vol. 39, No. 4, February, 1967, P. 26.

CONSIDERATIONS IN THE DEVELOPMENT OF TANTALUM BASE ALLOYS

By

R. W. Buckman, Jr. and R. C. Goodspeed

This paper reviews the alloying behavior of tantalum with discussion being restricted primarily to the effects of the alloying additions tungsten, hafnium, rhenium, and carbon. The effect of each of these additions on tensile strength, fabricability, and creep behavior is discussed. The status of current commercially available tantalum base alloys is presented and the potential strength limit of alloy compositions which retain room temperature ductility is proposed.

Westinghouse Astronuclear Laboratory
Pittsburgh, Pennsylvania 15236

INTRODUCTION

Applications for which tantalum base alloys are being considered can be divided into two categories; (a) short lived($\leq$ 10 hours) service at 2500-3500°F, typical of aerospace vehicle and rocket propulsion unit operation and (b) long life (>10,000 hours) service of space nuclear power conversion systems which will operate at 2000-3000°F. Those applications, which have provided the impetus for the development of tantalum base alloys, were discussed in part by Semmel in these proceedings and therefore will not be elucidated here.

Although this is a review paper, it is not the intent of the authors to present a detailed compilation of the experimental data generated by the various investigators in tantalum alloy development. Instead, an attempt will be made to place tantalum base alloy development in its proper perspective. Thus generally observed trends and limitations will be emphasized rather than specific property data which are available in excellent compilations elsewhere (1,2).

The observations presented are based primarily on data reported for the present generation alloy compositions and for those whose development is sufficiently advanced to be considered in the pilot plant stage. The topics which will be discussed are (1) the alloying behavior of tantalum emphasizing the effects of strengthening additions on the tensile properties, fabricability, and creep behavior; (2) the current status of tantalum base alloys, both the commercially available and the advanced experimental compositions; and finally (3) the potential strength limitations which will probably be achieved in useful tantalum alloy compositions. Although environmental interactions are critical to the application of tantalum base alloys, this is a subject within itself and is beyond the scope of this paper. This topic is also covered in some detail in this volume by Inouye, Chandler, and Hoffman.

ALLOYING BEHAVIOR

Although tantalum has a density of 0. 604 lbs/in^3, approximately twice that of columbium and molybdenum base alloys, it does possess unique low temperature ductility, a high melting point (5440°F/2996°C), and a moderately high elastic modulus (27 x 10^6 psi at RT). However, it is the extremely good low temperature ductility exhibited by tantalum which makes it an unique alloy base, and the retention of this feature must be the primary consideration in the development of tantalum base alloys.

Strengthening of tantalum has been achieved by solid solution and dispersed phase strengthening, but the former has received the greatest amount of attention. Although tantalum exhibits complete or extensive solubility with its near neighbor elements in the periodic table, the most effective substitutional solute strengtheners as determined by alloy development studies during the past ten years are tungsten, rhenium, and hafnium (3,4,5). A dispersed phase strengthening increment can be achieved by the addition of carbon, nitrogen, and oxygen with carbon being the most popular interstitial addition.   The dispersed second phase is normally a carbide, oxide, and/or nitride formed by the interaction of the interstitial addition with a reactive metal addition such as hafnium or zirconium.   However, in the case of carbon, a matrix carbide can be formed which also imparts a significant increment to elevated temperature strength.

Effects of Alloying Additions on Yield Strength

Alloying of tantalum to achieve high strength at elevated temperatures has been quite successful and an excellent compilation of the strength of a wide range of binary tantalum alloys at room temperature and at 2200°F has been prepared by Seigle (6).   This review discussed solution strengthening exclusively and did not deal with the concomitant effects of the alloying additions on low temperature ductility or fabricability.   As noted earlier, the most effective strengthening additions to tantalum are tungsten, hafnium, and rhenium.   Interstitial additions while not effective elevated temperature strengtheners, none-the-less are extremely potent strengtheners at room temperature.   Thus their effects on low temperature ductility and fabricability are extremely important.   For our purposes, the effects of solute additions on the strength of tantalum can best be described in terms of the increment-al strengthening contribution of each alloying element.   Considering that the yield strength of an alloy is comprised of the yield strength of pure tantalum plus the incremental contribution of each alloying element gives the following phenomeno-logical expression.

$$\sigma_{YS(alloy)} = \sigma_{YS(Ta)} + (\partial\Delta\sigma_{YS}/\partial C_1) C_1 + \ldots (\partial\Delta\sigma_{YS}/\partial C_i) C_i$$

The parameter $\partial\Delta\sigma_{YS}/\partial C$ is the stress increase per atom per cent addition and has the unit ksi/a/o and C is the concentration in atom per cent.   The parameter $\partial\Delta\sigma_{YS}/\partial C$ is plotted as a function of temperature for selected substitutional and interstitial solutes in Figure 1.   The $\partial\Delta\sigma_{YS}/\partial C$ values for the substitutional solute additions were calculated from data on alloys containing 7-17% W, up to 2-1/2% Hf, and up to 3% Re, and for the interstitial additions, up to 300 ppm carbon or nitrogen, and 500 ppm oxygen (1-3,5-10).   It should be emphasized that the strengthening index $\partial\Delta\sigma_{YS}/\partial C$ is strongly a function of concentration.   Thus the values in Figure 1 are only valid within the concentration limits noted above.   At room temperature, it is apparent that interstitials in solution have a much more potent effect on the room temperature yield strength than the substitutional additions.   The implications of this rapid hardening by the interstitial additions on fabricability and low tempera-ture ductility will be discussed later.   Strengthening of tantalum at 2200°F by interstitials in solid solution is ineffective and the increment observed by the carbon addition is most likely due to a carbide precipitate.   However, by adding hafnium or zirconium, the solubility of oxygen and nitrogen is significantly reduced and the formation of finely dispersed oxide, nitride, and/or carbide precipitates contributes a significant increment of strengthening.   The strengthening increment imparted by the nitride and/or carbide persists up to about 2700°F (3,9,11).   At 3000°F and above, dispersed phase strengthening by interstitial compounds has been found to be in-

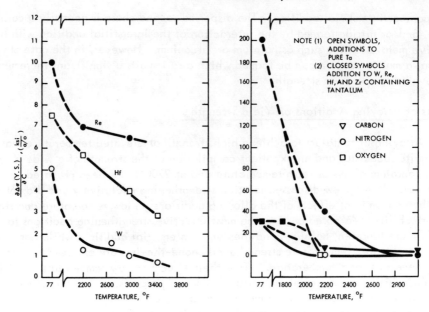

FIGURE 1 – Effect of Temperature on the Solute Strengthening of Substitutional and
Interstitial Elements in Tantalum

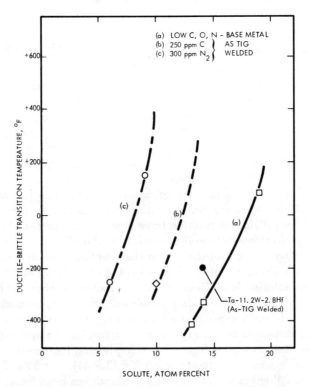

FIGURE 2 – Ductile-Brittle Transition Behavior of Tantalum Base Alloys
(Ref. 3, 5, 28, 34)

effective (3).   Strengthening by the substitutional solutes on the other hand are still quite effective, at least on a short time basis, at 3500°F which is 0.73 of the absolute melting temperature of pure tantalum.

For binary additions at the 5 atom percent level the strengthening indices $(\partial \Delta \sigma_{YS}/\partial C)$ at 2200°F, for rhenium, tungsten, and hafnium additions to tantalum are reported as 3500, 2100, and 2200 psi/atom percent addition respectively.   In Figure 1, both rhenium and hafnium appear to exhibit a disproportionate high rate of strengthening relative to the values exhibited for tungsten.   This most likely reflects the concentration dependence of strengthening since both hafnium and rhenium were at relatively dilute levels in the alloys used for evaluation.   The disproportionate rate of strengthening exhibited by hafnium may also be explained in terms of a dispersed phase contribution from the formation of interstitial compounds (6).   This may prove a reasonable explanation at temperatures below 2700°F.   However, at 3500°F interstitial compounds formed in alloys containing only residual carbon, oxygen, or nitrogen (< 100 ppm), are likely dissolved.   Thus at 3500°F a dispersed phase strength-ening explanation does not appear valid.   However, it is not the purpose here to discuss strengthening mechanisms, a subject which has already been covered in some detail in this volume by Wilcox.

Effect of Solute Additions on Fabricability

Although dramatic increases in the strength of tantalum are achieved by solute additions, the permissible amount which can be tolerated is limited by its effects on low temperature ductility, the most attractive characteristic of tantalum.   Since welding increases the ductile-brittle transition temperature (DBTT), tests on welded joints are an excellent measure of fabricability; that is, welded joints which show good low temperature ductility are indicative of a highly fabricable base metal.   However, as will be shown, base metal which exhibits good ductility does not always exhibit a low ductile-brittle transition temperature as welded.

The ductile-brittle transition temperature data plotted in Figure 2 are for 0.03-0.04 inch sheet tested in bending in the as-recrystallized or as-tungsten inert gas arc welded condition.   Schmidt et al (3) have shown that tantalum can tolerate up to 19 atom percent tungsten before the base metal bend ductile-brittle transition temperature is raised above room temperature.   They also reported that TIG welding increased the DBTT approximately 650°F which would limit a room temperature ductile weld to alloys containing less than 13 a/o tungsten.   Ammon and Begley (12) have shown that a Ta-11.2W-2.8Hf (14 atom percent solute) composition retained as-TIG welded bend ductility to -250°F.   However, they also reported that to achieve good as-welded ductility required maintaining the tungsten to hafnium concentration at about 4 to 1.   It is interesting to note here that the anomalous effect rhenium has on the ductility of molybdenum and tungsten is not observed in either tantalum or columbium since rhenium additions drastically increase the ductile-brittle transition temperature of the latter two.   The degrading effect of rhenium on low temperature ductility led Schmidt et al (3) to the conclusion that any elevated temperature advantage imparted by rhenium additions is nullified by the concomitant degradation

of low temperature ductility. Subsequent work has shown that at the 10 atom percent solute level, rhenium additions should be maintained at less than 3 a/o to retain
low temperature ductility and weldability (7,13).

The additions of carbon and nitrogen to an alloy matrix further limit the substitutional solute level before as-TIG welded ductility is seriously limited, reflecting
the solubility relationships of the additions with the matrix. Although carbon
additions of 1 atom percent to Ta-10 a/o(W+Mo+Zr) and Ta-10-12a/o (W+Hf) do not
seriously degrade the low temperature ductility of recrystallized sheet material,
during welding, carbon is dissolved and the cooling rate of the solidified metal is
sufficiently fast to retain a large amount of carbon in solid solution (3,14). The
potent low temperature strengthening of the tantalum matrix by carbon and nitrogen
in solid solution has been discussed in the previous section. By proper post weld
annealing treatments, low temperature ductility can be restored in carbon containing
alloys. This may prove impractical however in large complex fabricated structures.

Fortunately the trade-offs between elevated temperature strength and low
temperature ductility may not represent too complex a problem. The data of Schmidt
et al (3) plotted in Figure 3 show that increasing the solute level in tantalum above
13 a/o does not increase the elevated temperature tensile strength significantly
enough to absorb the drastic decrease in low temperature ductility. Thus for adequate
low temperature weld ductility, substitutional solute strengthened alloys should contain less than 12-14 a/o percent total alloy addition. If carbon is added at the
200-300 ppm level, the substitutional solute addition should be restricted to 10 a/o
or less. An equivalent amount of nitrogen would require a substitutional solute
level of less than 8a/o. Generally, the adverse effects nitrogen has on low temperature weld ductility will limit its usefulness as a strengthening addition in weldable
sheet alloys.

Effect of Alloy Additions on Creep Strength

Development of tantalum base alloys from the standpoint of improving resistance
to creep deformation has received attention only in the past 4 to 5 years. Creep is
a thermally activated process and the large body of experimental creep data indicates
that creep rate ($\dot{\varepsilon}$) is given by the general equation (15).

$$\dot{\varepsilon} = \sum_i A_i(\sigma,T,S)\, e^{-\Delta Hi(\sigma,T,S)/RT}$$

where A_i is the frequency factor and ΔHi the activation energy of one of a number of
deformation mechanisms. Both A_i and ΔHi may depend on stress (σ), temperature (T),
and structure (S). The structure term includes all lattice imperfections which influence deformation. Thus in the course of an alloy screening investigation which
encompasses wide compositional variations, it is particularly difficult to compare
quantitatively the effects of the various additions on creep behavior on the basis of
constant structure. An additional complication in studying the creep properties of
tantalum base alloys is of course the test environment interactions which can and do
occur at pressures on the order of 1×10^{-7} torr. For example, Inouye (16) has shown

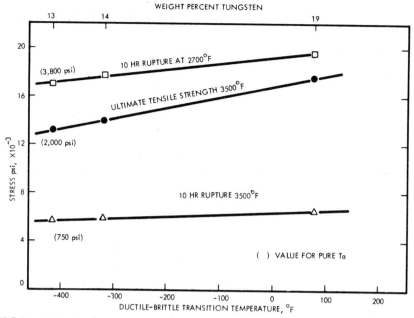

FIGURE 3 – 4T Bend Ductile-Brittle Transition Temperature of Tantalum Base Alloys (Ref. 3)

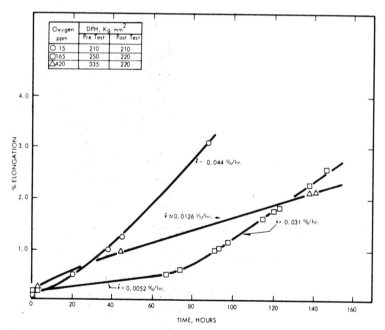

FIGURE 4 – Creep Behavior of Internally Oxidized Ta-8W-2Hf (T-111) at 2200°F and 21,000 psi (Ref. 17)

that oxygen is the primary contaminant during low pressure $(10^{-7}$ torr), elevated temperature exposure of tantalum and columbium. Buckman (17) has shown that oxygen contamination significantly improves the creep resistance of T-111 (Ta-8W-2Hf), a reactive metal containing tantalum base alloy, as illustrated by the creep curves in Figure 4. The strengthening mechanism which is reported as a coherent HfO_2 precipitate is similar to that observed by Bonesteel et al (18) for an internally oxidized Cb-1W-1Zr alloy. However strengthening of reactive metal tantalum alloys by oxide dispersions formed in situ is of little benefit at temperatures of $2200°F$ and above, as the coherency of the hafnium oxide precipitate is short lived. Thus in reviewing the creep data available, reactive-metal-containing tantalum alloy properties obtained at pressures greater than 1×10^{-7} torr were disregarded.

To compare the creep strengthening effects of the various alloy additions, a parameter similar to that used in describing short time tensile strength was utilized. Except now $\partial \Delta \sigma_c / \partial C$ is defined as the incremental change in stress due to an alloying element (on an atom percent basis) to cause 1% creep elongation in 100 hours at $2400°F$. The relative creep strengthening contributions for the individual substitutional and interstitial solute additions are illustrated in Figure 5. Data from which this figure was constructed were taken from the sources listed in Table 1. Where possible, in this comparison an attempt was made to use only creep data for material tested in the recrystallized condition with nominally similar grain sizes. Data for the material tested in the as-worked condition were used only at temperatures where recrystallization occurred during heating to the test temperature. All the creep data were first normalized using the Larson-Miller parameter with a value of 15 used for the constant in the expression $P = T_{oR}(C+t)$ where t is the time to 1% elongation in hours. The values for $\partial \Delta \sigma_c / \partial C$ in Figure 5 were all interpolated with the exception of that for the nitride which required extrapolation. Thus this value may be somewhat optimistic.

From Figure 5 it is immediately apparent that hafnium makes a negative contribution to the creep strength of tantalum alloys at $2400°F$. This behavior is similar to that reported by McAdam (26) for hafnium additions to columbium alloys. The theoretical factors which contribute to creep strength as summarized by Sherby (27) would lead one to predict this behavior for hafnium additions to tantalum alloys. Again, the values of $\partial \Delta \sigma_c / \partial C$ are only meant for qualitative comparison since they are a function of concentration as well as test temperature.

Although the physical properties of tungsten and rhenium are quite similar, the latter appears to exert a much more significant effect on the creep of tantalum alloys than does an equivalent amount of tungsten. Buckman and Goodspeed (28) have shown that the creep behavior of complex tantalum alloys is sensitive to minor rhenium additions as shown in Figure 6. The reason for this is not clear although it could be postulated from current theory (27) that the effect is due to an electronic contribution since little difference in elastic moduli and melting temperature exist between tungsten and rhenium. It is apparent that the improvement from rhenium is realized after only dilute additions (<1.5 a/o) and that further increases are

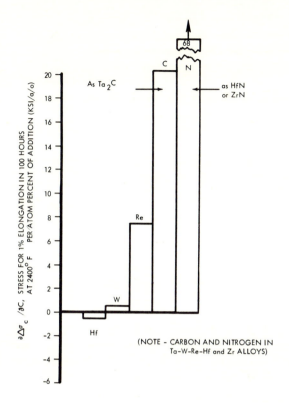

FIGURE 5 - Creep Strengthening Indices for Various Solutes in Tantalum Base Alloys

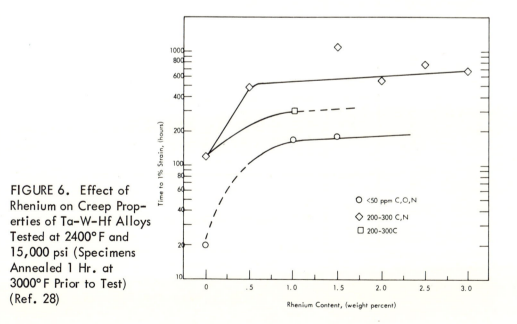

FIGURE 6. Effect of Rhenium on Creep Properties of Ta-W-Hf Alloys Tested at 2400° F and 15,000 psi (Specimens Annealed 1 Hr. at 3000° F Prior to Test) (Ref. 28)

TABLE 1 – Sources for Creep Data from Which Creep Strengthening Parameter was Determined

Element	Alloy Composition (w/o)	Sheet Thickness	Condition	Test Pressure (torr)	Reference
Ta	Unalloyed Ta	0.04"	Recrystallized	$<5 \times 10^{-5}$ torr	19
Ta	Unalloyed Ta	0.06"	As–Worked	$<2 \times 10^{-7}$ torr	20
W	Ta–10W	0.06"	As–Worked	$<2 \times 10^{-7}$ torr	20
W	Ta–10W	0.03"	Recrystallized	$<1 \times 10^{-8}$ torr	21
W	Ta–10W	0.04"	Recrystallized	$<1 \times 10^{-8}$ torr	22
Hf	Ta–8W–2Hf	0.06"	As–Worked	$<2 \times 10^{-7}$ torr	20
Hf	Ta–8W–2Hf	0.03"	Recrystallized	$<1 \times 10^{-8}$ torr	21
Hf	Ta–8W–2Hf	0.03"	Recrystallized	$<1 \times 10^{-8}$ torr	23
Re	Ta–8W–1Re–1Hf	0.04"	Recrystallized	$<1 \times 10^{-8}$ torr	24
C	Ta–8W–1Re–1Hf–0.025C	0.04"	Recrystallized	$<1 \times 10^{-8}$ torr	24
C	Ta–9W–1Hf–0.03C	0.04"	Recrystallized	$<1 \times 10^{-8}$ torr	25
N	Ta–7.1W–1.56Re–0.26Zr–0.02N	0.04"	Recrystallized	$<1 \times 10^{-8}$ torr	13

ineffective, at least under the test conditions which were investigated.

While short time strengthening by interstitial carbide and nitride compounds becomes less effective above 2400°F, their effect on creep properties at 2400°F is significant. In the creep resistant tantalum base alloys containing >200 ppm carbon and less than 1-1/2% hafnium, typical of the ASTAR-811C composition (Ta-8W-1Re-1Hf-0.025C), the phase relationships as proposed by Ammon and Harrod (11) and Buckman and Goodspeed (8) indicate only the presence of the dimetal carbide (Ta_2C) as the dispersed second phase. Although the strengthening parameter is expressed in units of "per atom percent addition", it is assumed that the carbon is present in the metal matrix as a dimetal carbide. Likewise it is assumed that the nitrogen is present in the alloy matrix as a reactive metal mononitride precipitate. Although the value of $\partial \Delta \sigma_c / \partial C$ for nitrogen may be somewhat high since it was determined by extrapolation, Buckman and Goodspeed (28) have shown that reactive metal nitrides are much more potent creep strengtheners than carbides. This is believed to be due to the fact that ZrN and HfN both form a coherent precipitate with the tantalum alloy matrix. Also the nitride precipitation kinetics are slow (28). Thus during a long time creep test precipitation could occur on dislocations thereby more effectively hindering further movement. This would coincide with the stress induced precipitation mechanism proposed by Chang (29).

Although no evidence of a coherent stage has been reported for precipitation of the dimetal carbide, the creep strength of carbide strengthened tantalum base alloys can be significantly altered by thermal treatment. This treatment which consists of a high temperature solution anneal results in a significant change in carbide morphology as well as grain size (28). The use of thermal treatments to control the final creep properties gives another degree of freedom to the designer since the strength level of carbide strengthened tantalum alloys can be controlled other than by alloy additions which must be restricted to a level which does not impair fabricability and weldability.

CURRENT STATUS

Tantalum base alloys designated commercially available by refractory metal alloy producers are listed in Table 2 (30,31). They are listed in chronological order of their development and include Ta-10W, T-111 (Ta-8W-2Hf) and T-222 (Ta-10W-2.5Hf-0.01C). These alloys have been made and are available in all the common mill shapes such as bar, sheet, strip, foil, and tubing. The remaining two alloys 473 (Ta-7W-3Re) and ASTAR-811C (Ta-8W-1Re-1Hf-0.025C) are the two most promising compositions in the advanced development stage and will probably reach commercial status within the next 5-10 years. With the exception of ASTAR-811C, these tantalum alloy compositions were developed on the basis of achieving the highest elevated temperature tensile strength without sacrificing fabricability and weldability (5,7, 12, 32,33) (See Figure 7 and Table 3). This combination of properties was of course dictated by the initial requirements of aerospace applications. Elevated temperature strengthening in Ta-10W, T-111, and 473 is achieved by solid solution additions.

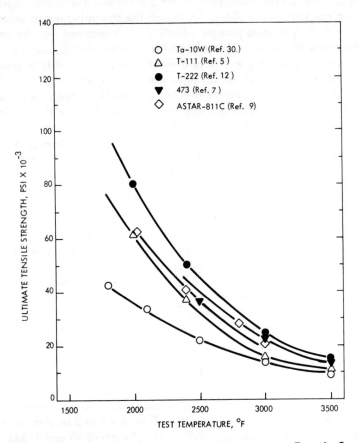

FIGURE 7 - Comparison of the Elevated Temperature Tensile Strength
of Tantalum Base Alloys

TABLE 2 – Current Status of Tantalum Base Alloys

Tantalum Base Alloys		
Composition, Weight Percent	Designation	Status
Ta–10W	Ta–10W	Commercial
Ta–8W–2Hf	T–111	Commercial
Ta–10W–2.5Hf–0.01C	T–222	Commercial
Ta–7W–3Re	473	Developmental
Ta–8W–1Re–1Hf–0.025C	ASTAR–811C	Developmental

TABLE 3 – Room Temperature Properties and Cryogenic Bend Ductility of Tantalum Base Alloys
(5,7,9,11,12,28,34)

Property	Alloy				
	Ta–10W	T–111	T–222	473	ASTAR–811C
Room Temperature* Tensile Strength Yield Strength Elongation	90,000 65,000 30	90,000 75,000 30	120,000 110,000 25	111,000 93,000 30	105,000 85,000 27
Bend** Transition Temperature (a) Recrystallized* Case Metal	$<-320^\circ$F	$<-320^\circ$F	$<-320^\circ$F	$<-320^\circ$F	$<-320^\circ$F
(b) As–TIG Welded	<-320 to -275°F	<-320 to -200°F	-320 to -100°F	$<$R. T. ***	-250 to -175°F

* Recrystallized at 2000°F–3000°F prior to testing.

** 1t bend factor.

*** 2t bend factor.

The hafnium addition used in T-111 was added to react with residual interstitial elements and is also necessary to confer corrosion resistance to alkali metals (34,36). T-222 exhibits the highest tensile strength over the temperature range of 2000-3500°F. This strength advantage is attributed not only to the increased substitutional solute level but also to dispersed phase strengthening imparted by the carbon addition (11).

ASTAR-811C was developed specifically with respect to time dependent deformation properties which are of vital concern in the operation of space nuclear power systems (37).   Thus the guidelines used in its development represent a departure in refractory metal alloy development.   By utilizing a combination of solid solution and dispersed phase strengthening, good creep properties were achieved without sacrifice of ductility or weldability.   The creep strength advantage of ASTAR-811C is evident from the following data on sheet recrystallized 1 hour at 3000°F prior to testing at 1 x 10^{-8} torr (8,9).   Referring to the tensile data in Figure 7 it is apparent that the improved creep properties of ASTAR-811C were obtained while still maintaining respectable short time tensile properties though somewhat lower than those of T-222.   This data serves to point out the danger associated with using short time properties for predicting long time behavior.

Alloy	Time to Elongate 1% at 2400°F and 15,000 psi
ASTAR-811C	260 Hours
T-222	80 Hours
T-111	20 Hours

POTENTIAL STRENGTH LIMITS

The discussion thus far has been restricted to the properties of the currently available and promising developmental tantalum alloy compositions. Let us now refer to Figures 8 and 9 where short time tensile and creep properties of tantalum base alloys are compared with those of the promising columbium, molybdenum, and tungsten base alloys. All strength values have been density compensated to provide a uniform basis for comparison. From short time tensile strength considerations (Figure 8) tantalum alloys do not show an advantage over columbium and molybdenum alloys until above 3000°F. The potential strength limit shown on Figure 8 was proposed by Schmidt et al (3) and represents an alloy composition which still exhibits room temperature base metal ductility. This point cannot be overemphasized for if the solute level is such that it results in raising the ductile-brittle transition temperature above room temperature, the advantage of tantalum as an alloy base is lost. Turning to Figure 9 where time-dependent deformation is the criterion, it becomes obvious that tantalum base alloys have a distinct advantage over both columbium and molybdenum base alloys at temperatures above 2200°F. The potential strength limit was estimated for an alloy composition which would exhibit base metal ductility at room temperature. It would be of a type similar to ASTAR-811C but would contain a total substitutional solute level on the order of 16 to 20 atom percent. It is most

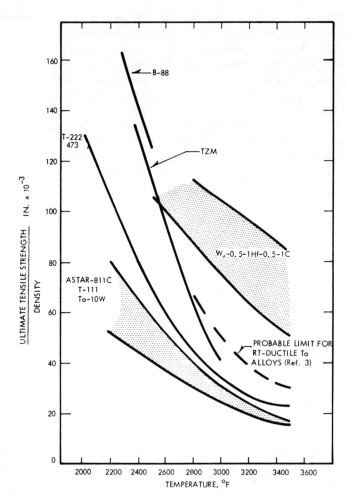

FIGURE 8 – Ultimate Strength/Density Plot for Refractory Metal Alloys
(Ref. 5, 7, 9, 12, 30, 38–41)

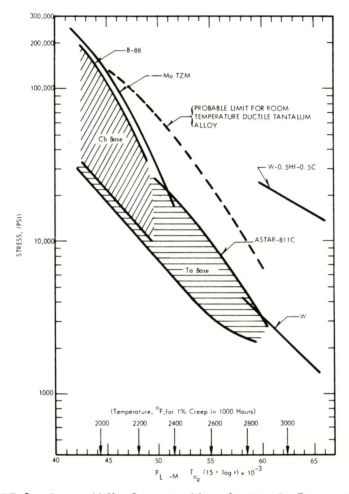

FIGURE 9 – Larson–Miller Parameter Normalization for Time to 1%
Creep for Refractory Metal Alloys (Ref. 9, 20, 21, 23, 39, 40)

likely that tantalum alloys of this solute level would find application primarily as bar and/or forging since sheet fabrication will most likely be difficult. As is obvious from both Figures 8 and 9 the tungsten alloys are the strongest refractory alloys. However, they are also exceedingly difficult to fabricate since working temperatures are in excess of 3500°F (38).

SUMMARY

Although tantalum has a high density, its unique low temperature ductility has made it an attractive alloy base. The retention of this feature must be the primary consideration in the development of tantalum base alloys. By proper control and selection of alloy additions, alloy compositions utilizing both solid solution and dispersed phase strengthening and having optimum combinations of strength and fabricability have been developed. Although much higher strength molybdenum and tungsten alloys are available tantalum alloys have the distinct advantage in that, with current refractory metal alloy technology, they can be fabricated readily into all mill shapes, formed into complex parts at room temperature, and joined by fusion welding into reliable hardware. Thus no matter how much strength is gained by alloying, if the resulting composition cannot be produced into usable and reliable hardware, it will remain as nothing more than a laboratory exercise.

REFERENCES

1. F. F. Schmidt, "Tantalum and Tantalum Alloys", DMIC Report 133, July 25, 1960.

2. F. F. Schmidt and H. R. Ogden, "The Engineering Properties of Tantalum and Tantalum Alloys", DMIC Report 189, September 13, 1963.

3. F. F. Schmidt, E. S. Bartlett, and H. R. Ogden, "Investigation of Tantalum and Its Alloys", Technical Documentary Report No. ASD-TDR-62-594, Part II, May, 1963.

4. A. L. Feild, Jr., R. L. Ammon, A. I. Lewis, and L. S. Richardson, "Fabrication and Properties of Tantalum Base Alloys", Metallurgical Society Conferences, Volume 18, High Temperature Materials II (Edited by G. M. Ault, W. F. Barclay, and H. P. Munger) pp 139-160.

5. R. L. Ammon and R. T. Begley, "Pilot Production and Evaluation of Tantalum Alloy Sheet", Summary Phase Report, WANL-PR-M-004, June 15, 1963.

6. L. L. Seigle, "Structural Considerations in Developing Refractory Metal Alloys", The Science and Technology of Selected Refractory Metals (Edited by N. E. Promisel), AGARD Conf. on Refractory Metals Held in Oslo, Norway, June 23-26, 1963, pp 63-93.

7. L. H. Amra and G. D. Oxx, Jr., "Tantalum Base Alloys with High Strength Above 3000°F", Metallurgical Society Conferences, Volume 34, High Temperature Refractory Metals, Part II (Edited by R. W. Fountain, J. Maltz, L. S. Richardson).

8. R. W. Buckman, Jr. and R. C. Goodspeed, "Development of Dispersion Strengthened Tantalum Base Alloy", Seventh Quarterly Report, WANL-PR-(Q)-008, NASA-CR-54894.

9. R. W. Buckman, Jr. and R. C. Goodspeed, "Development of Dispersion Strengthened Tantalum Base Alloy", Twelfth Quarterly Report, WANL-PR-(Q)-013, NASA-CR-72316.

10. R. A. Perkins and J. L. Lytton, "Effect of Processing Variables on the Structure and Properties of Refractory Metals", Technical Report No. AFML-TR-65-234, Part I, July, 1965.

11. R. L. Ammon and D. L. Harrod, "Strengthening Effects in Ta-W-Hf Alloys", Presented at AIME Symposium on the Physical Metallurgy of Refractory Metals, French Lick, Indiana, October 3-5, 1965, To be Published.

12.   R. L. Ammon and R. T. Begley, "Pilot Production and Evaluation of Tantalum
       Alloy Sheet", Summary Phase Report, Part II, WANL-PR-M-009, July 1, 1964.

13.   R. W. Buckman, Jr. , "Development of Dispersion Strengthened Tantalum Base
       Alloy", Fourth Quarterly Report, WANL-PR-(Q)-005, NASA-CR-54288.

14.   R. W. Buckman, Jr. and R. T. Begley, "Development of Dispersion Strengthened
       Tantalum Base Alloy", Third Quarterly Report, WANL-PR-(Q)-003, NASA-
       CR-54105.

15.   H. Conrad, "Experimental Evaluation of Creep and Stress Rupture", Mechanical
       Behavior of Materials at Elevated Temperatures, (Edited by J. E. Dorn) McGraw-
       Hill Book Co. , 1961, pp 168-169.

16.   H. Inouye, "The Contamination of Refractory Metals in Vacua Below 10^{-6}
       Torr", Refractory Metals and Alloys III: Applied Aspects – Part II, (Edited by
       R. I. Jaffee) Metallurgical Society Conferences, Vol. 30, pp 871-883.

17.   R. W. Buckman, Jr. , "Effect of Test Environment on the Mechanical Properties
       of Refractory Metal Alloys" Presented at 1967 Vacuum Metallurgy Conference,
       Barbizon Plaza Hotel, New York, N. Y. , June 13-15, 1967, To be Published.

18.   R. M. Bonesteel, J. L. Lytton, D. J. Rowcliffe, and T. E. Tietz, "Recovery and
       Internal Oxidation of Columbium and Columbium Alloys", AFML-TR-66-253,
       August, 1966.

19.   F. F. Schmidt et al, "Investigation of the Properties of Tantalum and Its Alloys",
       Battelle Memorial Institute, WADD-TR-59-13 (December 31, 1959).

20.   R. L. Stephenson, "Creep-Rupture Properties of Unalloyed Tantalum, Ta-10%W
       and T-111 Alloys", Oak Ridge National Laboratory, ORNL-TM-1994,
       December, 1967.

21.   R. H. Titran and R. W. Hall, "Ultra High Vacuum Creep Behavior of Columbium
       and Tantalum Alloys at 2000°F and 2200°F for Times Greater than 1000 Hours",
       NASA Technical Note, NASA-TND-3222, January, 1966.

22.   R. W. Buckman, Jr. , Unpublished Data. Ⓦ Astronuclear Laboratory.

23.   J. C. Sawyer and E. A. Steigerwald, "Generation of Long Time Creep Data of
       Refractory Alloys at Elevated Temperatures", Prepared for NASA by TRW, Inc.
       under Contract NAS 3-2545, ER-7203.

24.   R. W. Buckman, Jr. and R. C. Goodspeed, "Development of Dispersion
       Strengthened Tantalum Base Alloy", Thirteenth Quarterly Report, WANL-PR-
       (Q)-014, NASA-CR-72306.

25.   R. W. Buckman, Jr. , "Development of Dispersion Strengthened Tantalum Base Alloy", Fifth Quarterly Report, WANL-PR-(Q)-006, NASA-CR-54462.

26.   G. D. McAdam, "Substitutional Niobium Alloys of High Creep Strength", J. Inst. of Metals, 93, 1964-65, pp 559-564.

27.   O. D. Sherby, "Factors Affecting the High Temperature Strength of Polycrystalline Solids, "Acta Met, 10, 1962, pp 135-147.

28.   R. W. Buckman,Jr. and R. C. Goodspeed, "Development of Dispersion Strengthened Tantalum Base Alloy", Tenth Quarterly Report, WANL-PR-(Q)-011, NASA-CR-72093.

29.   W. H. Chang,"Strengthening of Refractory Metals", Refractory Metals and Alloys, AIME Met. Soc. Conference, Volume 11, 1961, pp 83-117.

30.   Anon, Columbium, Tantalum and Tungsten Alloys Technical Information, Vol. 3, Wah Chang, Albany, A Teledyne Co. , Albany, Oregon, 1968.

31.   Anon, Fansteel Product Bulletins.

32.   M. L. Torti,"Development of Tantalum-Tungsten Alloys for High Performance Propulsion System Components", Second Quarterly Report, Contract NOrd-18787.

33.   M. L. Torti,"Physical Properties and Fabrication Techniques for the Tantalum 10% Tungsten Alloy", High Temperature Materials II, Met. Soc. Conf. 18, pp 161-169.

34.   G. G. Lessmann and D. R. Stoner, "Welding Refractory Metal Alloys for Space System Applications", WANL-SP-009, November, 1965, Presented at 9th National SAMPE Symposium on Joining Materials for Aerospace Systems, November 15-17, 1965, Dayton, Ohio.

35.   R. T. Begley and R. W. Buckman, Jr. , "Tantalum Base Alloys for Space Nuclear Power Systems", Presented at ASME Winter Annual Meeting, Pittsburgh, Pa. , November 12-17, 1967. To be published.

36.   J. R. Stefano and E. E. Hoffman, "Corrosion Mechanisms in Refractory Metal-Alkali Metal Systems, The Science and Technology of Tungsten, Tantalum, Molybdenum, Niobium and Their Alloys, AGARD Conference on Refractory Metals held at Oslo University Centre, Oslo-Blinderns, Norway, June 23-26, 1963, Edited by N. E. Promisel.

37.   Anon, Space Power Systems Advanced Technology Conference, pp 169-200, NASA-SP-131, Lewis Research Center, Cleveland, Ohio, August 23-24, 1966.

38.   P. L. Raffo and W. D. Klopp, "Mechanical Properties of Solid Solution and
       Carbide Strengthened Arc Melted Tungsten Alloys", NASA Technical Note
       NASA-TN-D-3248, February, 1966.

39.   L. S. Rubenstein, "Effects of Composition and Heat Treatment on High Tempera-
       ture Strength of Arc Melted Tungsten-Hafnium-Carbon Alloys", NASA Technical
       Note, NASA-TND-4379.

40.   R. T. Begley, J. A. Cornie, and R. C. Goodspeed, "Development of Columbium
       Base Alloys", AFML-TR-67-116, November, 1967.

41.   F. F. Schmidt and H. R. Ogden, "The Engineering Properties of Molybdenum
       and Molybdenum Base Alloys", DMIC Report 190, September 20, 1963.

ADVANCED PROCESSING TECHNOLOGY AND HIGH
TEMPERATURE MECHANICAL PROPERTIES OF
TUNGSTEN BASE ALLOYS

A Review

H. G. Sell

Abstract

Advances made in the manufacturing of tungsten base
alloys by powder metallurgy, fusion and chemical vapor
deposition techniques are reviewed. The mechanical
properties of many alloys are discussed in terms of
the effect of alloying on the ductile-brittle transition temperature and on the high temperature tensile
and creep strength.

H. G. Sell is Manager of the Metals Research Section,
Westinghouse Lamp Division, Bloomfield, N. J.

Introduction

The total consumption of tungsten in the United States during 1966 was about 7200 tons. This quantity was employed to produce cemented tungsten carbide drill bits, a wide variety of steel alloys, heavy metal, incandescent lamps, dyes and paints, fluorescent phosphors, and many other products (1). In the same year, the country consumed approximately 170 million tons of steel, 3.8 million tons of aluminum, and 2.2 million tons of copper. Thus, on a volume of metal basis, tungsten represents a very small fraction of the other 3 metals cited; yet its importance in our society is vastly greater than is reflected in these statistics, for in many applications, no other material can take its place.

As a result of tungsten's unique combination of properties, it is not surprising that the metal is being investigated at a sustained high rate and that much effort is being devoted towards the development of useful tungsten base alloys.

This paper will concern itself primarily with the current status of the various tungsten alloy development efforts as they have been reported in the open literature. It will compare manufacturing methods employed for the alloys with those used for pure (unalloyed) tungsten and also the mechanical properties of the alloys with those of the pure metal. Furthermore, the paper will emphasize achievements in terms of specific objectives such as low temperature ductility or high temperature tensile and creep strength, although no attempt has been made to provide a complete bibliography. Graphs presented will show salient features of behavior in a somewhat abbreviated form, and the reader is referred to the pertinent references for specific or supplemental information. The paper will close with a discussion of the current survey and an outline of fruitful areas for further development work.

Since most of the pure tungsten and the tungsten base alloys are still produced by powder metallurgy techniques, this presentation is oriented towards powder metallurgy processing but does not include chemical extraction and refining methods.

Production of Tungsten Powders

In the most commonly used process (Fig. 1), pure tungsten metal powder is produced by reduction of tungsten oxide (WO_3) in hydrogen (2). Refinements of this process have been primarily concerned with improving the purity of the metal powder (3) or with controlling the particle size. Among various new approaches of powder production, two techniques—electrowinning and vapor phase reduction—have been shown to be feasible in recent years.

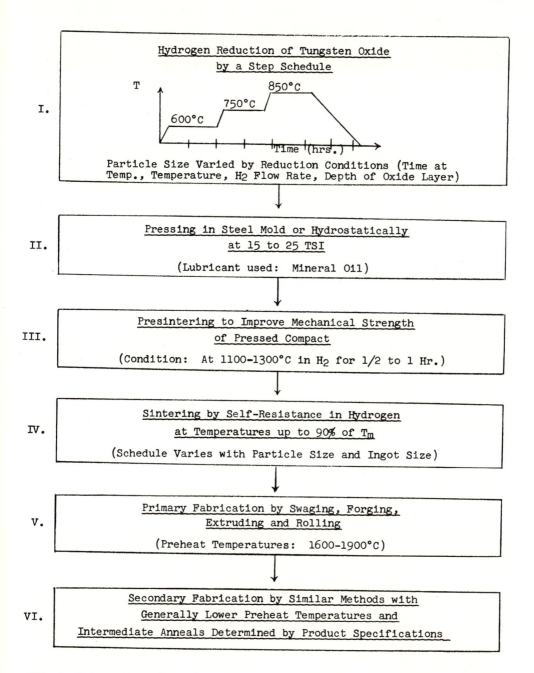

Fig. 1. Schematic outline of original powder metallurgy process for tungsten.

Electrowinning. In this electrochemical reduction process, powder can be produced directly from the ore (4, 5),
although WO_3 or ammonium paratungstate is preferred as a
starting material. The optimum electrolyte is a composition
of $Na_4P_2O_7$ (7 parts), NaCl (2 parts), and $Na_2B_4O_7$ (1 part)
which can be used indefinitely with WO_3 as the feed. For
best results, the electrolytic cell which is composed of a
graphite crucible (the anode) and a graphite rod (the cath-
ode) must be operated at approximately 1000°C. The electrowon "powder" is deposited in the form of acicular den-
drites (6). This powder is compared with powders obtained
by other techniques in Fig. 2. The particle size distribution, which cannot be readily compared with the distribution
of powders from other methods because of the odd shape of
the particles, can be controlled by varying the current density (5-300 amps/cm^2). The purity of the electrowon tungsten
powder is, in general, somewhat lower than that of powders
obtained by the standard powder metallurgical or the vapor
deposition processes*. Electrowinning of tungsten has
progressed on a relatively small scale and has not yet become a commercial process because it cannot compete economically with the standard process.

Vapor Phase Reduction. Significantly more effort has
been expended to produce ultrafine (~ 0.1μ) tungsten powders and larger tungsten granules (~100μ) by vapor phase
reactions. For the production of ultrafine powders, WCl_6
is reacted with H_2 to form tungsten powder. Lamprey et al.
(7) and Tress et al. (8) have reduced WCl_6 with H_2 at temperatures from ~500 to 900°C (Fig. 3). The resulting tungsten particles are about 0.05μ in size and as small as .01μ.
Neuenschwander (9) used the technique of reacting WCl_6 in a
hydrogen plasma at ~3000°C (Fig. 3). The resulting particles also range in size from 0.01μ to 0.06μ (Fig. 2).

Another technique of producing submicroscopic tungsten
powder is by arc vaporization (10). In this process, a tungsten anode is heated by an intense DC arc to ~7000°C. The
anode vaporizes, and, upon cooling, the vapors condense to
prismatic particles 0.01μ to 0.1μ in size.

---

* Impurity analyses will be sparingly given in the text, and
  no comparisons will be made in tabular form. Although it
  is recognized that impurities have significant effects on
  properties, attributing these effects to specific impurities and their concentrations fall too far into the realm
  of the speculative to be attempted in this paper. Also,
  these analyses are subject to change as different analytical techniques are used and improvements in a process are
  being made.

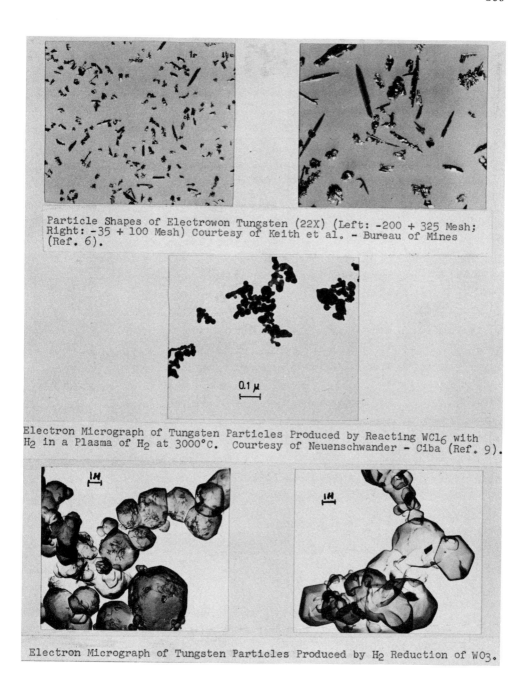

Particle Shapes of Electrowon Tungsten (22X) (Left: -200 + 325 Mesh;
Right: -35 + 100 Mesh) Courtesy of Keith et al. - Bureau of Mines
(Ref. 6).

0.1 μ

Electron Micrograph of Tungsten Particles Produced by Reacting WCl_6 with
H_2 in a Plasma of H_2 at 3000°C. Courtesy of Neuenschwander - Ciba (Ref. 9).

Electron Micrograph of Tungsten Particles Produced by H_2 Reduction of WO_3.

Fig. 2. Shapes and sizes of unalloyed tungsten powders produced by various
techniques.

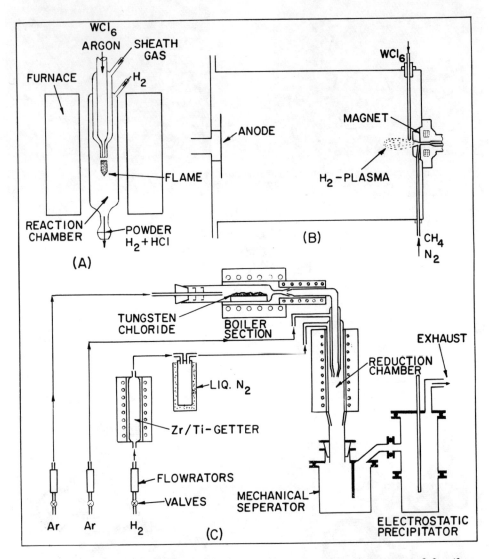

Fig. 3. Schematic drawing of various WCl_6 reduction systems used for the production of submicron size tungsten powders [A: Lamprey et al. (7), B: Neuenschwander (9) and C: Tress et al. (8)].

All vapor phase reduction techniques yielding submicron size particles are beset by the same problems: (1) Residual contamination by oxygen, chlorine and carbon, (2) the necessity for employing special collection techniques, e.g. electrostatic precipitation to achieve high collection efficiency, and (3) pyrophoricity. The oxygen and chlorine contamination can be reduced by heating at 900°C in H_2, but the carbon content is not affected by this treatment. The sporadic occurrence of pyrophoricity is not yet fully understood; although in general, these very fine powders can be handled with some care and powders with an oxygen level $\sim 1\%$ do not exhibit pyrophoricity.

On the other extreme of the particle size scale, tungsten granules have been produced ranging in sizes from 100μ to 10000μ (11, 12) by reduction of WF_6 (instead of WCl_6) with H_2 in a fluidized bed of particles (seeds) from tungsten powder of approximately 3μ Fisher Sub-Sieve Size. Material produced in this fashion has also found no commercial application although it has some interesting mechanical properties differing from those of standard tungsten.

Submicron size tungsten powders have been produced either as "another material" to extend the efficacy of a given "process", or with the objective of supplying powders which can be sintered more rapidly and for the production of finely dispersed second phase alloys by a powder metallurgy approach. A subsequent section will deal with some limitations of powders prepared in this fashion.

Preparation of Tungsten and Tungsten Alloy Powders for Consolidation

The preparation of powders, i.e. their treatment prior to consolidation, is required to facilitate purification during sintering in the case of unalloyed tungsten and to assure homogenization of a solute element or the retention of a desired dispersed second phase of hard particles in the case of alloys. Of course, it is equally important that the consolidated ingot be fabricable.

The criteria discussed in the following section are common to other types of powders as well as to tungsten powders. They will be presented in the order: (I) Powders for Pure Tungsten Compacts, (II) Powders for Dilute Solid Solution Alloy Compacts, and (III) Powders for Dispersed Second Phase Alloy Compacts.

I. Powders for Pure Tungsten Compacts. The most common problems encountered in the standard process (Fig. 1) are poor pressability, contamination by deleterious impurities, low sintered density, or heterogeneous grain growth. Remedial action must often be taken without firm knowledge of their causes.

There is little information available as to what controls the cold pressability of pure tungsten powders. Poster

(13) has shown that the particle morphology controls the
strength of a green compact; the more irregular the particles are, the greater is the degree of interlocking, and
thus, the stronger is the compact. He attributes the lack
of bonding of regular tungsten particles to their inherent
brittleness. He also showed that blends of 80% coarse (25µ)
to 20% fine (1.5µ) particles yielded the highest density.
Lewis et al. (14) found that oxide films affect the com-
pactibility of tungsten powders subjected to rolling. Lewis
and co-workers report the following results for powder pre-
treated (reduced) at 800°C in flowing H_2 and powder of the
same particle size distribution exposed to air at room temperature for 8 months (supposedly the powder was heavily oxidized).

Reduced Powder

Height of Powder in Roll Bite	Roll Load (Pounds)	Strip Thickness
3/4 inch	40,000	0.031 inch
1-1/4 inches	62,000	0.041 inch
2 inches	70,000	0.044 inch

Powder Stored in Air

1 inch	22,000	0.023 inch
2 inches	42,000	0.024 inch

To improve the purity of powders, one can resort to
acid washing techniques. Washing freshly reduced powders
sequentially in HF, HCl and deionized water has been found
to be very effective in removing metallic impurities (1, 15).
        Simple blending is also applied generally for the
purpose of achieving specific particle size distributions.
In practice, an empirically determined suitable particle
size distribution is reproduced by superimposition, i.e.
calculated or graphically determined weight fractions of
powders with known distributions are blended together. In
addition, particle size distributions are controlled by reduction conditions (Fig. 1).
        II. Powders for Solid Solution Alloys. The important
concern in these alloys is homogeneity of composition. If
the alloys are consolidated by arc or electron beam casting
instead of sintering, the elemental powders (usually specified to be of high purity) are blended and then hydrostatically compacted into a consumable electrode which is subsequently sintered (at least partially). Homogenization is
primarily left to the melting process, although the sintering employed may contribute to it.
        On the other hand, if the alloy is to be consoli-
dated by sintering at high temperatures, in which case

homogeneity is dependent on solid state diffusion processes, or by several other more advanced consolidation methods described below, ball milling with or without a grinding aid (16) is often employed subsequent to blending (17). The purpose is to achieve a smaller particle size and intimate mixing to minimize the diffusion distance required for homogenization. However, comminution by ball milling has limitations. For instance, in blends of powders individual constituents do not comminute to the same degree. Furthermore, the degree of comminution possible is a function of the specific grinding aid employed.

For consolidation methods such as hot compaction, gas pressure bonding and direct powder extrusion which are carried out at about .5 T_m of tungsten, pre-alloying of powders has been attempted. Maykuth et al. (18) co-reduced (in dry H_2) blends of ammonium paratungstate and ammonium perrhenate corresponding to W-5 w/o Re. X-ray analysis showed the major constituent to be unalloyed tungsten, but no unalloyed Re was detected; instead, a partially pre-alloyed powder (Chi phase - Re_3W) was found. These authors (18), in a modified approach, also tried to pre-alloy W-Re powders by re-reducing a blend of tungsten metal powder and ammonium perrhenate; however, a rhenium segregation still persisted.

A novel method of pre-alloying W-Re and W-Mo powders has been described by Smiley et al. (19). This method involves co-reduction of hexafluorides WF_6 and ReF_6 (or MoF_6) in a hydrogen-fluorine flame. Flame reduced alloy powders are of submicroscopic size (average size 340Å), but some size variation can be obtained by changing the gas feed system. Smiley and co-workers (19) have prepared W-17.3 w/o Mo and W-25 w/o Re powders which exhibited a single X-ray diffraction pattern. However, the powders are contaminated with fluorine (300-500 ppm) and to some extent with carbon. Treatment in hydrogen reduces the fluorine content to significantly lower levels. The carbon contamination depends on the purity of the raw materials and pickup in the reaction equipment.

III. <u>Powders for Dispersed Second Phase Alloys</u>. The experimental approaches in this area have been guided by concepts of dispersion strengthening. The aim is a submicron size dispersion of (hard) particles with an interparticle spacing of approximately 0.1μ.

From an alloy powder preparation point of view, basically 3 methods are employed to produce blends of submicron size powders: (1) Co-precipitation or co-reduction, i.e. by chemical methods, (2) dry or wet blending of submicron size elemental powders, and (3) comminution of alloy powder blends. All these methods have been tried singly or in combination. In their original work on W-ThO_2 alloys, Atkinson et al. (3) prepared W-ThO_2 powders by co-precipitation of H_2WO_4 and thorium oxalate from solutions of tungstate and thorium ions and by co-reducing powders of WO_3-ThO_2

in H_2. (Th$(NO_3)_4$ solution was slurried with WO_3 and the mixture fired at 800°C). In the first case, the precipitate was fired in air (800°C) and then co-reduced. Maykuth et al. (18) followed the same approach in the preparation of W-Re powders. They added thorium oxalate in an acidified solution to tungsten metal powder; and in a different approach, they added tungsten powder to a colloidal sol of thoria. The powders obtained by these various techniques differ primarily in the size of the thoria particles.

The second method, blending of submicron size powders, has been employed by White (20) for the preparation of a 68 W-20 Ta-12 Mo-5 v/o ThO_2 alloy and by King et al. (21), for a W-1 w/o ThO_2 alloy. In the latter work, submicron size thoria powder was blended into WO_3 powder followed by H_2 reduction.

The third method, milling, has been extensively employed by Morcom et al. (16). These authors ball milled co-reduced W-ThO_2 powders as well as powder blends of W-HfC, W-HfO_2-ZrO_2 and W-HfN-ZrN. A grinding aid of aluminum chloride was used. The effectiveness of milling is demonstrated in Fig. 4 for a W-$ThO2$ powder. A narrow distribution peaking at 0.2μ particle size was obtained.

Consolidation of Tungsten and Tungsten Alloys

Consolidation is considered to be the process of forming a solid compact of high strength and 85-100% density. Except for the chemical vapor deposition method (CVD) which can directly yield tungsten of nearly 100% density, all other methods start with powder. The following consolidation methods will be discussed:

    I.   Die Pressing and Sintering
    II.   Arc or Electron Beam Casting
   III.   Hot Pressing
    IV.   Extrusion
    V.   Chemical Vapor Deposition
   VI.   Slip Casting, Slip Extruding and Sintering

    I. Cold Pressing and Sintering. Powders are either pressed in a steel mold or hydrostatically, and the pressed compact is subsequently sintered in vacuum or in H_2. Particle size distribution, sintering temperatures and times, and impurities such as surface oxides are important factors which affect the rate of consolidation, the final density and the integrity of the ingot. Powders which are difficult to press are often made pressable by pressing aids such as camphor-ether solutions and others described in standard text books.

A new approach which is not commercially employed is activation sintering. In the established method, activators like Ni, Co, Fe, Ru, Pd or Pt in concentrations <1% are used (22, 23). Densities >85% have been obtained in 30 minutes

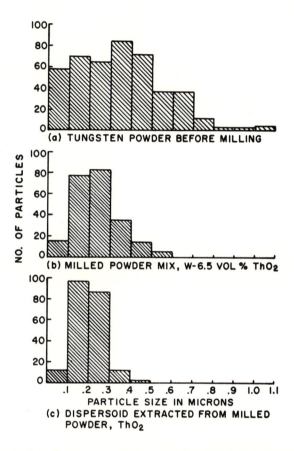

Fig. 4. Effect of milling on particle size distribution of a W-6.5 V/O ThO$_2$ powder.

406 H. G. SELL

by sintering in H_2 at temperatures as low as 1100°C. A new approach has been tried by Toth et al. (24) who have found some form of activation sintering to occur in the presence of bromine in the hydrogen atmosphere. The effect was most pronounced at 1400°C where the rate of densification was increased by 40%.

In a subsequent investigation, Toth et al. (25) again studied sintering activated by Ni and Pd and included in their investigation W-2 w/o ThO_2 powder. They confirmed the earlier results for pure tungsten (23). With W-ThO_2 powder, the sintering rates are slower, but the activators are still very effective. To appreciate the potential usefulness of activation sintering, one must keep in mind that the activated process is most effective in the temperature range 850 to 1200°C where no densification occurs at all in standard H_2 sintering. Morcom et al. (16) who attempted to activate the sintering of W-2 w/o ThO_2 compacts with Ni at 1700°C noted some increase in the final density, but the effect was by far less pronounced than observed by Toth et al. (25). This fact has been confirmed by Pugh et al. (26) who investigated activation sintering of tungsten with manganese ($\sim$5%), sulfur ($\sim$.3%), and yttrium ($\sim$.2%). The greatest effects were also observed in the temperature range 1250°C to 1400°C.

It is apparent that activation sintering of unalloyed tungsten, W-ThO_2 and possibly other alloys is most effective at about 1200°C. However, the application of this method in the consolidation of dispersed second phase alloys is discouraged because the activator retained after sintering tends to embrittle the ingot, not to mention chemical reactions which may occur between the activator and a dispersed second phase. The problem is exemplified by Sutherland's (27) unsuccessful attempts to hot roll (1320°C to 2200°C preheat temperature) tungsten which was sintered with Ni as activator. His materials had the following sintering history:

Activator Concentr.	Sintering Temp., °C	Sintering Time (hrs.)	Sintered Density (grams/cc)
0.04 w/o	1050	8	93
0.14 w/o	1230	1	93
0.14 w/o	1230	8	98

On hot rolling, the activation sintered tungsten cracked on the first rolling pass. Sutherland made tensile tests on 96% dense tungsten which had been sintered with 0.1 w/o Ni and reported the following properties:

Test Temp.	Tensile Strength (PSI)	% Elong.	% R.A.
1325°C	7650	0	0
1540°C	1780	0	0

An additional point to be made is that even though the sintering rate is enhanced at significantly lower temperatures, actual total sintering times are longer than those required in standard high temperature sintering.

II. Vacuum Arc and Electron Beam Casting. Following the initial success achieved in vacuum DC and AC arc melting (28, 29) and in electron beam melting (30) of larger size (2-1/2" diameter x 8" long and larger) tungsten ingots, both techniques were rapidly applied to alloys (Table 1). In pure tungsten, the coarse grain structure and residual porosity are undesirable in regards to primary breakdown. In solid solution alloys, homogeneity of composition is an additional very important factor, i.e. the degree to which coring may occur on freezing. Finally, there is the class of precipitation hardened alloys, primarily by a dispersed second phase of a carbide. In these alloys, one is concerned with intergranular precipitation of the second phase which can cause embrittlement.

As Table 1 shows, both melting techniques are employed, but arc casting appears to be preferred over electron beam melting. A major problem encountered in skull or centrifugal casting is that the fluidity differs amongst alloys. This problem has often prevented the casting of sound billets. A good example is the centrifugal casting of the alloy 68 W-20 Ta-12 Mo (38) where a W-30 Mo alloy could be cast readily but not the W-Ta-Mo alloy. Nevertheless, with respect to the development of solid solution and precipitation hardened alloys, consolidation by either of these techniques has become a well established alternative to consolidation by sintering.

III. Hot Pressing. A major problem in the development of dispersion strengthened alloys which must be consolidated by sintering (because the dispersoids are not generally stable at the melting temperature of tungsten) is agglomeration. Hot compaction, as distinct from high temperature sintering (2900°C), was investigated as a means of consolidating powder compacts at temperatures (1500-2200°C) where agglomeration was not supposed to occur. White (39, 40) was the first to show that billets with greater than 90% density can be hot pressed in a graphite mold. Specifically, he pressed a W-2 w/o ThO_2 billet (2.2" diameter x 2.5" long) at 6000 PSI and 2200°C which was later extruded (41). Morcom et al. (16), using a single acting press, hot pressed many billets (1.2" diameter x 1.5" long) of a variety of dispersed second phase powders to densities >95%, also at about 6000 PSI but some-

Table 1. Some Tungsten Base Alloys Consolidated
by Arc- or Electron-Beam-Melting

Alloys	Ref.	Method of Casting
W-3Re-0.5Hf/0.5Zr	(31)	C.A.M.*
W-3Re-0.5Cb-0.25Hf/0.25Zr	(31)	C.A.M.
W-0.6Cb, 0.6Cb-0.12Zr/0.04Ti	(32)	C.A.M.
W-6Mo-6Cb	(33)	C.A.M.
W-20Ta-12Mo	(31, 33)	C.S.C.**/C.A.M.
W-15Mo	(31)	C.S.C.
W-(1-26%)Re	(34)	C.A.M./E.B.M.***
W-Hf, Ta-Hf, Re-Hf, Cb-Hf, Ta-Re	(35)	C.A.M.
W-C, Cb-C, Ta-C, Ta-Re-C, Hf-C	(35)	C.A.M.
W-Hf-C	(36)	C.A.M.
W-Zr-C/B	(37)	C.A.M.
W-Hf-C/B	(37)	C.A.M.
W-Cb-C	(37)	C.A.M.
W-Cb-Zr/Hf-C	(37)	C.A.M.

  * Consumable Arc
 ** Centrifugal Skull Casting
*** Electron Beam Melting

what lower temperatures (1800°C to 2000°C) than employed by White.

Much more elaborate than hot pressing in graphite molds is isostatic compaction by gas pressure bonding (42). Pre-compacted powders are vacuum sealed in a container (Mo) and placed inside of an autoclave in which they are heated to about 1600°C under a pressure of He at ~10,000 PSI. Ingots from large CVD-tungsten granules (12) and from W-Re-ThO_2 powders (18) have been prepared by this technique.

Hot compaction in graphite is occasionally employed, but the technique has not become a production method. The initial objective of preventing agglomeration has not been achieved, and economically, the method cannot compete with the standard process. Isostatic compaction has remained a laboratory curiosity for tungsten although it is now used extensively in consolidating ceramics and cermets.

IV. <u>Extrusion</u>. Consolidation in one step from a powder to a fully densified ingot is very appealing. The most direct method is some form of high rate compaction. This method has been studied by Reinhardt et al. (43) using, among others, W-ThO_2 powders. The results have not been encouraging because the small pellets could not be produced free of cracks or delaminations. Dynapak extrusion without precompaction of pure tungsten and W-ThO_2 powders encapsulated in a Mo container was attempted by Sell et al. (41). Short lengths of 99% dense rods were obtained, but no further effort was made to develop this technique. More recently, Goodspeed et al. (44) have repeatedly extruded, on a conventional press and on a Dynapak, hydrostatically compacted W-2 w/o ThO_2 billets ($<$ 60% density) into sound sheet and round bars at as low as 1350°C extrusion temperature (Table 2). The billets were encapsulated in Mo. Extrusion of large hydrostatically compacted billets without a liner on a conventional extrusion press has not yet been demonstrated. For the extrusions with a liner, Goodspeed et al. have determined 1750°C as the optimum extrusion temperature for billets having ~60% of theoretical density.

V. <u>Chemical Vapor Deposition (CVD)</u>. Pure tungsten is now routinely produced in tubes, rods and slabs on substrates of W, Mo, Cu or Ni. The preferred chemical reaction is $WF_6(g) + 3H_2(g) \longrightarrow W(s) + 6HF(g)$ (45, 46, 47) instead of the reaction $WCl_6 + 3H_2$ which has also been investigated (48). The latter reaction is more difficult to control because of the higher melting and boiling points of WCl_6 (WF_6: M.P. = 2.5°C, B.P. = 19.5°C; WCl_6: M.P. = 270°C, B.P. = 347°C) and also because of oxychlorides, the formation of which is critically dependent on the temperature distribution in the reaction chamber if O_2 is present in the reaction chamber.

The major shortcomings of CVD-tungsten are its columnar grain structure and contamination with fluorine. Holman et al. (49) achieved a grain refinement by rubbing and brushing the deposit as it forms. Deposition rates as a function

Table 2. Extrusion Parameters and Results of Dynapak and Conventionally Extruded (44) Molybdenum Clad Billets of Hydrostatically Pressed W-3.8 v/o ThO2

Extr. Type (a)	Extr. Temp. (°C)	Reduc. Ratio	Extr. Dimensions (b) (in.) L x W x t	% of Theoretical Density				Comments
				Initial	Final			
					Nose	Center	Tail	
SB,C	1750	6.3/1	10-3/4 x 1.52 x 0.42	59	99.7	99.6	98.6	Molybdenum can split in 2 places near nose
SB,C	1650	5.4/1	13-3/16 x 1.52 x 0.52	59	99.5	98.0	98.0	Excellent surface
SB,C	1650	5.7/1	11-3/8 x 1.47 x 0.50	59	99.4	98.3	98.2	Excellent surface
RB,D	2100	5.5/1	5-7/8 x 0.42 dia.	60	----	99.5	----	Some tearing of clad
RB,D	2100	7.5/1	8-3/4 x 0.39 dia.	58	----	100.0	----	Some tearing of clad
SB,D	2100	4/1	3-1/2 x 1 x 0.25	56	97.4	----	97.9	Some tearing of clad
SB,D	1750	4/1	3-3/4 x 1 x 0.25	59	99.7	----	99.8	Excellent surface
SB,D	1750	4/1	3-3/8 x 1 x 0.25	60	99.6	----	99.7	Excellent surface
SB,D	1600	4/1	2-5/8 x 1 x 0.25	57	99.1	----	99.5	Excellent surface
SB,D	1350	4/1	4-1/8 x 1 x 0.26	57	98.7	----	98.7	Excellent surface

(a) RB = Round Bar; SB = Sheet Bar; D = Dynapak; C = Conventional

(b) After removal of cladding.

of reaction temperature and H_2/WF_6 ratios have also been widely investigated, but authors disagree on the most favorable conditions under which deposited layers of high integrity form. They agree, however, on the deleterious effects of oxychlorides and oxyfluorides. Excessive contamination by fluorine (>20 ppm) causes the formation of gas bubbles at high temperatures which in turn affect the mechanical properties (50, 51).

Tungsten alloys have also been produced by CVD-techniques. Holman et al. (49) prepared various W-Re alloys containing up to 30% Re from premixed WF_6 and ReF_6 liquids. These alloys in concentration of Re $>18\%$ have a finer grain structure and contain a new phase referred to as federite (A-15, W_3Re). Donaldson et al. (52) attempted to co-deposit W-Re alloys from WF_6 and ReF_6 pre-mixed in the gas/vapor phases to different ratios of H_2/MF_6 and separately introduced into the reaction chamber, but non-uniform deposits of heterogeneous composition resulted. In an earlier work, Donaldson and Kenworthy (53) successfully co-deposited a W-Mo alloy. Ternary alloys of W-Mo-Re were deposited by Fairchild (54), but the deposit had a highly non-uniform composition. The preparation of a dispersed second phase alloy by a CVD process has been attempted by Landingham et al. (55). The following reactions were involved:

$$WCl_6 + 3H_2 \longrightarrow W + 6HCl$$

$$2HfCl_4 + H_2 + 2NH_3 \longrightarrow 2HfN + 8HCl$$

While the decomposition of tungsten and hafnium nitride was successfully accomplished, the deposit was composed of layers rich either in W, W_2N, or HfN. However, the tungsten nitride was readily eliminated by heating up to $900^\circ C$.

The technology of chemical vapor deposition of pure tungsten is well advanced, and rods, tubes, and slabs are now prepared with little difficulty. The gradient in structure (fine grained near the substrate, coarse-columnar towards the outside surface) is not desirable but has not posed an insurmountable problem in fabrication. The preparation of alloys by CVD-techniques is still in the early stages of development, but these techniques have the potential of becoming major competitors to the preparation of alloys (especially dispersed second phase alloys) by powder metallurgy processes.

VI. <u>Slip Casting, Slip Extruding and Sintering</u>. The use of slip casting in lieu of pressing has been beset by the problem that it was not possible to produce consistently high density, crack-free, castings or sintered forms. In slip casting, the requirement is for a vehicle which combines the properties of keeping the powder uniformly suspended and of possessing high fluidity for ease of elimination (drainage into the mold). Organic vehicles singly or in combinations have not met these requirements. However, in a new approach, Stoddard et al. (56) have had excellent results with powders coated by milling with oxide or carbon. Oxide coated powders are mixed with H_2O (100 grams of powder to 6-7 milliliters of H_2O) and the slurry cast into a plaster mold. Wall thicknesses as thin as 1/16" have been produced in one minute and the casts were sintered to densities exceeding 95%.

Milner et al. (57) have extruded W-30 Re-30 Mo slip prepared by first blending the elemental metal powders together with a binder of 3% methyl cellulose and then mixing the blend with water and a glycerin plasticizer. They were able to produce 0.3" diameter rods and tubing 0.13" O.D. and 0.08" I.D. Tubing was drawn from this slip up to 68% reduction in area at room temperature and at temperatures from 300°C to 600°C. Using the same technique, Milner (58) also produced W-25% Re rods. Tubes or rods sintered after drawing have exhibited attractive mechanical properties in comparison with wrought alloys of the same composition.

Fabrication

Among the various possible fabrication techniques, extrusion as a mode of primary breakdown of solid billets and tube blanks has been thoroughly investigated. The development of secondary fabrication methods for tungsten alloys, especially rolling of alloy sheet, has been lagging. In contrast, the development of secondary tube fabrication methods of unalloyed tungsten and W-Re alloys has made good progress. Also, an impressive set of fabrication data has been generated for unalloyed tungsten, and this information will be briefly summarized first; then the fabrication of alloys will be discussed with tube fabrication being considered separately at the end of this section.

Unalloyed Tungsten

The fabricability and the formability of unalloyed tungsten have been thoroughly investigated under the MAB's Refractory Metal Sheet Rolling Program (59). The material yield at finished sizes (0.250, 0.100, 0.060, 0.020 and 0.010 in. thickness) was significantly higher with powder metallurgy tungsten than with arc cast tungsten, although the finished size of the powder metallurgy sheet was smaller (18 x 24 inches, versus 24 x 72 inches) because the sintered powder metallurgy ingot was smaller than the arc cast billet (by design of the experiments). The results of the formability investigations are summarized in Table 3. A broad range of forming techniques have been evaluated and were found to be applicable to tungsten sheet with relatively little difficulty.

Table 3. Minimum Preheat Temperatures (°F) Required
to Form Various Gages of Tungsten Sheet
Produced by Powder Metallurgy   (58)

Fabrication Method	Gage Thickness (mils)			
	10	20	30	40
Square Shearing	310	420	900	1200
Rotary Shearing	350	510	950	1250
Blanking	400-700	480-800	700-1000	1050-1500
Brake Forming	400	500	700	1050
Perforating (1/8" I.D.)	200-500	400-600	400-700	700-1100
Perforating (1/2" I.D.)	250-475	450-650	600-900	800-1200
Joggling	350	390	675	850
Corrugating	500*	700**	---	---
Roll Forming	225	400	650	800
Drawing	---	---	1250-1500	1600-1850
Expl. Forming	---	---	----	1470-1650

 * R = 0.11 in.
** R = 0.18 in.

Tungsten Alloys

Fabricability has been and still is the most important
problem with many tungsten base alloys and a rather sharp
separation is often made between primary and secondary fabrication. While this separation may be desirable from the
viewpoint of a fabrication schedule, it has led to the re-
porting of much property data on extruded alloys, especially
arc cast alloys. Since obviously the requirements are for
information on strength properties and ductility of finished
products (sheet), the data on extruded alloy bars are only
of limited value and may even be misleading.

In primary fabrication, extrusion has become the preferred
technique of producing bar stock for subsequent swaging or
rolling. For this purpose, billets are generally canned in
molybdenum. Canning practically eliminates surface tearing,
and the material yield is increased. To improve lubricity,
extrusion blanks have initially been coated with glass. However, more and more experimenters depend on the naturally
forming oxides (MoO_3 or WO_3) for lubrication.

Extrusion temperatures (preheat) and extrusion ratios for
a number of alloys are listed in Table 4. Groups 1 and 2 of
the W-Hf-C alloys were consolidated by arc casting (37, 36)
and the third group by powder metallurgy (60). Both Groups 1
and 2 were extruded clad in Mo, but Group 1 could not be
swaged. Group 2 was swaged and rolled, and Group 3 was also
rolled into sheet. Three differences are noted in the fabrication technique between Group 1 which could not be worked
after extrusion and Group 2 which could: (1) The extrusion
ratio of Group 1 was low (4:1 vs. normally 6 to 8), (2) the
Mo cladding was removed after extrusion, and the bars were
recladded for swaging, and (3) the swaging temperature was
low ($<1650°C$). It is not known to what extent these 3 factors affected the poor fabricability of the Group 1 alloys;
but combined with possible inhomogeneity in composition of
the arc cast billets, they may have meant the difference between failure and success. Another alloy, No. 5 in Table 4,
which had a non-uniform composition, also could not be swaged.

The powder metallurgy alloys (Nos. 6 and 7) have been fabricated into sheet. In the case of alloy No. 6, gas pressure
bonded sheet bars were pack clad in Mo (18) and alloy No. 7
was extruded (Mo clad) into sheet bars and subsequently forged
or rolled (44) without the cladding.

In general, the picture that has evolved is not too encouraging. Among all alloys produced, only alloys containing
Re and/or Mo in solid solution and the W-2 w/o ThO_2 alloy
have yielded sheet of respectable size. In the case of all
other alloys, secondary fabrication was either unsuccessful,
or the size (2-4" wide x 8-16" long) of the sheet was only
sufficient in quantity to determine some mechanical properties.

Table 4.   Primary Fabrication Parameters for
Various Tungsten Base Alloys

Material	Extr. Temp. (°F)	Ratio	Ref.
1. W-(0.27-0.96) Hf[a] -(0.017-0.056) C	4000[b]	4.2:1	(37)
2. W-(0.23-1.76) Hf[a] -(0.21-0.94) C	4000[b]	8:1	(36)
3. W-(0.015-2.00) Hf -(0.01-0.70) C	3270-* 3540	-----	(60)
4. W-25 Re[a]	3650	4:1	(67)
5. W-20 Ta-12 Mo	3750[b]	5.5:1	(33)
6. W-5 Re-22ThO$_2$[c]	2900[b]	-----	(18)
7. W-2 ThO$_2$	2700[b]	6:1	(44)
8. W-27 Re-1 Cb-0.1 C	3750[b]	5:1	(100)

* Forged or Swaged

[a] Arc Melted

[b] Clad in Mo

[c] Powder Met. - Rolled

One major factor which probably contributes to the difficulties encountered is lack of homogeneity. However, this is not well substantiated in many cases. Interstitial pickup may also be a serious problem.

Tubing

The situation is much more encouraging in the development of tubing where, by necessity, extrusion is employed as the primary fabrication method. Two different techniques have evolved: (1) The filled billet technique, and (2) the floating mandrel technique. Schematics of the billet designs are shown in Fig. 5, and some fabrication parameters are compared in Table 5.

Filled Billet. Pure tungsten and W-25% Re tubing was successfully extruded and re-extruded by Isserow et al. (61) using the filled billet approach. Following primary fabrication, tubing of both materials was drawn to 0.25" O.D. x 0.020" wall thickness with a moving mandrel. A similar approach was taken by Burt et al. (62). Besides pure tungsten and W-25% Re, these authors extruded a tungsten sleeve made from CVD-tungsten granules and also a sleeve of the alloy W-30% Mo-30% Re.

Floating Mandrel. The floating mandrel technique has been used by McDonald et al. (63) for the extrusion of larger diameter heavy wall tungsten tubing. This technique requires no cladding of the billet; and in the case of unalloyed tungsten, tube blanks can be extruded at rather low temperatures (supposedly because of self-lubrication by surface oxidation; molten oxide-WO_3 forms at 1473°C). McDonald et al. (63) extruded tungsten blanks 3" O.D. x 5" long with a 1-1/8" bore from 1750°C preheat temperature at a ratio 5.88:1. Blankenship et al. (64) have shown that the floating mandrel technique can also be used to extrude small diameter tungsten tubing. They have produced tungsten tubing 0.5" and 0.375" in diameter with a wall thickness of approximately 0.020" (Table 5) from starting tube blanks ~1.3" O.D. x 0.3" wall thickness which were contained in duplex billets (Fig. 5). These remarkable extrusions were carried out in the temperature range 1750°C to 2200°C and at the high ratio of 17:1.

The floating mandrel technique was recently applied to W-25% Re (in addition to unalloyed tungsten) by McDonald et al. (64). The extrusions were again performed without lubrication. However, in the case of the W-25% Re alloy, a layer of pure tungsten was first chemically vapor deposited onto the extrusion billet because an oxide did not form as readily on the alloy surface as on a pure tungsten surface.

In the work mentioned here, the tube blanks with one exception were prepared by powder metallurgy and not by arc casting. The consensus appears to be that this is the most suitable starting material, in particular, since Blankenship et al. (64) were able to extrude tubing from 65% dense tube

Table 5. Tungsten and Tungsten Alloy Tube Shell Extrusion Data

Material	Unalloyed Tungsten Sintered [a]	W-25 Re [a]	W-30% Mo -30% Re [a]	Unalloyed Tungsten Sintered [b]	Unalloyed Tungsten Wrought [c]	W-25% Re Arc Cast [c]
Method	F.B.*	F.B.*	F.B.*	Float.Mandrel	Float.Mandrel	Float.Mandrel
Jacket	304-SS	Mo	304-SS	Mo	None	
Core	Mo	Mo	304-SS	Mo Tubing	None	
Temp., °C	1150	1400	1100	1750-2200	1650	2100
Ratio	4:1	4:1	7.4:1	17:1	5:1	4.9:1
O.D.(in.)	0.625	0.625	-----	.5 and ~.375	1.621	1.639
Wall(in.)	0.062	0.062	-----	~0.020	~0.32	~0.33

(a) Burt et al. (Ref. 61)

(b) Blankenship et al. (Ref. 63)

(c) McDonald et al. (Ref. 64)

* F.B. - Filled Billet (Fig. 5)

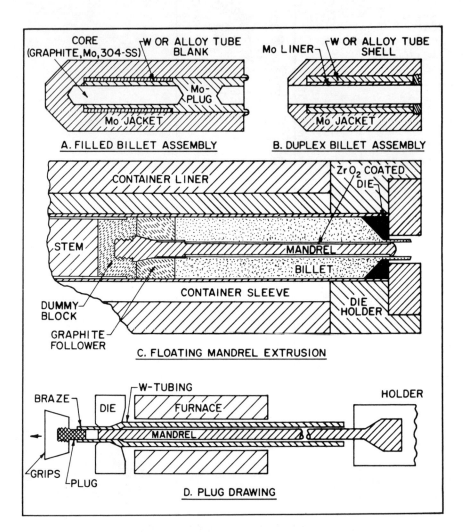

Fig. 5. Schematics of billet assemblies, floating mandrel extrusion and plug drawing techniques employed in fabrication of tungsten and tungsten alloy tubing.

blanks. However, a new non-powder metallurgy method appears
to be as promising for the preparation of tube blanks; namely,
by CVD. Martin et al. (66) have readily extruded tungsten
tube blanks prepared by CVD.

Primary fabrication of tungsten and tungsten alloy tubing
by extrusion and re-extrusion is well developed, and good
progress has been made in the development of secondary fabrication techniques where plug drawing, also shown schematically
in Fig. 5, has been successfully employed for the fabrication
of thin wall small tubing. Concerning the preparation of tube
blanks, the slip extrusion technique developed by Milner et
al. (57) may have an advantage over the conventional preparation of blanks by pressing and sintering.

Mechanical Properties

The evaluation of the mechanical properties presented a
difficult task. Although an effort towards standardization
of test procedures for alloy sheet has been made (68), not
every investigator has strictly adhered to these procedures.
Frequently, the thermal-mechanical history of test specimens
is poorly described and test conditions are insufficiently
specified. However, the data are numerous and general trends
are observed. These will be compared and discussed in the
following sequence:

I. Low Temperature Tensile Properties, II. Elevated Temperature Tensile Properties, and III. Creep Properties.

I. Low Temperature Tensile Properties. The main interest
in the low temperature mechanical properties of tungsten is
its ductile-brittle transition (DBTT), which occurs above room
temperature, and the various factors which either lower or
raise the transition temperature or affect the mode of fracture. Many development programs were undertaken with the objective of improving the low temperature ductility by alloying. To appreciate the extent to which these efforts have
been successful, the behavior of unalloyed tungsten of different processing origin will be compared with that of
alloys for the recrystallized condition.

Unalloyed tungsten as well as tungsten base alloys become increasingly more ductile below the DBTT of the recrystallized metal with increasing amounts of deformation (R.T.
being about the lower limit). The recrystallized condition
is, therefore, the most adverse condition for low temperature
ductility, but it does not define an exact boundary because
factors such as the surface roughness of the test specimen,
the strain rate, purity, and grain size can have a pronounced
influence on the DBTT. These effects have been thoroughly
discussed by Seigle et al. (69). Results subsequently published indicate that neither the effects of impurities nor
the effects of grain size on the mode of fracture or on the
DBTT are unambiguously established.

Unalloyed Tungsten: A summary of the DBTT's reported in the literature for polycrystalline tungsten is presented in Fig. 6. Grain sizes vary from material to material but with few exceptions lie around 50μ, although recrystallization treatments differ markedly. In comparing impurity levels in the various materials, one notes that interstitial impurity levels are on the average by a factor of 2-3 higher in PM-tungsten than in AC, EB-, and CVD-tungsten (83). Metallic impurity levels are about the same in PM- and CVD-tungsten but are by a factor of 5 to 10 lower in AC-tungsten and still less in EB-tungsten.

Orehotsky et al. (70) who have investigated the DBTT of tungsten wires fabricated from rod purified by electron beam floating zone melting found that the DBTT was significantly lowered in this material. They concluded that the DBTT decreases with increasing purity. On the other hand, Witzke et al. (30) have found a higher DBTT in high purity EB-tungsten than in PM-tungsten of comparable grain size.

In comparing the results of Clark (77) and Stephens (78) on decarburized and carburized pure tungsten, respectively, it is noted that the DBTT increases with increasing carbon content essentially independent of grain size (Table 6).

Considering the mode of fracture, Clark found a greater percentage of transgranular fracture in the low carbon containing tungsten (2 ppm) tested in bending (4T), while Stephens observed a greater percentage of transgranular fracture in carbon dosed tungsten (>36 ppm) tested in tension. Unambiguous evidence for an impurity effect was reported by Stephens (78) for oxygen. Dosing with oxygen increased the DBTT and the specimens failed completely by intergranular fracture at temperatures at and above the transition temperature where unalloyed tungsten normally exhibits a mixed mode of fracture.

In regard to the effect of grain size on the DBTT, Seigle et al. (69) and Jaffee et al. (71) have reported that the DBTT increases with increasing grain size, while Gilbert (72) and Raffo et al. (73) observed it to decrease with increasing grain size. Klopp et al. (82) found in triple electron beam melted tungsten that the DBTT was significantly higher than in AC-tungsten with an order of magnitude smaller grain size. On the other hand, Farrell et al. (74) and Koo (75) reported the DBTT to be independent of grain size but dependent on heat treatment (impurity distribution).

These apparently contradictory observations emphasize that the brittleness of tungsten is controlled by a complex mechanism which involves both grain boundaries and impurities. However, if grain size differences are great, grain size overrides the effect of impurity content and distribution (e.g. fine grained PM-tungsten vs. large grain EB-tungsten). On the other hand, if grain size differences are small, heat treatments, i.e. distribution of impurities, affect predominantly the DBTT of unalloyed tungsten.

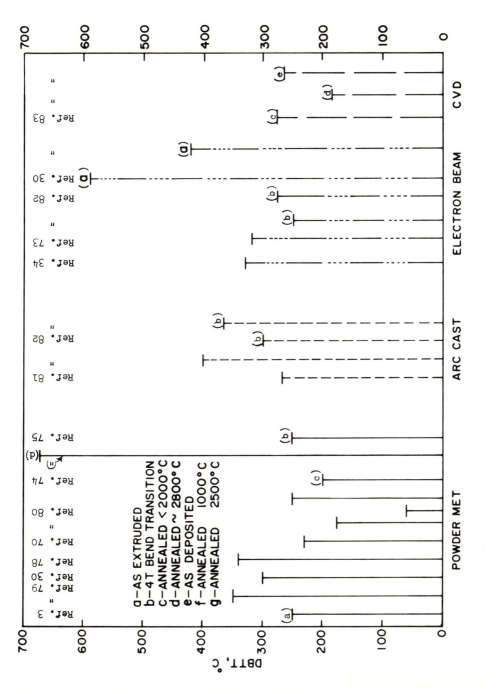

Fig. 6. Ductile–brittle transition temperatures of unalloyed tungsten.

Table 6.   DBTT in Carburized (Stephens - Ref. 78)
   and Decarburized (Clark - Ref. 77) Tungsten

Tensile Tr. Temp.* (78)			4T Bend Tr. Temp. (77)		
C (ppm)	Grain Size (μ)	DBTT °C	C (ppm)	Grain Size (μ)	DBTT °C
8	25	232	2	25	136
36	25	368	7	19	178
45	30	390	17	23	210
60	60	416	41	22	218

* 0.005" min^{-1} to yield; 0.05" min^{-1} to fracture.

Tungsten Alloys: DBTT's for a variety of alloys have been summarized in Fig. 7, where the position of the arrow on a column indicates the lowest DBTT temperature determined. Unalloyed PM-tungsten has been included in this figure for comparison (the dashed part of the column indicates the range of DBTT's reported with the exception of the extreme values (70, 75)).

In all the solid solution alloys investigated by Clark (77) and Raffo et al. (73), the DBTT increases slightly with increasing grain size, but as in unalloyed tungsten, the annealing temperature exerts a greater effect on the DBTT. This is attributed to a redistribution of solutes upon annealing. For instance, in W-(2-5%) Hf and W-0.15% Ru alloys, Clark (77) reports that the DBTT reaches a maximum at an intermediate annealing temperature and then decreases with higher annealing temperatures. Such maxima have not been observed in unalloyed tungsten.

The well established effect of rhenium in the concentration range 24% to 27% Re on the DBTT has not been equalled by alloying with other solid solution elements. This effect of Re is more pronounced in AC- or EB-alloys than in an alloy made by powder metallurgy (67). The latter alloy in the recrystallized condition has a DBTT significantly above room temperature. Clark (77) reports also a strong grain size effect in an EB-melted W-25% Re alloy. Both Klopp et al. (34) and Clark noted that these alloys initially deform by twinning. Dickinson et al. (84) claim lower DBTT's for an alloy of W-1% Re-(.1-.3%) Mn (nominal composition). This effect is thought to result from deoxidation by Mn forming MnO_2 which volatilizes during sintering.

For dispersed second phase alloys, Hahn et al. (76) have recently proposed that second phase particles may be instrumental in lowering the DBTT. This hypothesis is not supported by the evidence on tungsten base dispersed second phase alloys. Of special interest in this regard are the W-ThO_2 alloys, where somewhat lower DBTTs (3, 79) were reported. In subsequent investigations, it was determined that those alloys were tested in the stress relieved condition. As King (85) has recently shown, temperatures in excess of 2700°C and times longer than 10 hours are required to recrystallize a W-2 w/o ThO_2 alloy with a fine dispersion. The DBTT of completely recrystallized W-2 w/o ThO_2 has not yet been determined.

The results on the two recrystallized dispersed second phase alloys, included in Fig. 7, present no unambiguous evidence for a direct particle effect improving the ductility. In the W-.67 B alloy (dispersion of W_2B) investigated by Raffo et al. (73), the DBTT is higher than in unalloyed W. On the other hand, in the W-5% Re-2.2% ThO_2 alloys studied by Maykuth et al. (18), the lowest DBTT is lower than the DBTT's of unalloyed tungsten with the exception of that found by Orehotsky et al. (70) for electron beam floating zone melted material. However, the alloy with the lowest DBTT had the finest grain

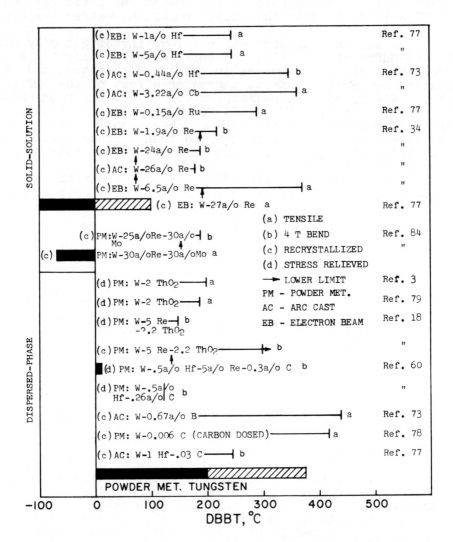

Fig. 7. Ductile to brittle transition temperature (DBTT), °C, of various tungsten base alloys compared with powder met. tungsten.

size, the thoria dispersion apparently having been responsible for the grain refinement. Since the DBTT showed a grain size dependence in these alloys, it is not certain whether the lower DBTT is a direct consequence of the thoria dispersion.

Except for high concentrations of Re in solid solution, alloying has not resulted in a major improvement of the low temperature ductility of tungsten. In dilute solid solution alloys, grain size is less influential than in dispersed second phase alloys in affecting the DBTT. There is a further indication that heat treatments affect the DBTT of unalloyed tungsten to a greater extent than the DBTT of the alloys for which data have been reported.

II. Elevated Temperature Tensile Properties. The most recent comprehensive compilation of mechanical property data on unalloyed tungsten and tungsten base alloys was prepared by Schmidt et al. (86). The information presented in this summary pertained primarily to materials consolidated by sintering. In the meantime, the mechanical properties of AC- and EB-melted tungsten have been investigated by Klopp and co-workers (81, 82) and by Witzke et al. (30), and the high temperature tensile properties of CVD-tungsten by Taylor et al. (87).

To present a more complete evaluation, the more recent information on unalloyed tungsten will be compared with earlier data on PM-tungsten (Figs. 8, 9, 10 and 11). Subsequently, the tensile properties of tungsten base alloys will be reviewed. The evaluation will again be limited to recrystallized materials tested at slow strain rates (0.02-0.05 min^{-1}) because this is the condition most clearly defined. Tests at higher strain rates (21, 88) have shown that the ultimate stress and the ductility increase with increasing strain rate over a wide temperature range (1200-3000°C).

Unalloyed Tungsten: The yield stress of the various materials, depicted in Fig. 8, decreases as a function of increasing temperature in the order PM-tungsten, AC-tungsten, and EB-tungsten. The spread in data for PM-tungsten (the wide band) supposedly reflects different impurity levels in the source materials. This also holds true for the comparison of the ultimate stress in Fig. 9, where the EB-tungsten (30, 82) has the lowest strength of all materials. The ultimate strengths of the CVD-tungsten (87) falls within the band. On the other hand, AC-tungsten competes with PM-tungsten for the highest ultimate strength values.

Ductility information is presented in terms of % R.A. in Fig. 10 and in terms of % El. in Fig. 11. In terms of % R.A. (Fig. 10), the most ductile tungsten is EB- and AC-tungsten. The ductility of PM-tungsten is strongly temperature dependent and drastically decreases from $\sim$90% R.A. at 1000°C to $\sim$10% R.A. at 2400°C. The least ductile material is CVD-tungsten.

In terms of % El. (Fig. 11), AC-tungsten again has the highest ductility but EB-tungsten is, surprisingly, less

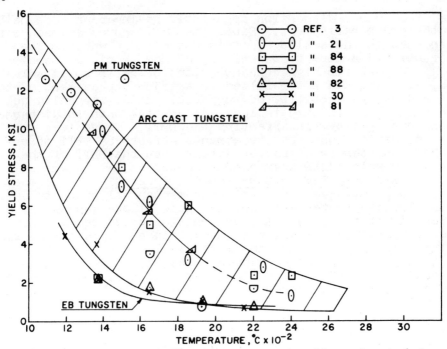

Fig. 8. Yield stress of unalloyed tungsten produced by various techniques.

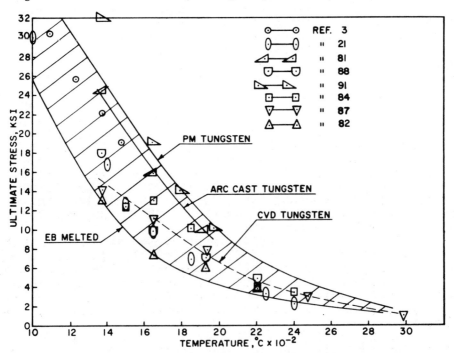

Fig. 9. Ultimate stress of unalloyed tungsten produced by various techniques.

ductile than PM-tungsten which shows a maximum at about 1500°C followed by a steady decline to approximately 10% elongation at 2400°C. CVD-tungsten's % El. amounts to about 20% up to 2000°C and then increases to $\sim$30% at 2600°C.

Any correlation of the tensile properties of unalloyed tungsten with impurities, grain size or recrystallization treatment can only be qualitative. In regard to purity, a correlation is indicated between total impurity content and tensile and yield strength. The EB-melted tungsten which is the purest tungsten has the lowest strength. However, no correlation has yet been established between the yield strength and the concentration of a specific impurity.

The parameter which appears to exert the greatest influence is grain size. A comparison of the reported data shows that the grain size was smallest in PM-tungsten (50-100μ), larger in AC-tungsten (50-200μ), and largest in EB-tungsten ($\sim$300μ). A grain size effect has been reported by Klopp et al. (89) and by Taylor et al. (90). The experimental evidence points toward a Petch type relationship between yield strength (or ultimate strength) and grain size. However, since in the latter two investigations the grain size was varied by annealing, impurity redistribution cannot be ruled out as a contributing factor.

The major parameters which influence the ductility are void growth, stress induced grain growth, and impurity distribution. The loss in ductility of the PM-tungsten with increasing test temperature is clearly related to the greater rate of void formation resulting from grain boundary sliding, while stress induced grain growth is inhibited in this material because of the relatively high impurity level. On the other hand, in the purer AC- and EB-tungsten stress induced grain growth occurs to a more significant degree and the rate of cavitation is correspondingly reduced. CVD-tungsten is a very special case because in this material bubbles are either pre-existent or form at relatively low annealing temperatures, and these bubbles undergo rapid coalescence under stress (51). These bubbles have been attributed to fluorine. They lodge primarily in grain boundaries and have a restraining effect on stress induced grain boundary migration. Their growth leads to intercrystalline failure.

Tungsten Base Alloys: The development of tungsten base alloys has followed the same lines of approach used with other refractory metals like solid solution and/or dispersed second phase alloying, but the effort has not been concentrated on a few selected alloy systems as, for instance, in the development of Nb-base or Mo-base alloys. Instead, a wide variety of alloy systems and alloy compositions were attempted and, when fabricable material was obtained, investigated (frequently on a screening basis by making a few high temperature tensile tests). In order to avoid extensive tabulations of little merit, only those alloys were included in this review which either exhibited exceptional properties or

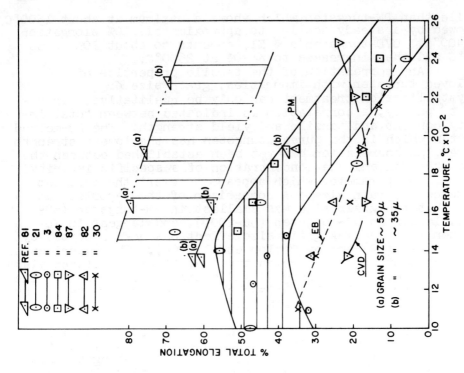

Fig. 11. Ductility (tot. elongation) of unalloyed tungsten produced by various techniques.

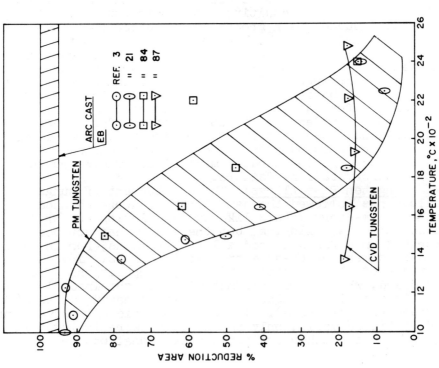

Fig. 10. Ductility (reduction of area) of unalloyed tungsten produced by various techniques.

for which a larger body of data has been reported.

The goal of most of the alloy development efforts has been an alloy with high strength in the temperature range 3500 to 5000°F (1930°C to 2760°C). The guideline for the strength levels sought was laid down by the Materials Advisory Board, Sheet Rolling Panel (68). Several of the alloys which are included in this review have surpassed these guidelines.

The yield stress of the alloys chosen for this review are compared in Fig. 12 and their ultimate stress in Fig. 13. The alloys included in this comparison are listed below.

Solid Solution Alloys

    W-25 Re (34)
    W-1.7 a/o Hf (35)
    W-20 a/o Ta-12 a/o Mo (92)

Precipitation Hardened Alloys

    W-0.5 Hf-0.26 C (60)
    W-0.52 Hf-0.09 C (35)
    W-0.48 Hf-0.5 C (36)
    W-0.58 Cb-0.085 C (73)
    W-12.6 Cb-0.29 V-0.12 Zr-0.07 C (33)

Dispersed Second Phase Alloys

    W-2 ThO_2 (79)
    W-25 Re-2 ThO_2 (91)
    W-5 Re-2.2 ThO_2 (18)

The fact stands out in Fig. 12 that all alloys have a higher yield stress than unalloyed PM-tungsten over the temperature range for which a comparison can be made. This is also true for the ultimate stress with the exception of the W-Re alloys which were found to have strength equal to or even lower than unalloyed PM-tungsten at and above 2400°C, Fig. 13 (79). Two bands are indicated in Fig. 13, one small band for the W-Re alloys and a broader band for the strength levels of the solid solution alloys (lower limit) and the precipitation hardened alloys with a carbide second phase (upper limit). The dispersed second phase alloys (W-2 ThO_2, W-(5)25Re-(2.2)2ThO_2 are shown separately.

The distinction between alloys lies primarily in the temperature dependence of their yield stress and their ultimate stress. It is quite apparent that the temperature dependence of the yield and ultimate stress of solid-solution alloys is, in general, greater than that of the dispersed second phase alloys. The temperature dependence of the yield stress in these alloys is the more pronounced the greater the concentration of the solid solution(s). The behavior of the

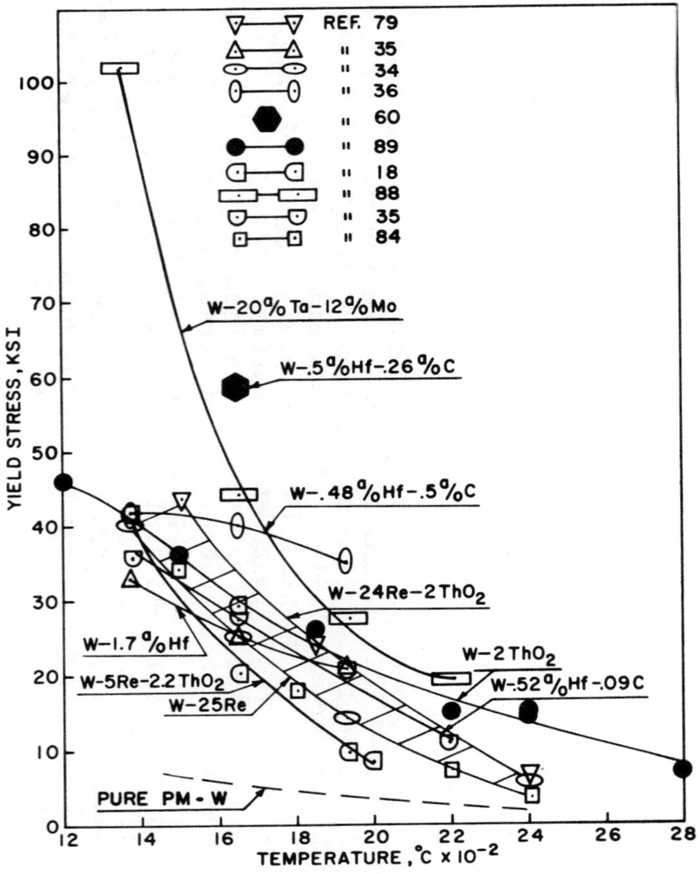

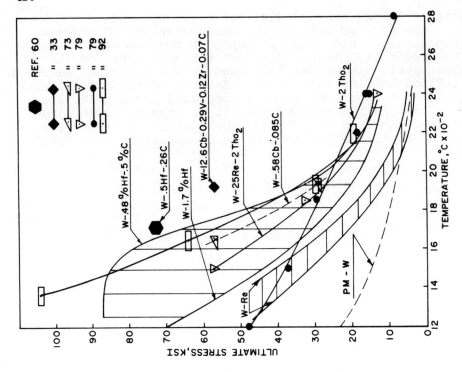

Fig. 13. Ultimate stress of various tungsten base alloys compared with powder met. tungsten.

Fig. 12. Yield stress of various tungsten base alloys and powder met. tungsten.

precipitation hardened alloys is less well defined because
again treatments can greatly influence the properties of
this class of alloys.

In regard to ductility (Figs. 14 and 15), the alloys
containing high Re concentrations have relatively low ductility over a wide temperature range except for an EB-melted
W-24% Re alloy which exhibits extreme ductility above 1700°C.
The precipitation hardened alloys show indications of maxima
and minima. In terms of % R.A. (Fig. 14) all alloys are more
ductile than PM-tungsten above 2100°C and most are as ductile
as PM-tungsten below this temperature. In terms of % El.
(Fig. 15), PM-tungsten is more ductile below 1500°C.

Among all the alloys investigated, three alloys have
outstanding short time high temperature strength properties:
The solid solution alloy W-20 Ta-12 Mo (92), the precipitation hardened alloys W-.5 Hf-.26 C (60) and W-.48 Hf-.5 C
(36), and the dispersed second phase alloy W-2 w/o ThO2
(79). The latter alloy has superior strength properties
above 2200°C (samples annealed for 1/2 hour at 2400°C).

III. _Creep Properties_. The creep behavior of unalloyed
PM-tungsten rod was investigated initially by Green (93) and
later by Sell et al. (79). AC- and EB-tungsten rod were
studied by Klopp et al. (81, 82). A very extensive investigation of the creep properties of AC-tungsten sheet was conducted by Flagella (94), and the creep behavior of CVD-tungsten was studied by McCoy (95). In the case of unalloyed
tungsten, the temperature range 1482°-3000°C was covered.
In the majority of the investigations, tests were made at
more than 2 temperatures. In the case of alloys, data were
usually obtained at one temperature; data at 2 temperatures
are the exception.

For the purpose of this review, the published creep
results were subjected to a computerized analysis, using the
following creep equation (96):

$$\dot{\varepsilon} = A\sigma^{n_c - Q/RT}$$

where $\dot{\varepsilon}$ is the steady state creep rate, A is a constant, σ is
the applied stress (usually the dead weight load), n is the
stress exponent, and the other symbols have their usual meaning. The computer program was written to yield the temperature compensated creep rate as a function of the applied stress
for an activation energy of volume self-diffusion of ~155Kcal
(97). Included in the computer program was a regression
analysis of the data to determine the optimum activation
energy.

The results of this computation for a constant activation energy of ~155Kcal/mole were plotted in Figs. 16 and
17 for unalloyed tungsten and a selection of tungsten alloys,
respectively. The results of the regression analyses of
optimizing the activation energy are summarized in Tables 7
and 8.

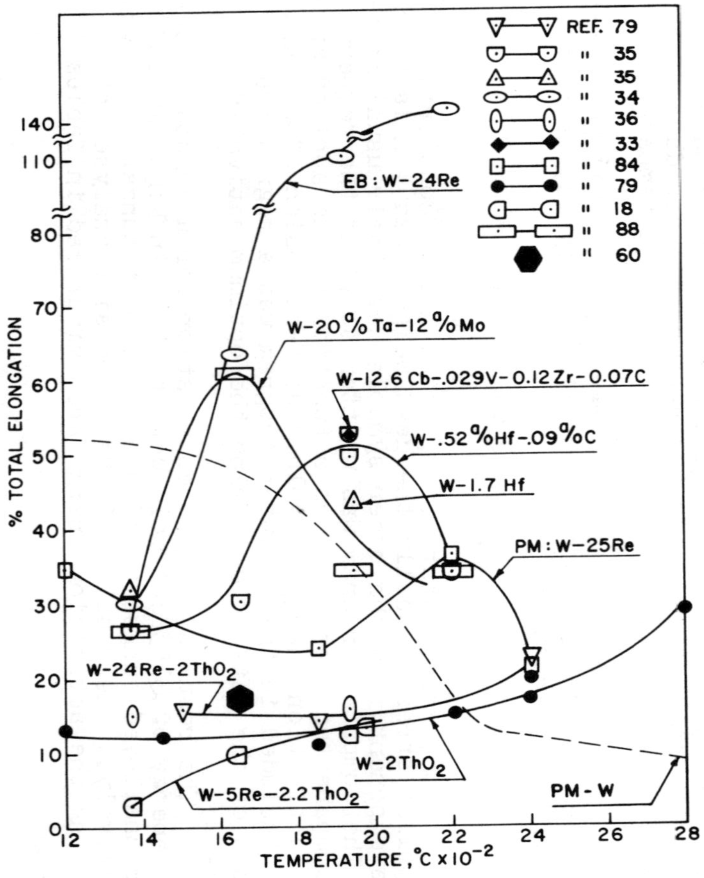

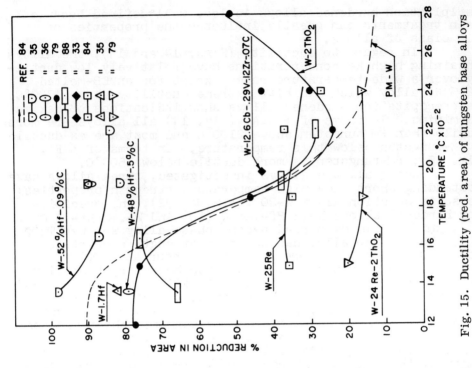

Fig. 15. Ductility (red. area) of tungsten base alloys compared with powder met. tungsten.

Fig. 14. Ductility (tot. elongation) of tungsten base alloys compared with powder met. tungsten.

Table 7. Summary of Creep Test Results on Pure Tungsten Presented in Fig. 16 Optimized with Respect to Activation Energy

Material	Temperature Range, °C	Points	Activation Energy cal/Mole	Slope* (n)	Ref.
PM-W (Rod)	1482-2200	4	112270	6.34	(79)
AC-W (Rod)	1648-2204	3	102440	5.58	(81)
EB-W (Rod)	1579-1998	5	80110	6.69	(82)
PM-W (Rod)	2250-2800	4	97800	5.53	(93)
AC-W (Sheet)	1600-3000	8	106290	3.58	(94)
CVD-W (Sheet)	1650-2200	2	140460	4.98	(95)
PM-W (Sheet)	1650-2200	2	63200	3.75	(95)

* The slopes in Fig. 16 correspond to an activation energy of Q = 155 Kcal/Mole.

Table 8. Compilation of Some Creep Test Results on Tungsten Alloys; Activation Energy Q Optimized for Data Available***

Material	Temperature Range, °C	Points	Activation Energy cal/Mole	Slope (n)	Ref.
W-2% ThO2	1371-1482	2	197890	6.68	(3)
W-2% ThO2	1650-2200	2	93830	3.8	(95)
W-2.5% Re(a)	1648-1926	2	96600	4.89	(34)
W-2.8% Re(a)	1648-1926	2	114709	5.84	(34)
W-4.5% Re(a)	1648-1926	2	95102	4.23	(34)
W-3% Re*	1648-1926	2	108061	3.81	(34)
W-5.1% Re*	1648-1926	2	105515	5.69	(34)
W-26% Re*	1648-1926	2	109596	4.66	(34)
W-25% Re**	1650-2200	2	122880	6.73	(99)
W-25% Re*	1650-2200	2	118479	4.04	(99)
W-2 ThO2	2100-2200	2	107352	5.51	(79)

(a) EB-Sheet

* AC-Sheet

** PM-Sheet

*** The slopes in Fig. 17 correspond to an activation energy of Q = 155 Kcal/Mole.

The data plotted in Figs. 16 and 17 cover only a limited range, but they can readily be extrapolated. This extrapolation is justified for all unalloyed tungsten curves (Fig. 16) but not for all the alloy curves included in Fig. 17. For instance, since the solid solution alloys W-1.7 a/o Hf and W-0.88 a/o Ta-0.3 a/o Hf (35) and the precipitation hardened alloys W-1.76 a/o Hf-0.72 a/o C and W-0.48 a/o Hf-0.72 a/o C (36) were only tested at one temperature, their calculated slopes are in doubt. However, the position of their curves in the log-log plot (Fig. 17) is significant in that it shows the creep strength of those alloys relative to other alloys.

For the same reason, alloys for which creep data were only reported at one temperature were omitted in Table 8.

Unalloyed Tungsten: It is clearly apparent from Fig. 16 that PM-tungsten has a higher creep strength than either AC- or EB-tungsten. The creep strength of CVD (TCD)--tungsten (95) falls in between the PM- and the AC-tungsten.

Most remarkable are the results of the regression analysis. If the data of references (95) and (82) are disregarded, the remaining data on PM- and AC-tungsten are characterized by an average activation energy of $\bar{Q}$ = 104.7 Kcal/mole and an average slope of $\bar{n}$ = 5.2. It will be recognized that the value for $\bar{Q}$ = 104.7 Kcal/mole is lower than expected. However, this value confirms recent considerations by Neumann (98) who has suggested that an activation energy of this magnitude is in line with a volume diffusion mechanism involving single vacancies. The single vacancy mechanism applies for temperatures $\leq$ 0.8 T_m. Above 0.8 T_m volume diffusion supposedly proceeds by a di-vacancy mechanism for which the activation energy is approximately 155 Kcal/mole. This will not be further discussed here.

Tungsten Alloys: The alloys selected for comparison from a multitude of compositions show important trends in regard to alloying effects. Clearly discernable in Fig. 17 is the lower creep strength of the W-Re alloys (99, 34) in comparison with PM-tungsten. Noteworthy, too, are the results on two W-2 w/o ThO_2 alloys (95, 79). The higher creep strength of one of these alloys (79) reflects the effect of the much finer thoria dispersion. This alloy also competes favorably with the carbide precipitation alloys which overage rapidly above 1900°C (36). The creep strength of the dilute solid solution alloys W-Hf (35) and W-Hf-Ta (35) are superior to the creep strength of unalloyed PM-tungsten but possess significantly lower strength in the medium temperature region than either the carbide precipitation hardened alloys or the W-2 w/o ThO_2 alloy with a fine dispersion.

Results of the regression analysis for a number of alloys are presented in Table 8. Only alloys have been included in this table for which data have been determined at more than one temperature. This excludes, unfortunately, the carbide precipitation hardened alloys. If one eliminates

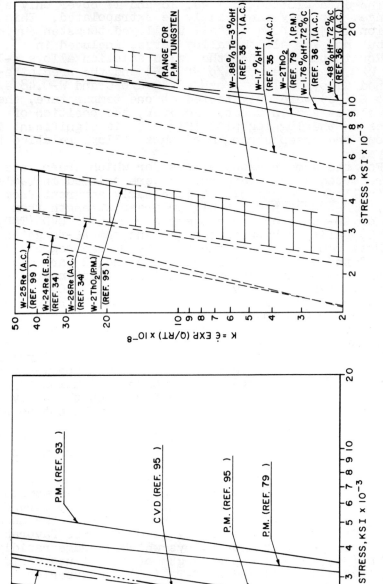

Fig. 17. Temperature compensated creep of various tungsten alloys compared with unalloyed tungsten (Q = 155 KCAL).

Fig. 16. Temperature compensated creep rates of unalloyed tungsten produced by various techniques (Q = 155 KCAL).

from consideration the first alloy listed in Table 8 and
averages the data for the remaining 10 alloys, one calcu-
lates an average activation energy of $\bar{Q}$ = 107.2 Kcal/mole
and an average slope of $\bar{n}$ = 5.02. This result suggests
that the creep mechanism in the alloys (including the W-ThO$_2$
dispersed second phase alloys) is the same as in unalloyed
tungsten.

In general, the creep data on alloys are not sufficiently complete to make other than qualitative comparisons.
In particular, the aging characteristics of the precipitation
hardened alloys must be more thoroughly investigated. The
alloy that has currently the best potential for high temperature ($\succ$2000°C) creep applications is W-2% ThO$_2$ (79).

Summary and Conclusions

A research metallurgist, who was investigating the nucleation of cavities on grain boundaries, when asked why he
used tungsten for this basic study,--there was no obvious
connection of his work on tungsten with the interests of his
employer--, replied: "It's a wonderful metal to work with
because it is brittle at room temperature and so easy to
fracture for replication". This brittleness and other features of tungsten and tungsten alloys (high density, oxidation), which have essentially inhibited their broader commercial use, have made them preferred materials for basic and
applied research. A separate report could readily be devoted
to an elaboration on the role of tungsten in basic research.
In the present review, an attempt was made to assess the
state of the tungsten alloy technology, and to determine how
successful numerous development efforts have been in achieving high temperature strength and in improving low temperature ductility. Since a concluding paragraph was included
at the end of each section, only a general summary on the
alloy development effort is given below.

More thoroughly characterized than most of the many
alloys investigated were the alloys W-Re, W-Hf-C, and W-ThO$_2$
and to a lesser extent the alloys W-Hf-Re-C and W-Re-ThO$_2$.
It will be noted that these 3 classes of alloys have relatively simple compositions and, with the exception of the
W-Re-ThO$_2$ alloy, can be readily fabricated. The W-Re, W-Hf-
C and W-Hf-Re-C alloys have been consolidated by fusion as
well as by sintering and the W-ThO$_2$ and W-Re-ThO$_2$ alloys
only by sintering. The manufacturing technique for the W-ThO$_2$
alloys is well developed and fine dispersions of thoria
($\sim$0.3μm) have been reproduced. The process is simplified
because consolidation and primary breakdown can be performed
simultaneously. For this purpose, a billet is hydrostatically
pressed to $\sim$60% density, canned in Mo, and extruded at 1700°C.

All these alloys possess favorable mechanical properties
In regard to tensile and long time creep strength at high tem-

peratures, the oxide dispersed second phase alloy W-2 w/o ThO_2 is superior to the other classes of alloys for use above 1900°C because thoria is chemically stable up to temperatures approaching its melting point. In the temperature region below 1900°C, the oxide dispersed second phase alloys, as distinct from the precipitation hardened W-Hf-C alloys, in general, have lower tensile strength but still superior long time creep strength. The higher tensile strengths of the W-Hf-C alloys is a consequence of a fine carbide dispersion obtained by heat treating and aging. However, above 1900°C overaging rapidly deteriorates their strength, so that at or slightly below this temperature they will not be useful for long time applications. Optimum properties were obtained with equal volume fractions ($\sim$ 0.5 v/o) of Hf and C. This is a lower volume fraction than is required for the $W-ThO_2$ alloys to obtain optimum strengthening. The tensile and creep strength in the $W-ThO_2$ alloys reaches a maximum with about 4 v/o thoria. Higher volume fractions of thoria do not increase the strength but rather impair the fabricability.

Solid solution strengthening at and above 1900°C has gained interest only with respect to Re because of its ductilizing effect. Significant strengthening is observed when the Re concentration exceeds 10%. However, the solid solution effect does not persist to temperatures in excess of $\sim$1900°C. In particular, the creep strength above 1650°C is lower than that of unalloyed tungsten.

A major improvement in the low temperature ductility of tungsten has been achieved by solid solution alloying with Re in the highest possible concentrations (24-27% Re). This improvement is greater in EB- or AC-alloys and less in PM-alloys. A less pronounced effect on the DBTT was found for 3-5% Re in dilute solid solution. This has led to the preparation of ternary $W-Re-ThO_2$ and quartenary W-Re-Hf-C alloys, respectively, with $\sim$ 5% Re.

In $W-Re-ThO_2$ alloys, the Re-addition has effectively lowered the DBTT's, but the process of making these alloys is less controlled than for the binary $W-ThO_2$ alloys. Also, the $W-Re-ThO_2$ alloys were not very fabricable. However, in this class of alloys, solid solution strengthening and dispersed second phase strengthening were found to be additive effects. The same was not observed in the W-Hf-Re-C alloys where Re did not enhance the strength. The Re addition in the latter alloys was, however, effective in lowering the DBTT for the stress relieved condition.

In conclusion, one cannot fail to be impressed by the progress that has been made in the manufacturing technology of tungsten alloys. Many alloys have been produced, and some, with exceptional high temperature strength properties. While the parameters which control these properties are reasonably well understood, none of the alloys has been completely characterized with respect to fabrication schedules and long time temperature effects. The alloy most likely to satisfy high temperature strength requirements is the alloy W-2 w/o ThO_2. An alloy which combines room temperature ductility with high temperature strength must still be developed.

Acknowledgment

The author wants to thank his co-worker, Mr. G. W. King, for assisting in the compilation and discussion of the mechanical property data.

REFERENCES

1. Tungsten, R. F. Stevens, Jr., Bureau of Mines Minerals Yearbook 1966, Vol. 1, Superintendent of Documents, Washington, D.C.

2. Tungsten, A. N. Zelikman, O. E. Krein, and G. V. Samsonov, "Metallurgy of Rare Metals", 2nd Edition, Ed. L. V. Belyaerskaya, 1-57 (1964) (Translated from Russian, NASATT F-359, U. S. Dept. of Commerce, Clearinghouse for Tech. and Sc. Inf.).

3. Physical Metallurgy of Tungsten and Tungsten Base Alloys, R. H. Atkinson, et al. (Westinghouse Lamp Div., Bloomfield, N. J.), WADD-TR-60-37, March, 1960.

4. Electrowinning Molybdenum and Tungsten, D. H. Baker, Jr., U. S. Bureau of Mines, Reno Metallurgy Research Center, Nev.), J. Metals, 9, 873-76 (1964).

5. Electrolytic Production of Tungsten and Molybdenum, H. L. Slatin (to Timax Associates, New York, N. Y.), Canada No. 713033 (July 6, 1965).

6. Evaluation of Electrowon Tungsten, G. H. Keith, B. D. Jones, and E. A. Rowe (U. S. Bureau of Mines, Reno Metallurgy Research Center, Nev.), R.I. No. 6578, 13 pp, (1965).

7. Ultrafine Tungsten and Molybdenum Powders, H. Lamprey and R. L. Ripley, J. Electro. Chem. Soc. 109, 8, 713-16, (1962).

8. Preparation of Submicron Tungsten Powder by Hydrogen Reduction of Tungsten Hexachloride, J. E. Tress, T. T. Campbell, and F. E. Block (U. S. Bureau of Mines, Albany Metallurgy Research Center, Oreg.), R.I. No. 6835, 14 pp, (1966).

9. Herstellung und Charakterisierung von Ultrafeinen Karbiden, Nitriden und Metallen, E. Neuenschwander (Ciba Ltd., Basel, Switzerland), J. Less-Common Metals, 11, 365-75, (1966).

10. Some Characteristics of Arc Vaporized Submicron Particulate, D. J. Holmgren, J. O. Gibson, and C. Sheer, "Ultrafine Particles", W. E. Kuhn Editor, John Wiley and Sons, Inc., 129-45, (1963).

11. Fluoride Tungsten, J. H. Oxley, E. A. Beidler, J. M. Blocher, Jr., C. J. Lyons, R. S. Park, and J. H. Pearson, "Metals for the Space Age", Plansee Proceedings 1964, Ed. F. Benesovsky, Metallwerk Plansee A. G., Reutte/ Tyrol Springer-Publisher, 278-300, (1965).

12. Reduction of Tungsten Hexafluoride to Form Improved Tungsten Particles, J. M. Blocher, Jr. and J. H. Pearson (to Allied Chemical Corp., N. Y.), U. S. No. 3234007 (February 8, 1966).

13. Factors Affecting the Compaction of Tungsten Powders, A. R. Poster, (Sylvania Electric Products, Inc., Towanda, Pa.), Powder Metallurgy No. 9, 301-15, (1962).

14. Some Observations on the Effect of Atmospheric Humidity on the Flow of Metal Powder, H. D. Lewis, T. J. Ready, and H. H. Hausner (Los Alamos Scientific Laboratory; Metals, Ceramics and Materials Div., Los Alamos, N. M.) LAMS No. 2773, 22 pp, (October, 1962).

15. Tungsten and Tungsten Base Alloys, R. H. Atkinson, G. H. Keith, and R. C. Koo (Westinghouse Lamp Div., Bloomfield, N. J.), Refractory Metals and Alloys, Metallurgical Soc. Conferences, Vol. 11, Eds. M. Semchyshen and J. J. Harwood, Intersc. Publ., 319-55, (1961).

16. Stability of Selected Submicron Refractory Dispersoids in Tungsten, W. R. Morcom and N. F. Cerulli, "Modern Development in Powder Metallurgy", Vol. II, Ed. H. H. Hausner, Plenum Press, 203-15, (1966).

17. The Production of Submicron Metal Powders by Ball Milling with Grinding Aids, M. Quatinetz, R. J. Schafer and C. R. Smeal (NASA, Lewis Research Center, Cleveland, Ohio), Ultrafine Particles, John Wiley and Sons, N. Y., 271-96, (1963).

18. Further Development of Ductile Tungsten-Base Sheet Alloy, D. J. Maykuth, K. R. Grube, H. R. Ogden, A. Gilbert, R. I. Jaffee, and J. M. Blocher, Jr. (Battelle Memorial Institute, Columbus, Ohio), Bureau of Naval Weapons Contract N600(19) 61982, Final Report, 66 pp., (October 31, 1965).

19. Preparation of Refractory Metal Powders with Unusual Properties, S. H. Smiley, C. D. Brater and H. L. Kaufman, J. Metals, 17, 6, 605-10, (1965).

20. Alloy and Dispersion Strengthening by Powder Metallurgy,
     J. E. White (formerly with Aerospace Corp., El Segundo,
     Calif.), J. Metals, 17, 6, 587-93, (1965).

21. The Effect of Thoria on the Elevated Temperature Tensile
     Properties of Recrystallized High-Purity Tungsten, G. W.
     King and H. G. Sell (Advanced Dev. Dept., Westinghouse
     Lamp Div., Bloomfield, N. J.), Trans. A.I.M.E., 233, 6,
     1104-13, (1965).

22. Über die Beeinflussung des Sinterverhaltens von Wolfram,
     J. Vacek (Forschungsinstitut fur Pulvermetallurgie,
     Vestec bei Prag, CSR) Plansee Berichte fur Pulvermetallurgie,
     7, 6-17, (1959).

23. The Activation Sintering of Tungsten with Group VIII Ele-
     ments, H. W. Hayden and J. H. Brophy (Now with Inter-
     national Nickel Co., Tuxedo, N. Y.), J. Electrochem. Soc.,
     110, 7, 806-10, (1963).

24. Activation Sintering of Tungsten, I. J. Toth, N. A.
     Lockington and L. W. Derry (Battersea College of Techn.,
     London, Gr. Britain), J. Less-Common Metals, 9, 157-67,
     (1965).

25. The Kinetics of Metallic Activation Sintering of Tungsten,
     I. J. Toth and N. A. Lockington (see Ref. 24), J. Less-
     Common Metals, 12, 353-65, (1967).

26. Powder-Metallurgical Tungsten-Base Alloy and Methods of
     Making Same, J. W. Pugh, D. T. Hurd and L. H. Amra (to
     General Electric Co., Nela Park, Cleveland, Ohio),
     Canada No. 777,071, (January 30, 1968).

27. Tungsten Sintered with Nickel Activator, E. C. Sutherland
     (formerly with NASA, Lewis Research Center, Cleveland,
     Ohio), Unpublished Research (Information communicated
     5/3/68 by R. W. Hall, Chief, Refractory Metals Branch,
     NASA Lewis Research Center).

28. Consumable Electrode Arc Melting of Refractory Metals,
     S. S. Noesen and R. M. Parke, Vacuum Metallurgy, Ed. R. F.
     Bunshah, Reinhold Publ. Corp., 162-71, (1958).

29. Arc-Melted Tungsten and Tungsten Alloys, F. A. Foyle, High
     Temperature Materials, Vol. 18, Part 2, Metallurgical Soc.
     Conferences, Eds. G. M. Ault, et al., Intersc. Publ.,
     109-24, (1963).

30. Preliminary Investigation of Melting, Extruding and Mechanical Properties of Electron-Beam Melted Tungsten, W. R. Witzke, E. C. Sutherland and G. K. Watson (NASA, Lewis Research Center, Cleveland, Ohio), NASA TN D-1707, 41 pp, (May, 1963).

31. The Primary Working of Refractory Metals, D. R. Carnahan and V. DePierre (Air Force Materials Laboratory, Wright-Patterson Air Force Base, Ohio), AFML-TR-64-387, Part II, 139 pp, (October, 1965).

32. Arc Melting, Working and Properties of Tungsten + 0.6 Columbium Alloys, S. Inouye and G. Saul (Air Force Materials Laboratory, Wright-Patterson Air Force Base, Ohio), AFML-TR-65-401, 25 pp, (January, 1966).

33. Research on Workable Refractory Alloys of Tungsten, Tantalum, Molybdenum and Columbium, R. C. Westgren, V. R. Thomson and V. C. Petersen (Crucible Steel Co. of America, Pittsburgh, Pa.), WADD-TR-61-134, Part II, April, 98 pp, (April, 1963).

34. Mechanical Properties of Dilute Tungsten-Rhenium Alloys, W. D. Klopp, W. R. Witzke and P. L. Raffo (NASA, Lewis Research Center, Cleveland, Ohio), NASA TN-D-3483, 32 pp, (September, 1966).

35. Mechanical Properties of Solid-Solution and Carbide Strengthened Arc-Melted Tungsten Alloys, P. L. Raffo and W. D. Klopp (NASA, Lewis Research Center, Cleveland, Ohio), NASA TN-D-3248, 35 pp, (February, 1966).

36. Effects of Composition and Heat Treatment on High-Temperature Strength of Arc-Melted Tungsten-Hafnium-Carbon Alloys, L. S. Rubinstein (formerly with NASA, Lewis Research Center, Cleveland, Ohio), NASA TN-D-4379, (February, 1968).

37. Development and Evaluation of Tungsten Base Alloys, M. Semchyshen, R. Q. Barr and E. Kalns (Climax Molybdenum Co., Ann Arbor, Michigan), Final Report B.N.W. No. NOw 64-0057-C, 94 pp, (August 10, 1965).

38. Centrifugal Casting of Tungsten, E. D. Calvert and R. A. Beall, J. Metals, 1, 1, 38-46, (January, 1966).

39. Refractory Metal Alloy Hot Pressed from Elemental Powders, J. E. White, Trans. ASM, 57, 3, 756-94, (1965).

40. Alloy and Dispersion Strengthening by Powder Metallurgy,
     J. E. White, J. Metals, <u>17</u>, 6, 587-94, (1965).

41. Development of Dispersion Strengthened Tungsten Base
     Alloys, H. G. Sell, W. R. Morcom and G. W. King (Westing-
     house Lamp Division, Bloomfield, N. J.), AFML-TR-65-407,
     Part I, 89 pp, (November, 1965).

42. Advances in Hot Isostatic Processing, M. J. Ryan and
     A. C. MacMillan, Battelle Tech. Rev., <u>17</u>, 2, 15-20,
     (1968).

43. Cold Consolidation of Metal Plus Dispersoid Blends for
     Examination by Electron Microscopy, G. Reinhardt, W. S.
     Cremens and J. W. Weeton (NASA, Lewis Research Center,
     Cleveland, Ohio), NASA TN-D-3511, 34 pp, (July, 1966).

44. Development of High-Strength Tungsten-Thoria Alloy Sheet,
     R. C. Goodspeed and G. W. King (Westinghouse Astronuclear
     Laboratories, Large, Pa., and Westinghouse Lamp Division,
     Bloomfield, N. J.), Third Quarterly Progress Report, Con-
     tract AF33(615)-67-C-1282, Westinghouse ANL-PR-(WW)-003,
     31 pp, (January 20, 1968).

45. Effect of Certain Process Variables on Vapor Deposited
     Tungsten, F. W. Hoertel (Rolla Metallurgy Research Center,
     Rolla, Mo.), Bureau of Mines, RI No. 6731, 15 pp, (1966).

46. Vapor Deposition of Tungsten by Hydrogen Reduction of
     Tungsten Hexafluoride, J. F. Berkeley, A. Brenner and
     W. E. Reid, Jr. (National Bureau of Standard, Washington,
     D.C.), J. Electroch. Soc., <u>114</u>, 6, 561-68, (1967).

47. Preparation and Evaluation of Vapor-Deposited Tungsten,
     R. L. Heestand, J. I. Federer and C. F. Leitten, Jr.
     (Oak Ridge National Laboratory, Oak Ridge, Tenn.), USAEC
     Report ORNL-3662, 32 pp, (August, 1964).

48. Preparation of Vapor Deposited Tungsten at Atmospheric
     Pressure, E. J. Mehalchick and M. B. MacInnis (Sylvania
     Electric Products Inc., Towanda, Pa.), Electroch. Techn.,
     <u>6</u>, 1-2, 66-69, (1968).

49. CVD-Tungsten and Tungsten-Rhenium Alloys for Structural
     Applications, W. R. Holman and F. J. Huegel (Lawrence
     Radiation Laboratory, Livermore, Calif.), Proceedings of
     Conf. on Chem. Vapor Deposition of Refractory Metals,
     Alloys and Compounds, Ed. by A. C. Schaffhauser, Publ.
     by Am. Nuclear Society, Interstate Printers, Danville,
     Illinois (1967).

50. The Growth of Grain Boundary Gas Bubbles in Chemically Vapor Deposited Tungsten, K. Farrell, J. T. Houston and A. C. Schaffhauser (Metals and Ceramics Div., Oak Ridge National Laboratory, Oak Ridge, Tenn.), Proceedings of Conf. on Chem. Vapor Deposition of Refractory Metals, Alloys and Compounds, Ed. by A. C. Schaffhauser (see Ref. 49).

51. Mechanical Behavior of CVD-Tungsten at Elevated Temperatures, H. E. McCoy and J. O. Stiegler (see Ref. 49), 391-425.

52. A Preliminary Study of Vapor Deposition of Rhenium and Rhenium-Tungsten, J. G. Donaldson, F. W. Hoertel and A. A. Cochran (Rolla Metallurgy Research Center, Bureau of Mines, Rolla, Mo.), J. Less-Common Metals, 14, 93-101, (1968).

53. Vapor Deposition of Molybdenum-Tungsten Alloys, J. G. Donaldson and H. Kenworthy (see Ref. 52), U. S. Bureau of Mines, RI No. 6853, 12 pp, (March, 1966).

54. Chemical Vapor Deposition of Tungsten-Molybdenum-Rhenium Ternary Alloys, C. I. Fairchild (Los Alamos Scientific Laboratory, Los Alamos, N. M.), Proceedings Conf. on CVD of Refractory Metals, Alloys and Compounds, Ed. by A. C. Schaffhauser (see Ref. 49).

55. Dispersion Strengthening of Tungsten by Co-Vapor Deposition, R. L. Landingham and J. H. Austin (Lawrence Radiation Laboratory, Livermore, Calif.), UCRL Report No. 50209, 27 pp, (February 24, 1967).

56. Tungsten Slip Casting Method, S. D. Stoddard and D. E. Nuckolls, (Los Alamos Scientific Laboratory, Los Alamos, N. M.), U. S. 3,322,536, (May 30, 1967).

57. Properties of Sintered W-Mo-Re Alloys, A. Milner, D. D. Rowenhorst and T. R. Bergstrom (3 M Company Central Research Laboratories, St. Paul, Minn.), J. Less-Common Metals, 13, 488-92, (1967).

58. Tensile Evaluation of Unworked Tungsten-25% Rhenium Alloys Consolidated by Sintering Elemental Powders, A. Milner (see Ref. 57), Powder Metallurgy, 11, 21, (1968) - To be published.

59. Summary of Contractor Results in Support of the Refractory Metals Sheet Rolling Program, D. J. Maykuth, DMIC Report No. 231, 22-47, (December 1, 1966).

60. Powder Metallurgy Tungsten Alloys Based on Carbide Dis-
     persions, S. Friedman and C. D. Dickinson (G. T. and E.
     Laboratories, Bayside, N. Y.), TR-65-3513, 18 pp,
     (November, 1965).

61. Fabrication of Tungsten and Tungsten-25% Rhenium Tubing,
     S. Isserow, R. G. Jenkins, J. G. Hunt and G. I. Friedman
     (Whittaker Corporation, Nuclear Metals Div., West Concord,
     Mass.), Nuclear Applications, 2, 8, 304-07, (1966).

62. Development of Techniques for Fabrication of Small Diam-
     eter Thin-Wall Tungsten and Tungsten Alloy Tubing, W. R.
     Burt, Jr., D. C. Brillhart and R. M. Mayfield (Argonne
     National Laboratory, Metallurgy Div., Argonne, Ill.),
     ANL-7151, Interim Progress Report, Period January 1, 1964,
     through December 31, 1965, 36 pp, (July, 1966).

63. Method for Extruding Molybdenum and Tungsten, R. E.
     McDonald and C. F. Leitten, Jr. (To the United States of
     America), U. S. No. 3,350,907, (November 7, 1967).

64. Extrusion of 1/2- and 3/8-Inch Diameter, Thin Wall Tung-
     sten Tubing Using the Floating-Mandrel Technique, C. P.
     Blankenship and C. A. Gyorgak (NASA, Lewis Research
     Center, Cleveland, Ohio), NASA TN-D-3772, 25 pp,
     (December, 1966).

65. Floating Mandrel Extrusion of Tungsten and Tungsten Alloy
     Tubing, R. E. McDonald and G. A. Reimann (Oak Ridge Na-
     tional Laboratories, Metals and Ceramics Div., Oak Ridge,
     Tenn.), ORNL-4210, 25 pp, (February, 1968).

66. Application of Chemical Vapor Deposition to the Production
     of Tungsten Tubing, W. R. Martin, R. L. Heestand and R. E.
     McDonald (see Ref. 65), ORNL-TM-1889, 13 pp, (July, 1967).

67. Some Properties of Tungsten-Rhenium Alloys, B. F. Kieffer,
     G. S. Root and S. A. Worcester (Wah Chang Albany, Albany,
     Oregon), Metals for the Space Age, Plansee Proceedings
     1964, Ed. F. Benesovsky, Springer Publ., 571-76, (1965).

68. Evaluation Test Methods for Refractory Metal Sheet Mate-
     rial, Materials Advisory Board, National Academy of
     Sciences, National Research Council, Report MAB-192-M,
     38 pp, (April 22, 1963).

69. Effect of Mechanical and Structural Variables on the Duc-
    tile-Brittle Transition in Refractory Metals, L. I.
    Seigle and C. D. Dickinson (G. T. and E Laboratories,
    Inc., Bayside, N. Y.), Refractory Metals and Alloys II,
    Metallurgical Society Conferences, 17, Ed. by M. Sem-
    chyshen et al., Intersc. Publ., 65-116, (1963).

70. The Effect of Zone Purification on the Transition Temper-
    ature of Polycrystalline Tungsten, J. L. Orehotsky and
    R. Steinitz (G. T. and E. Laboratories, Inc., Bayside,
    N. Y.), Trans. A.I.M.E., 224, 556-60, (1962).

71. Effects of Dispersions on the Recrystallization and the
    Ductile-Brittle Transition Behavior of Tungsten, R. I.
    Jaffee, B. C. Allen and D. J. Maykuth (Battelle Memorial
    Inst., Columbus, Ohio), Powder Metallurgy in the Nuclear
    Age, 4th Plansee Seminar, Ed. by F. Benesovsky, Springer
    Publ., 770-98 (1962).

72. A Fractographic Study on Tungsten and Dilute Tungsten-
    Rhenium Alloys, A. Gilbert (Battelle Memorial Institute,
    Columbus, Ohio), J. Less-Common Metals, 10, 328-43,
    (1966).

73. Mechanical Properties of Arc-Melted and Electron Beam
    Melted Tungsten Base Alloys, P. L. Raffo, W. D. Klopp
    and W. R. Witzke (NASA, Lewis Research Center, Cleveland,
    Ohio), NASA TN-D-2561, 20 pp, (January, 1965).

74. Recrystallization, Grain Growth and the Ductile-Brittle
    Transition in Tungsten Sheet, K. Farrell, A. C. Schaff-
    hauser and J. O. Stiegler (Oak Ridge National Laboratory,
    Oak Ridge, Tenn.), J. Less-Common Metals, 13, 141-55,
    (1967).

75. Observations on the Ductility of Polycrystalline Tung-
    sten as Affected by Annealing, R. C. Koo (Westinghouse
    Lamp Div., Bloomfield, N. J.), Trans. A.I.M.E., 227,
    280-82, (February, 1963).

76. Effects of Second Phase Particles on Ductility, G. T.
    Hahn and A. R. Rosenfield (Battelle Memorial Inst.,
    Columbus, Ohio), AFML-TR-65-409, 77 pp, (January, 1966).

77. Flow and Fracture of Tungsten and Its Alloys: Wrought,
    Recrystallized and Welded Conditions, J. W. Clark
    (General Electric Co., Flight Propulsion Lab., Evendale,
    Ohio), ASD-TDR-63-420, 85 pp, (April, 1963).

78. Effects of Interstitial Impurities on the Low Tempera-
    ture Tensile Properties of Tungsten, J. R. Stephens
    (NASA, Lewis Research Center, Cleveland, Ohio), NASA
    TN-D-2287, 14 pp, (June, 1964).

79. Development of Dispersion Strengthened Tungsten Base
    Alloys, H. G. Sell, W. R. Morcom and G. W. King (Westing-
    house Lamp Division, Bloomfield, N. J.), AFML-TR-65-407,
    Part II, 112 pp, (November, 1966).

80. Properties of Tungsten and Tungsten Base Alloys, E. L.
    Harmon (Union Carbide Metals Co., Niagara Falls, N. Y.),
    Investigation of the Properties of Tungsten and Its
    Alloys, WADD-TR-60-144, 82 pp, (May, 1960).

81. Effect of Purity and Structure on Recrystallization Grain
    Growth, Ductility, Tensile and Creep Properties of Arc-
    Melted Tungsten, W. D. Klopp and P. Raffo (NASA, Lewis
    Research Center, Cleveland, Ohio), NASA-TN-D-2503, 46 pp,
    (November, 1964).

82. Mechanical Properties and Recrystallization Behavior of
    Electron-Beam-Melted Tungsten Compared with Arc-Melted
    Tungsten, W. D. Klopp and W. R. Witzke (see Ref. 81),
    NASA-TN-D-3232, 34 pp, (January, 1966).

83. Low Temperature Ductility and Strength of Thermochemically
    Deposited Tungsten and Effects of Heat Treatment, A. C.
    Schaffhauser (Metals and Ceramics Div. Oak Ridge National
    Laboratory, Oak Ridge, Tenn.), Summary of 11th Refractory
    Composites Working Group Meeting, AFML-TR-66-179, 15 pp,
    (July, 1966).

84. Ductile Tungsten Alloys, C. D. Dickinson and E. N. Mazza
    (To G. T. and E. Laboratories, Inc.), U. S. No. 3,359,082
    (December 19, 1967).

85. An Investigation of the Yield Strength of a Dispersion
    Hardened W-3.8 v/o ThO$_2$ Alloy, G. W. King (Westinghouse
    Lamp Division, Bloomfield, N. J.), To be published in
    Trans. AIME.

86. The Engineering Properties of Tungsten and Tungsten Base
    Alloys, F. F. Schmidt and H. R. Ogden (Battelle Memorial
    Institute, Columbus, Ohio), DMIC Report 191, 128 pp,
    (September 27, 1963).

87. Tensile Properties of Pyrolytic Tungsten from 1370°C to 2980°C in Vacuum, J. L. Taylor and D. H. Boone, (Jet Propulsion Laboratory, Pasadena, California), J. Less-Common Metals, 6, 157-64 (1964).

88. Effect of Strain Rate on Mechanical Properties of Wrought Sintered Tungsten at Temperatures Above 2500°F, P. F. Sikora and R. W. Hall, (NASA, Lewis Research Center, Cleveland, Ohio), NASA TN-D-1094, 24 pp, (October, 1961).

89. Effects of Grain Size on Tensile and Creep Properties of Arc-Melted and Electron-Beam Melted Tungsten at 2250°C to 4140°F, W. D. Klopp, W. R. Witzke and P. L. Raffo, (NASA, Lewis Research Center, Cleveland, Ohio), Trans. AIME, 233, 1860-66, (October, 1965).

90. Effect of Grain Size and Impurities on Tensile Strength and Ductility of Tungsten From 2500 to 5000°F in Vacuum, J. L. Taylor, D. H. Boone and O. W. Simmons (see Ref. 87), JPL Technical Report No. 32-632, 9 pp, (July 1, 1964).

91. The High Temperature Tensile Properties of Tungsten and Tungsten-Rhenium Alloys Containing a Dispersed Second Phase, G. W. King, W. R. Morcom and H. G. Sell (Westinghouse Lamp Division, Bloomfield, N. J.), Refractory Metals and Alloys IV, Vol. 1, The Metallurgical Society of AIME Conf. Volumes, Ed. R. I. Jaffee et al., Gordon and Breach (1968).

92. Evaluation of High Strength Tungsten Base Alloys, C. R. Cook (Equipment Laboratories, TRW Inc., Cleveland, Ohio), AFML-TR-65-397, 115 pp, (November, 1965).

93. Short-Time Creep-Rupture Behavior of Tungsten at 2250°C to 2800°C, W. V. Green, (Los Alamos Scientific Laboratory, Los Alamos, N. M.), Trans. AIME, 215, 6, 1057-60, (December, 1959).

94. High Temperature Creep-Rupture Behavior of Unalloyed Tungsten, P. N. Flagella, (Nuclear Technology Department, Nuclear Energy Division, G. E. Company, Cincinnati, Ohio), GEMP-543, AEC Contract AT(40-1)-2847, 24 pp, (August 31, 1967).

95. Creep-Rupture Properties of Tungsten and Tungsten-Base Alloys, H. E. McCoy (Metals and Ceramics Division, Oak Ridge National Laboratory, Oak Ridge, Tennessee), Report No. ORNL-3992, 48 pp, (August, 1966).

96. Steady State Creep of Crystals, J. Weertman, (Metallurgy Dept., Northwestern University, Evanston, Illinois), J. Appl. Phys., 28, 10, 1185-89, (October, 1957).

97. Diffusion of Tungsten and Rhenium Tracers in Tungsten,
     R. L. Andelin, J. D. Knight and M. Kahn (Los Alamos
     Scientific Laboratory, Los Alamos, N. M.), Trans. AIME,
     233, 19-24, (January, 1965).

98. Diffusions-und Transportvorgange in und an Wolfram,
     G. M. Neumann (Osram Studiengesellschaft, Augsburg,
     Germany), 6th Plansee Seminar, Reutte, Tyrol, (June 24-
     28, 1968). To be published.

99. Creep Rupture Properties of Refractory Alloys, A Com-
     pilation by Metals and Ceramics Division, Oak Ridge
     National Laboratory, ORNL-Drawings 67-6748/49, (July,
     1967).

100. A Study of the Influence of Heat Treatment on the Micro-
     structure and Properties of Refractory Alloys, V. W.
     Chang, (General Electric Company, Evendale, Ohio), ASD-
     TDR-62-211, Part III, 124 pp, (August, 1964).

THE LESS COMMON REFRACTORY METALS
(RHENIUM, TECHNETIUM, HAFNIUM, NOBLE METALS)

Joseph Maltz

Abstract

Previous reviews of the status of the less-common refractory metals are cited and supplemented with discussions of
newer developments. The physical metallurgy of rhenium and
of technetium, an artificial element which has only recently
become available in sufficient amount for metallurgical
studies to be possible, are surveyed. New data on phase
equilibria, mechanical and physical properties and availability are included. Special attention is also given to
the roles of hafnium-tantalum alloys and of iridium as
oxidation-resistant coatings suitable for special situations.
Insofar as the potential usage of the less common refractory metals in applications which depend upon their high
melting point is concerned, high cost and low availability
preclude any radical changes in the near future. Instead,
an orderly growth is indicated as research pinpoints the
occasional unique benefits which can justify their very
high cost.

National Aeronautics and Space Administration, Washington,
D. C. 20546

Introduction

If tungsten, tantalum, molybdenum, columbium, vanadium and chromium are arbitrarily classified as the "major" refractory metals, seven additional metallic elements deserve consideration as "minor" ones. In order of melting point these are rhenium, osmium, iridium, ruthenium, hafnium, technetium and rhodium.

Some physical properties of these seven metals are summarized in Table 1 and their position in the periodic table in Figure 1. Figure 2, taken from Jaffee et al[1] and modified to include data for technetium, illustrates the periodicity of melting point, peaking at the VIA group. Interestingly enough, the slight anomaly shown by manganese is duplicated by technetium.

The "minor" position of these refractory metals is based upon some very practical considerations, including availability and cost. It is true that the production of several of them has increased very impressively in the past decade when expressed on a relative basis: several orders of magnitude for technetium, at least one order of magnitude for rhenium, a factor of two for hafnium (following a sharper increase the decade before), and a sizeable percentage for the refractory noble metals. Prices have not followed as clear a pattern: technetium has dropped sharply from a very high level and rhenium has dropped modestly, while the others have remained stable or increased. Looked at realistically, the most striking of these changes represent the beginnings of meaningful price and production structures for commodities which were formerly produced on too small a scale to find a real position in the marketplace. From another viewpoint, they reflect a transition from a position where the major customer is a basic chemist or physicist to one where the customer is an engineer exploring government-supported advanced technology. Hafnium is an exception: its production and price are geared to the zirconium market and its usage involves actual production requirements in nuclear reactors. And, of course, the refractory noble metals have established themselves in limited but important commercial applications since their discovery in the early nineteenth century.

There is no reason to expect the minor refractory metals to move up to "major" status in the near future or for their prices to continue to drop sharply. Hafnium is already out of the "precious metal" category and technetium still has room for movement. This latter is discussed more fully in the section devoted to that metal. As for the platinum metals, Vinogradov's estimate of their occurrence in a 16 kilometer thickness of the solid portion of the earth's

TABLE 1 - PHYSICAL PROPERTIES

Metal	Re	Os	Ir	Ru	Hf	Tc	Rh
M.Pt., °C	3180	3050	2443	2310	2222	2170	1960
B.Pt., °C, est.	5900	5500	5300	3900	5400	----	4500
Density, 20°C							
gm. per c.c.	21.04	22.61	22.6	12.45	13.09	11.49	12.41
lb. per cubic in.	0.756	0.815	0.813	0.441	0.473	0.415	0.447
Atomic No.	75	76	77	44	72	43	45
Atomic Radius, Å[a]	1.373	1.350	1.355	1.336	1.585	1.361	1.342
Atomic Wt.	186.2	190.2	192.2	101.1	178.5	99	102.9
Atomic volume cc. per gram atom	8.85	8.43	8.54	8.29	13.04	8.62	8.27
Young's Modulus static, psix10[6]	67.5	81	75	60	20	57[b]	46.2
Dynamic	----	----	76.5	68	----	----	55
Lattice	HCP	HCP	FCC	HCP	HCP/BCC	HCP	FCC
a_o, Å	2.761	2.734	3.839	2.706	3.188	2.741	3.804
c_o, Å	4.458	4.320	----	4.281	5.042	4.398	----
c/a	1.615	1.580	----	1.582	1.581	1.604	----
Thermal conductivity cal./cm. sec. °C, 0-100°C	0.11	0.21	0.35	0.25	0.053	0.12	0.36
Elect. resistivity microhm cm, 20°C	19.8	9.5	5.3	7.6	35.1	18.5	4.7
Spec. heat at 20°C cal./g-°C	.033	.031	.031	.057	.035	----	.059
Superconducting critical temp,°K	1.7	0.7	0.14	0.49	0.37	7.8	----
First ionization potential, volts	7.9	8.7	9.	7.7	7.3	7.3	7.7
Linear expansion coef. microinches/in.-°C	6.6 (20-200°C)	6.1	6.8	9.1	5.8	5.9 (20-200°C)	8.3
Work function, electron volts	4.8[c]	4.5[d]	5.4[c]	4.5[d]	3.6[c]	4.4(est)	4.8[d]
Magnetic susceptibility, cgs units x 10[-6], 20°C	68	9.9	26	43	75	270	111
Thermal neutron capture cross section, barns	85	15.3	440	2.56	115	22	149
Vapor pressure, mm. Hg	1.18×10^{-6} at 2225°C	10^{-6} at 2160°C	10^{-6} at 1810°C	10^{-6} at 1720°C	10^{-6}(est) at 2000°C	----	10^{6} at 1470°C

(a) Ref. (78)

(b) Ref. (72). No experimental details.

(c) thermionic

(d) by contact potential method

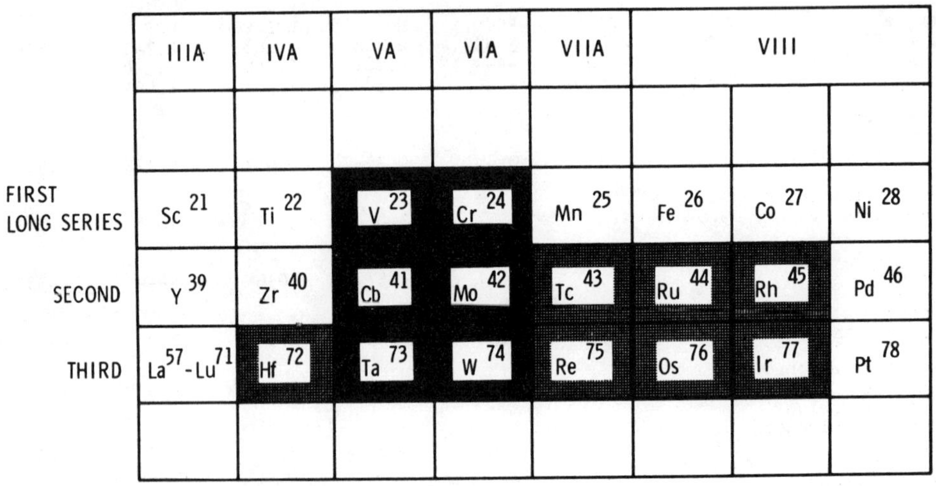

Fig. 1. The refractory metals.

crust, shown in Figure 3, and the absence of concentrated
ore deposits lead one to believe that they will continue to
be sold and priced by the ounce. The author does not intend
that this prognosis be entirely pessimistic. Rare and high-
priced materials are finding important niches in our technology. Obviously these are places where "a little goes a
long way" -- where a minute amount of the material performs
a vital function which less expensive materials cannot perform nearly as well. Most of the potential applications
discussed in this paper -- protective coatings, additions to
structural alloys, superconductors, etc. -- do not quite
meet these guidelines. That is why they are potential.
Nevertheless, they come close enough to be of scientific and
technological interest.

The remainder of this review will deal mainly with certain
aspects of the technology of rhenium, technetium, hafnium
and iridium. In a number of cases definitive reviews prepared a few years ago are cited. The present paper will not
attempt to repeat these in detail but will concentrate on
the newer developments. The other three metals will be
given lesser attention, since the more recent data do not
appear to presage any major change in their positions as
refractory structural metals.

Rhenium

A comprehensive state of the art review[1] in a previous conference in this series and the proceedings of a 1960 symposium[2] provide the base for a discussion of the physical
metallurgy of rhenium. Several factors combine to focus
attention upon this metallic element: the second highest
melting point among the metals, about midway between the
melting points of tungsten and tantalum; a very high modulus
of elasticity; a hexagonal close packed cubic structure which
provides different and sometimes more favorable patterns of
behavior than those encountered with the more common body
centered cubic refractory metals; relative inertness to carbon and water vapor at high temperatures; a very high work
hardening coefficient; and good superconducting properties.

Cost and scarcity keep rhenium from a place among the major
refractory metals. To illustrate the factor of cost: the
addition of only three percent of rhenium more than doubles
the price of commercial tungsten wire.

In spite of this handicap, rhenium and rhenium-containing
alloys are used in electrical contacts, high temperature
thermocouple wire and sheathing, thermionic emitter parts,
brazing filler metals for refractory metal alloys and special-
purpose filaments. Considerable rhenium is consumed in the
production of developmental alloys to support further study

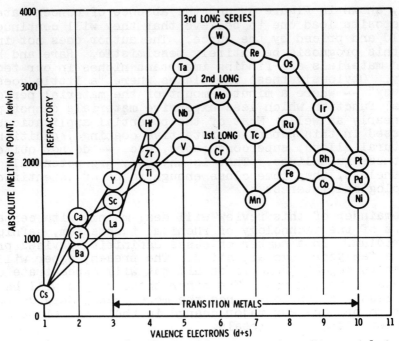

Fig. 2. Variation of melting point among the transition metals in the long series of the periodic table.

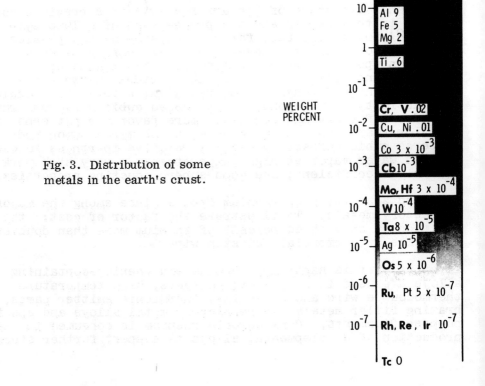

Fig. 3. Distribution of some metals in the earth's crust.

of the well-known "rhenium effect" -- the drastic lowering of
the ductile-to-brittle transition temperature which accompanies
the addition of rhenium near the limit of solid solubility
(approximately 50 weight percent) to molybdenum, tungsten and
chromium.(3)

Peacock has written the most recent of several monographs
dealing with the chemistry and alloying behavior of rhenium(4).
Phase diagrams of the binary systems of rhenium with aluminum,
carbon, columbium, cobalt, chromium, iron, hafnium, manganese,
molybdenum, nickel, platinum, silicon, tantalum, titanium,
uranium, vanadium, tungsten, yttrium and zirconium have been
collected by Elliot(5), along with more fragmentary data on
a number of other systems. Savitskii and Tylkina(6) have reported on the extensive Soviet work on phase diagrams, mechanical properties, fabricability and corrosion resistance of
rhenium, including alloys with titanium, molybdenum, tantalum,
columbium, cobalt, nickel, chromium, zirconium, manganese,
tungsten, vanadium, yttrium, platinum, hafnium and the Re-W-Mo
ternary system.

The binary equilibrium diagrams of rhenium with the refractory
metals tungsten, molybdenum, chromium, columbium and tantalum
are of special interest. Their most characteristic feature
is the extensive solid solubility of rhenium in the terminal
phase accompanied by the ductility improvement of the Group
VI-A metals previously mentioned. Beyond the solubility limit
sigma phase is formed by an eutectic or peritectic reaction
and the ductility drops sharply. Niemiec(7) has proposed a
system of classification of the structures of the rhenium and
technetium alloys with other transition metals, based upon
molar concentration of components, electron concentration,
location in the periodic table and ratio of atomic radii.

Previous reviews of rhenium have dealt extensively with the
"rhenium effect" mentioned earlier and with the properties of
the alloys of rhenium which show this effect. The volume of
literature in this area make this no longer practicable.
Several of the earlier papers in this conference dealt with
some of these alloys. A critical review confined entirely
to this phenomenon is in preparation(8).

Small additions of rhenium to the major refractory metals
appear to be beneficial, though less spectacular in their
effect than the solubility-limit additions. The benefits are
of several kinds: (1) solid solution strengthening, (2) reduction in hardness accompanied by an improvement in ductility,
(3) resistance to embrittlement by hydrogen and (4) resistance
to oxidation or corrosion. These are discussed, as follows:

(1) Rhenium is among the most effective solid solution strengtheners of columbium and tantalum(9, 10) and has been used for
this purpose in a number of developmental alloys(14).

Tarasov et al[9] list chromium, rhenium, titanium, zirconium, tungsten and molybdenum as the most effective strengtheners, in order of decreasing effectiveness, of columbium at room temperature. At 1100°C (2012°F) the order is slightly changed to chromium, rhenium, zirconium, molybdenum, tungsten, and tantalum. McAdam [10] places tungsten first at 1200°C (2192°F), followed by osmium, rhenium and chromium.

(2) The first additions of rhenium to the Group VIA metals chromium, molybdenum and tungsten produce a reduction in hardness, sometimes accompanied by a reduction in ductile-to-brittle transition temperature, which has been called the "solution-softening effect". This has sometimes been equated to the high rhenium ductilizing effect. More recent findings indicate that the two should be separated[8]. Klopp, Witzke and Raffo[15] have observed a ductile-to-brittle bend transition temperature of -75°F (-60°C) in worked sheet of electron-beam melted tungsten containing as little as 1.9% rhenium. "Doped" tungsten to which 3% of rhenium has been added is marketed commercially as a wire alloy. A hardness minimum accompanied by a maximum in ductility after annealing is reported near this rhenium level[16].

(3) The effect of rhenium additions upon the compatibility of tantalum, columbium and their alloys with hydrogen at 1500°F (816°C) was reported by Stephens and Garlick[17]. The subject is reviewed at length by Chandler and Walter in a paper in this volume, which also includes some data on hydrogen compatibility with tungsten-rhenium and molybdenum-rhenium.

(4) A reduction in scaling rate of columbium-zirconium alloys in oxygen has been noted[18] with the addition of up to 10 atomic percent of rhenium. Rhenium, at levels below 1% is also reported[19] to be the most effective of ten addition elements in improving the resistance of 18% Cr-14% Ni stainless steel to pitting corrosion. The authors explain this in terms of the greater stability of the passive state, which is related in turn to the structure of the surface film.

Mechanical properties of low and medium-rhenium alloys with the major refractory metals are fairly well documented but very few investigations of the mechanical properties of unalloyed rhenium or rhenium-base alloys have been reported. Room temperature tensile properties of cold-worked 0.005" thick rhenium sheet[11], of 0.007 - 0.010" thick sheet[12] and of 0.050 - 0.065" diameter wire[13] are grouped in Figure 4. The high strength and extraordinarily high rate of work hardening are apparent in this figure. Short time tensile strengths at elevated temperatures have been documented only for wire[13] under a limited range of conditions. They are reproduced in Figure 5 without change.

Until recently the only available creep rupture information was in the form of a few data points[12] at temperatures up

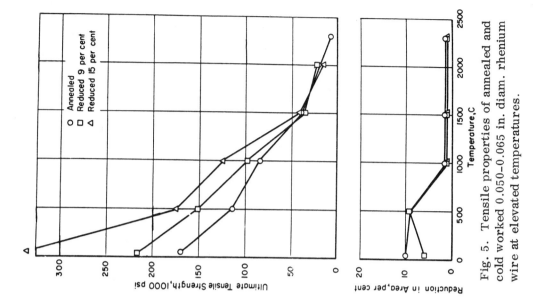

Fig. 5. Tensile properties of annealed and cold worked 0.050–0.065 in. diam. rhenium wire at elevated temperatures.

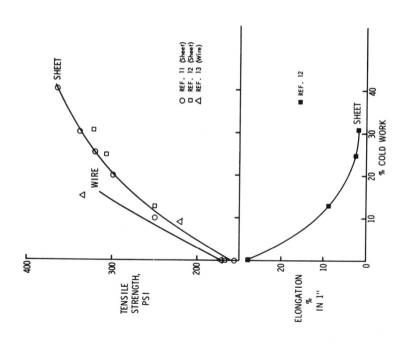

Fig. 4. Effect of cold work on room temperature tensile properties of rhenium.

to 2000°C. These were obtained on 0.050" wire in a helium-
5% hydrogen atmosphere containing some residual oxygen. More
complete tests have been reported[20] on 0.020" sheet from
1600° to 2800°C (2912° to 5072°F). Some of the curves are
reproduced in Figure 6. A comparative curve for tungsten at
2800°C is included to illustrate the authors' conclusion that,
within the temperature range studied, the short-time rupture
life for rhenium is greater than that for arc-cast tungsten.
In more prolonged tests the reverse is true. Four of the
older data points are included for comparison.

The physical properties of rhenium have been catalogued in
various reference works[21, 22] and are listed in Table 1.

Powell, Tye and Woodman have recently re-determined the thermal conductivity[23] and give a figure of approximately 0.11
cal./square cm./cm./sec./°C between room temperature and 500°K.
This value is comparable to those for columbium and tantalum
but less than half those for molybdenum and tungsten. The
previously quoted value, 0.17, was based upon measurements
at subzero temperatures. The same authors have extended the
range of electrical resistivity data down to 1.85°K.

They report a residual resistivity ratio of 4500 for high
purity rhenium. Near room temperature their figures for
resistivity are in good agreement with the normally quoted
value of 19.3 microhm-cm.

Rhenium is of interest for thermionic emitters because of the
high value of its work function, 4.8 eV. Recent determinations
[24] on single crystals indicate that a value as high as 5.5
eV is obtainable on the (0001) orientation.

Vapor deposition of rhenium and co-deposition of rhenium and
tungsten by hydrogen reduction of mixed halides have attracted
considerable attention because of the potential of the process for producing thermionic emitter parts and other space
hardware.[25] Coherent, high-purity rhenium deposits have
been successfully produced. Co-deposits, however, tend to
lack uniformity. It is likely that these difficulties will
eventually be overcome.

Technetium

Technetium, atomic number 43, is especially interesting because of the newness of its appearance on the metallurgical
scene. Although the existence of "ekamanganese" and its
position in the periodic table were correctly predicted by
Mandeleev, persistent searches have resulted in no confirmed
reports of its recovery from natural sources. This puts it
in a position among the cis-uranium elements which is shared
only by the much heavier promethium atom. It now appears that

all the isotopes of technetium are radioactive. Three isotopes with masses of 97, 98 and 99 have half lives in excess
of 10^5 years and correspondingly low levels of radiation but
none possesses a half life or a position in a natural decay
chain which allows it to occur naturally.[26]

Technetium was first produced artifically in trace amounts
by bombardment of molybdenum in the Berkeley cyclotron[27];
this was followed by isolation of milligram quantities from
reactor-irradiated uranium. In keeping with this history,
the name assigned to it derives from a Greek word meaning
"artificial". Obviously these first sources could not be
expected to yield quantities of commercial importance. The
real change came with the separation of the first gram of
technetium from reactor waste products in 1952. More recently
a kilogram quantity of technetium -99 metal powder has been
prepared at Hanford[28] by anion exchange extraction, conversion to pure ammonium pertechnate and hydrogen reduction.
A 100-gram ingot of metal 3/8" in diameter by $2\frac{1}{2}$" long has
been cast at Oak Ridge[29]. It is estimated that production
could be increased at least two orders of magnitude in the
near future and to the level of tons within a few years if
the demand so warranted. This would place the technetium
supply in nearly the same category as that of rhenium. The
low density and atomic weight are additional favorable factors
when one considers the corresponding figures for rhenium
(table 1).

The price of technetium in such a supply situation is a highly
artificial one. It can be bought for $55 per gram ($1500 per
ounce), compared to $100-$200 per ounce for some of the
platinum metals or $40 for rhenium. This in itself is a
drop to one fiftieth of the price a decade ago. Moreover,
design estimates indicate that a pilot plant can now be built
to produce technetium for $1 to $3 per gram, the exact figure
depending on quantity and the degree to which other metals
are simultaneously recovered. This would put the metal in a
nearly competitive position relative to rhenium.

The handling of Tc^{99} does not appear to represent a major
hazard. The weak β radiation (0.3 MeV) is stopped by laboratory glassware. Handling of larger amounts may require
secondary shielding from the weak gamma radiation produced
by reaction with the primary container. There is no problem
of concentration within the body; about half the Tc is eliminated in one day. As a first approximation the radiation
level of technetium is comparable to that of thorium but the
hazards associated with handling it are somewhat less[30].

The chemistry of technetium has recently been reviewed by
Peacock[4]. In the normal unexcited state its atoms, like
those of rhenium and manganese, have five d- and two s-type
valency electrons. However, rhenium and technetium resemble

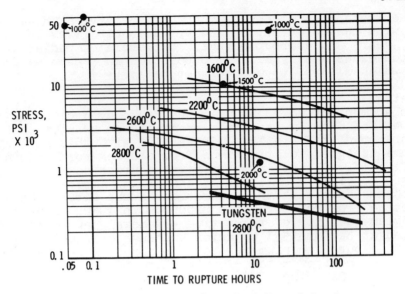

Fig. 6. Stress-rupture characteristics of rhenium.

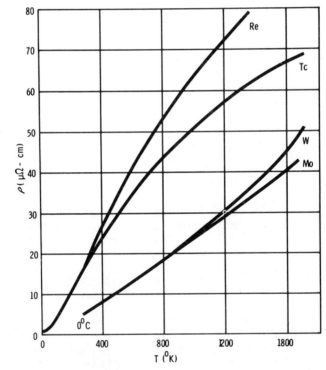

Fig. 7. Electrical resistivity of technetium and
neighboring transition elements.

one another much more closely than either resembles manganese.
Both exhibit valencies all the way from zero to seven, with
four and seven the most common for each. Both form dioxides
and heptoxides. The high temperature oxidation characteristics
of technetium can be expected to be poor. TcO_7 melts at 120°C
(248°F) and boils at 311°C (592°F). The corresponding figures
for Re_2O_7 are 297°C (570°F) and 363°C (685°F). Pertechnate
ion is a most effective corrosion inhibitor[31]. When kept in
a solution containing as little as 5 to 50 ppm TcO_4, mild
carbon steels do not corrode at temperatures up to at least
250°C.

Technetium, like rhenium, crystallizes in the close-packed
hexagonal system. Complete solid solubility of the two elements
between 1200° and 1800°C has been reported[32] on the basis
of X-ray analysis of sintered and quenched samples. This is
not unexpected in view of the similarity in atomic size and
electronic structure. The ionic radius of each element is
0.56 Å. There have been a considerable number of surveys of
phase relationships in other technetium-bearing binary systems.
Peacock notes binaries of technetium with molybdenum, zirconium,
columbium, scandium, titanium, hafnium, vanadium, chromium,
tungsten, manganese, iron, cobalt, nickel, the platinum metals,
aluminum and the rare earths. More recent studies extend this
list to include zinc[33], silicon[34] and thorium[35] and include additional data for technetium-columbium[36].

Few determinations of the physical properties of technetium
have been published. Koch and Love[37] have measured the
electrical resistivity of arc-melted technetium -99 between
7.5°K (-266°C) and 1700°K (1427°C, 2600°F) (Figure 7). Below
room temperature the values are comparable to those for rhenium.
At higher temperatures they drop somewhat below the rhenium
values but are roughly twice as high as the resistivities of
tungsten or molybdenum. Thermal conductivity of electron
beam melted technetium has been determined by Baker[28] from
thermal diffusivity measurements and found to be 0.12 - 0.13
calories/sq.cm./cm./°C/sec. over the range 25°C - 575°C
(77°F - 1067°F).

Technetium has attracted interest as a hard superconductor,
with a transition temperature second only to that of columbium.
Values of 7.73°K[38] and 7.92°K[39] have been reported. Still
higher transition temperatures have been noted for its binary
alloys: 12.9°K for Cb-Tc[40], 11.3°K for Tc-V[41, 42] and
7.0°K for W-Tc[43].

Mechanical property data for technetium and its alloys are
nearly nonexistent. Some are being obtained on a current
study[30] of technetium and tungsten-technetium alloys which
has, as its primary objective, a determination of the degree
to which technetium may act as a rhenium analog in its effect
upon the ductility of the Group VIA metals.

Fig. 8. Electron beam melted technetium.

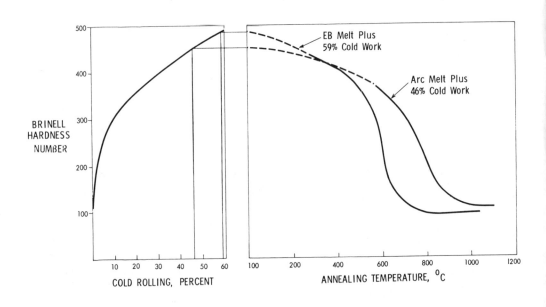

Fig. 9. Work hardening and annealing of technetium.

Unalloyed technetium shares with rhenium an extreme propensity
for deformation twinning and a very large work-hardening coefficient. As-cast arc or electron beam melted technetium
exhibits a heavily twinned microstructure (Figure 8) presumably
induced by cooling stresses. Arc melted technetium has been
cold rolled to 46 per cent reduction in area and electron beam
melted technetium to 59 per cent reduction in area prior to
severe edge cracking. The corresponding figure for commercial
rhenium is about 35 per cent. The progress of work hardening
is shown on the left-hand side of Figure 9. No significant
difference between the hardness of the arc melted and that of
the electron beam melted metal was noted at equal percentages
of cold work. Upon reheating, the 59% worked material annealed
at a somewhat lower temperature than that which had been deformed the lesser amount, as shown on the right-hand side of
the figure.

A single tensile test has been conducted by the same investigators on arc melted unalloyed technetium sheet cold worked
58% to a thickness of 0.036". The $\frac{1}{4}$" gauge length specimen
was deformed at a rate of 0.004 inch per inch per minute and
fractured at a load corresponding to an ultimate strength of
197,000 pounds per square inch.

Hardness figures for arc melted tungsten-technetium alloys
are shown in Figure 10. Lowered hardness in the range 20 to
50 at. % Tc is attributed by the investigators to the increasing
twinning deformation with technetium additions to the body
centered cubic tungsten solid solution. This is accompanied
by a gradual increase in the capacity of 30 gram arc-cast alloy
buttons to deform at room temperature under the repeated blows
of a 6000 lb. pneumatic hammer, up to the limit of the single
phase field (Figure 11). Beyond 50 at. % Tc the usual brittle
sigma phase appears.

Electron beam remelted and canned alloy buttons were also press
forged to a 60% reduction at 1700°C. No significant increase
in hot workability was noted, beyond a small improvement re-
latable to the grain refining effect of the first additions
of technetium to tungsten.

Bend tests of electron beam melted tungsten-technetium sheet
are in progress but no results are available at this time.

Hafnium

Although hafnium certainly qualifies as a refractory metal on
the basis of its 4032°F (2222°C) melting point, it has not
customarily been treated as one. This is due partly to the
relationship of hafnium to the less refractory Group IVA metals
titanium and zirconium, and partly to the fact that its principal use, as a control material for water-cooled nuclear
reactors, does not depend upon its high melting point.

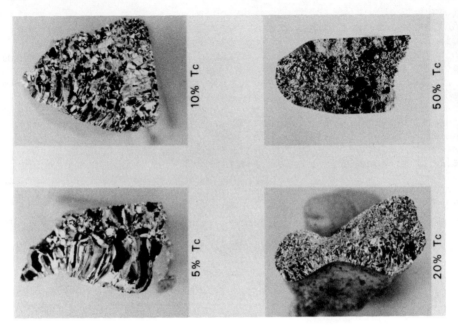

Fig. 11. Fracture surfaces of tungsten-technetium alloys showing cast structure and cold deformation.

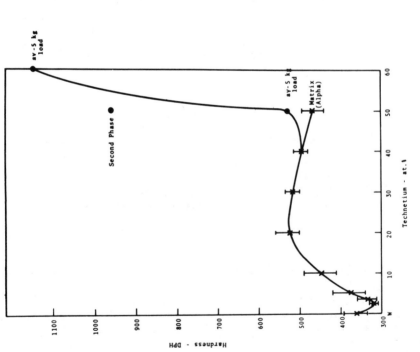

Fig. 10. Microhardness of as-cast tungsten technetium alloys.

Hafnium and zirconium normally occur together in nature in a
ratio of one part of hafnium to fifty of zirconium, so that
the supply and price of hafnium are intimately tied to the
market for zirconium. The difficulty of separating hafnium
from zirconium, together with an extreme difference in neutron
absorption cross-section, have further tied the history and
technology of the two metals together. A recent estimate [44] placed the U. S. demand for zirconium in 1966
at 2.5 million pounds, corresponding to a potential hafnium
supply of 50,000 pounds. The current price of zirconium-
bearing hafnium is about $75 per pound in sponge form or $150
per pound as powder.

There is a large volume of high-quality literature dealing
with extraction, separation, reduction, analysis, control rod
fabrication, aqueous corrosion and related matters. This will
not be dealt with here. Those interested are referred to the
classic review by Thomas and Hayes[45] and to coverage of more
recent developments in the Nuclear Science Abstracts.

About a decade ago hafnium began to assume some importance as
a potent strengthener of the other refractory metals, especially when combined with the interstitial elements in the
correct proportions to form dispersed carbides and nitrides
after suitable heat treatment. This function is analogous
to the empirically discovered role of titanium and zirconium
as strengtheners for molybdenum. Tantalum-tungsten-hafnium
alloys were the first to achieve some measure of commercial
acceptance[14]. This was followed by the use of hafnium in
molybdenum alloys and columbium alloys and, most recently, in
tungsten alloys[46]. The development and properties of these
low-hafnium refractory metal alloys are part of the subject
matter of the papers of Buckman and Goodspeed, and Sell in
this volume and will also not be further discussed here.
Instead, this review will concentrate on a new class
of alloys based on hafnium with approximately 20 percent
tantalum. These alloys are of interest because of their relatively slow rate of oxidation in air at temperatures between
2200°F and 4000°F. Reports of this development have appeared
only within the past three years. This work has been pursued
by Marnoch[47], Hill and Rausch[48] and Wimber and associates
[49].

Hafnium metal, thanks to a highly refractory oxide, is among
the best of the refractory metals in oxidation resistance at
1200°C (2192°F). At that temperature its static oxidation
weight change rate is comparable to that of chromium and less
than a thousandth that of molybdenum or rhenium. Limited data
are available at higher temperatures[50]. However, it appears
that hafnium does not remain protective at temperatures of the
order of 1700°C (3092°F) and above, since HfO_2 does not form
a dense, adherent film. Moreover, both hafnium metal and
hafnium oxide undergo allotropic transformations near 1700°C.

These are accompanied by volume changes and thermal expansion mismatch so that breakaway and essentially linear oxidation conditions exist.

Marnoch studied complex oxides based upon hafnia in an effort to improve performance above 3000°F. He reasoned that the lower vapor pressure of Ta_2O_5 compared to that of SiO_2 made the former a potentially useful glass former at temperatures too high for the latter to function. The Hf-Ta combination, then, might perform similarly to a silica-protected refractory oxide but in a higher temperature regime. An added bonus was the possibility of growing the desired complex oxide from a solid solution alloy of the elemental metals, by-passing some of the problems of coating application and adherence. In addition, the thermal expansion of Hf-10Ta is nearly the same as that of some of the substrate alloys, such as Ta-10W, for which protection is needed.

The actual mechanisms of protection have proven to be more complex than Marnoch expected but very favorable nevertheless. As Metcalf and Stetson point out in this volume, tantalum oxide suppresses the transformation of HfO_2 and minimizes the effect of thermal shock. Hafnium also acts as a powerful interstitial sink and prevents oxygen embrittlement of an alloy substrate. This interstitial sink effect has been demonstrated at 3500°F - 3800°F (1927 - 2093°C) and also at 2500°F (1371°C).

The binary Hf-Ta equilibrium diagram based upon the work of Kato and associates (51, 52) is shown in Figure 12. Body centered cubic β hafnium exhibits complete solid solubility with tantalum between the solidus and the temperature at which it transforms to the hexagonal α structure. Tantalum additions up to 24% sharply depress the temperature of this transformation and extend the single-phase β field down to 1050°C (1922°F), where an eutectoid reaction takes place. In the high-tantalum regime there is a monotectoid decomposition of β into hafnium-rich and tantalum-rich cubic phases.

An important result of this eutectoid decomposition mechanism is the lameller structure of alloys near the Hf-24Ta composition level, with the lamellar oriented at right angles to the free surface. Oxidation then proceeds along preferred crystallographic planes and provides mechanical continuity between the completely oxidized scale and the partially oxidized subscale.

Hf-Cb alloys behave much like alloys in the Hf-Ta system, the principal difference being the absence of a monotectoid reaction. Columbium, like tantalum, stabilizes β-hafnium down to approximately 2000°F (53).

Hill and Rausch(69) studied Hf-Ta binary alloys and more complex modifications thereof for the purpose of optimizing

oxidation resistance. Screening tests in static air at 2500°F
(1371°C) and in an oxygen-hydrogen torch at 3150°F (1732°C)
were used for this purpose. Some tests were also conducted
at 3500°F (1927°C). In the binary alloy series the maximum
oxidation resistance was obtained in the 20 - 30% Ta range.
Additions of 2-5% of tungsten, zirconium, yttrium, titanium
and columbium provided no significant improvement over the
binary alloys. Wimber[49] has added nickel up to 3% to this
list. There was also no major change attributable to the
residual zirconium content of about 1.5% normally present in
commercial hafnium. Ternary additions of molybdenum improved
the cold workability but also produced no significant improvement in resistance to oxidation under the test conditions.

Hill[54] has investigated the utility of modified Hf-Ta and
Hf-Cb alloys for structural use in oxidizing environments in
a lower temperature regime. Preliminary screening tests were
conducted in static air in a furnace at 2500°F (1371°C).
These were followed by extended oxidation tests, cyclic oxidation tests and low temperature oxidation studies of the most
promising compositions.

The screening tests of binary alloys indicated a minimum oxidation rate at the 20-25% Ta or 10-15 weight % Cb levels.
The best performance corresponded to 30 hours of life at 2500°F.
Breakaway oxidation commenced at 10-20 hours. Among the
ternary additions boron, aluminum, chromium, platinum, silicon
and iridium were found to decrease the rate of oxidation of
the best Hf-Ta binary alloys and to improve the density of the
scale. Vanadium, nickel, tungsten, molybdenum, tin, gold and
yttrium were not effective. Ternary additions to a Hf-13 weight
% Cb or Hf-15 weight % Cb base resulted in some improvement
but the weight gains were considerably higher than those observed in the best Hf-Ta base alloys. The best performance
of all was obtained with alloys in the Hf-Ta-Cr-Si, Hf-Ta-Cr-B
and Hf-Ta-Ir-Al-Si systems. In each case these survived 450
hours at 2500°F in static air.

Cyclic exposures of one hour at 2500°F followed by air cooling
to room temperature resulted in weight gains comparable to
those obtained under static conditions. However, low temperature oxidation tests in static air at 1200° - 1800°F (649° -
982°C) demonstrated that the alloys with highest oxidation
resistance were subject to accelerated degradation in this
range. There were several possible reasons for this, including
an inability of the protective oxide to be self healing and
an expansion mismatch associated with the α-β transformation
of hafnium.

It should be emphasised that eutectoid hafnium-tantalum does
not provide the same type of oxygen barrier as silicides or
the noble metals. It is essentially a system based upon the
moderately good oxidation resistance of hafnium and upon its

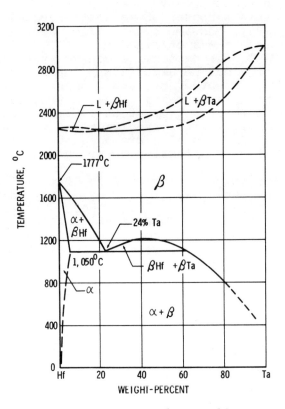

Fig. 12. Hafnium–tantalum equilibrium
diagram.

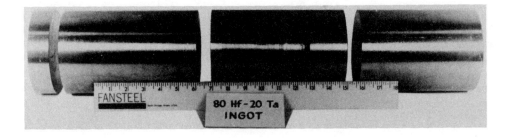

Fig. 13

ability to delay embrittlement of a substrate while it acts as a sink for oxygen. The role of tantalum is to let hafnium do these jobs efficiently by suppressing certain allotropic transformations and by producing the mechanically favorable laminated structure. If applied as a relatively thick cladding, it can offer minutes of protection near 4000°F or hours near 3000°F.

Small uncooled liquid rocket engines provide an example of an application for which this may be an adequate degree of protection. Mash, Donlevy and Bauer[55] have investigated this application by producing nozzle inserts of Hf-20Ta clad on a Ta-10W substrate.

Hf-20Ta alloy ingots ranging up to 250 lbs. in weight were produced by electron beam melting blended elemental powders. Figure 13 shows such an ingot after machine conditioning. Both press forging and extrusion were used for ingot breakdown, the former processing proving to be the more efficient on the basis of yield. The press forged sheet bar was hot rolled to heavy plate, assembled into a five layer sandwich with a Ta-10W core and unalloyed tantalum interleafs by edge welding in argon, and hot rolled to sheet at 2450°F (1343°C). The sheet is shown in Figure 14. The roll-bonded clad sheet was spun to a nozzle configuration in air at 1800 - 2100°F (982° - 1149°C). Finished nozzles, fitted with Ta-10W flanges, are shown in Figure 15.

Several such nozzles have been subjected to cyclic firing tests simulating engine restarts and reaching peak conditions of 800 seconds at a 5000°F flame temperature. The Hf-20Ta cladding did not fail, even in this extreme environment.

In summary, hafnium-tantalum binary alloys have been produced as ingot, plate, clad sheet, duplex tubing and in component configurations. When applied as a cladding, they offer protection against oxygen for a limited but significant time at temperatures above 3000°F. Ternary and quaternary additions may improve the workability somewhat but have little effect on service performance in this range. Tantalum-hafnium alloys also offer useful potential in a lower temperature regime slightly beyond the capability of nickel- or cobalt-base superalloys. Here they might be used in massive form, or as cladding. Here, also, modifications of the basic alloy are likely to be useful to improve oxidation resistance and to increase mechanical strength.

The Platinum-Group Refractory Metals

Unlike hafnium, rhenium or technetium the platinum-group refractory metals have been known for about a century and a half and have established secure positions in commerce as components

Fig. 14. Clad sheet – Hf-20 Ta on tantalum alloy.

Fig. 15. Nozzles hot spun from Ta-10W alloy roll clad with Hf-20Ta.

of alloys used for catalysts, corrosion-resistant equipment, thermocouples, electrical contacts, heater elements and jewelry.

The platinum metals comprise two triads at the ends of the second and third long periods of the periodic table. It is generally agreed that the number of s and d electrons per atom effectively participating in metallic bonding does not continue to increase in following the long periods beyond the VIA elements, although the disposition of the additional electrons is still a matter of dispute. The decrease in effective valence is demonstrated by the decrease in melting point beyond the VIA metals in Figure 2. On the other hand elastic modulus, which can also be considered an index of cohesive bond strength, rises to a peak value in the VIIIA elements. Osmium has the highest value of Young's modulus of any metal, following by the flanking elements iridium and rhenium. Ruthenium and rhodium occupy correspondingly high positions in the previous long period. Beyond the VIIIA elements the drop-off in melting point is great enough to keep platinum and palladium off the list of refractory metals and to place rhodium at the lower edge of the list.

The rate of work hardening follows a similar pattern[1] except that, as might be expected, the hexagonal metals work harden more rapidly than corresponding metals of cubic symmetry. Darling[72] relates the fact that osmium has never been effectively worked either hot or cold to the highest known rate of work hardening. Ruthenium has the next highest rate and can only be hot worked with great difficulty. The pattern is continued through rhenium, technetium and iridium, by which time workability is no longer a major problem.

The chemical and physical properties of the noble refractory metals and their alloying behavior with one another are well documented. Information on the constitution of the alloys of the four metals has accumulated steadily, a large proportion of the effort in recent years going into studies of their alloys with the rare earths and the Groups IVA to VIIA transition metals. The progress of this effort may be judged by Table 2, which compares the number of binary systems catalogued by Hansen as of 1955-1957 with the number found by Elliot[5] in compiling a supplement some seven years later. This work is continuing, of course. For example, the International Nickel Company listed nineteen binary iridium-base systems which were known well enough by 1964 to classify by the extent of solid solubility in the terminal phase (Table 3). Even a rapid review of the most recent literature provides evidence of effort on an undiminished scale by laboratories such as those of Savitskii and associates in the USSR.

Hume-Rothery[73] has interpreted the alloying behavior of the platinum metals in terms of electronic structure. He has

TABLE 2 - AVAILABILITY OF BINARY CONSTITUTION DIAGRAMS

	Hansen (1955-57)			Elliott (1962)		
	Diagram	Some Data	Total	Diagram	Some Data	Total
Iridium	5	29	34	8	53	61
Osmium	3	27	30	6	46	52
Rhodium	8	28	36	15	43	58
Ruthenium	3	31	34	12	41	53
Total	19	115	134	41	183	224

TABLE 3 - STATUS OF IRIDIUM PHASE DIAGRAMS

SOLID SOLUBILITY IN IRIDIUM WEIGHT PERCENT			
Less Than 2%	More Than 2%	Complete Solubility	Limited Data
	Cb		Co	As	Ge	Se
Ag	Cr	Re	Fe	B	O	Si
Au	Mn	Ru	Ni	Be	P	Sn
Bi	Mo	Ta	Pt	C	Pb	Te
Mg	Os	Ti	Rh	Cu	S	Th
	Pd	W			Sb	V
						Zr

shown that the alloys of greatest interest, those of the
early transition metals with the platinum metals, tend to
follow a sequence of following types of structure as one goes
across the equilibrium diagram:

$$bcc \quad \sigma \quad \chi \quad hcp \quad fcc$$

and that the location of a phase boundry can be correlated
with Average Group Number (AGN). For this purpose Group
Numbers 8, 9 and 10 are arbitrarily allotted to the elements
of Groups VIIIA, VIIIB and VIIIC respectively. Thus an
equiatomic alloy of tungsten and osmium has an AGN of 7.

Available data on mechanical properties were collected by
Jaffee and associates in 1960[1] and supplemented in 1962[56].
Brookes and Harris reported on iridium in 1961[57]. With a
few additions which are mentioned below, these papers represent the current state of knowledge. Data on yield strength,
tensile strength, hardness, ductility and stress to rupture
have been obtained by several investigators for unalloyed
iridium up to 2000°C (3632°F). A lesser amount of information is available for rhodium and ruthenium to 1500°C (2732°F).
The difficulty of preparing test specimens has discouraged
the collection of data, other than a few hardness values,
for osmium.

The tensile properties of these metals compare very favorably
with those of the more common refractory metals. Iridium and
ruthenium are stronger than columbium, tantalum or molybdenum
at all test temperatures keeping in mind the limited appli-
cability of comparisons of materials at different levels of
cold work. Below about 1250°C (2282°F) they are also stronger
than hot finished tungsten. On a homologous temperature
basis iridium, rhodium and ruthenium all show an advantage
relative to tungsten over a wide temperature range. A compensation for density naturally favors ruthenium.

Douglass and Jaffee[56] noted that high temperature tensile
properties of rhodium and ruthenium are sensitive to atmospheric
contamination, strengths being significantly higher in a
vacuum of 10^{-6} mm. Hg than in air. No such effect was noted
for iridium. Reinacher, on the other hand, found a considerable effect for both iridium[74] as well as rhodium[75].
Reinacher's observations were based upon comparisons of bare
wire with platinum-sheathed wire, both tested in air, and the
interpretation of the results is complicated by sheath-sub-
strate interactions as well as by passage of oxygen through
the sheathing.

There have been a few attempts to improve the high temperature strength of the refractory noble metals by alloying.
Geach, Knapton and Woolf[76] found that a single-phase alloy
consisting of 45 atomic percent of molybdenum dissolved in
ruthenium was much more fabricable than ruthenium, had higher
hot hardness than tungsten at all temperatures up to 1220°C

and retained the corrosion resistance of unalloyed ruthenium.
Tensile properties were not determined. Limited data have
also been reported on iridium strengthened with additions of
zirconium[58] and tungsten[58, 77]. Dispersion strengthening
with oxides[59] and carbides is an especially interesting
technique which has been applied to platinum and, judging by
the patent literature, to some of the refractory platinum
metals[61, 62].

In spite of these encouraging results the limited supply and
high price of the refractory noble metals severely restrict
their use in concentrated form for major structural components.
The Bureau of Mines, for example, gives the 1966 imports of
iridium to the United States for consumption as 8161 troy
ounces. If to this are added 3979 troy ounces of new production and 402 troy ounces of secondary metal recovery within the United States, the total is equivalent to only four
hundred square feet of 0.020" sheet.

Iridium Coatings

The use of the oxidation resistant platinum metals as coatings for the major refractory metals rather than in bulk
offers one means of extending the available supply. Research
in this direction has involved all of the noble metals but
has concentrated, in the recent past, on iridium. Osmium
and ruthenium are too brittle and far too oxidation-prone for
use as coating materials (Figure 16). Platinum and palladium
have excellent resistance to oxidation but their lower melting points provide a limit to the practicable service temperature. Rhodium most nearly resembles iridium but its melting
point is lower by some five hundred degrees Centigrade. This
becomes a more serious objection when one considers the possible lowering of the melting point by possible interaction
with a substrate. Table 4 lists the minimum melting points
taken from the applicable binary systems and demonstrates
that iridium is the most practicable choice for protecting
columbium above 3000°F (1649°C), tantalum above 3400°F
(1871°C) or tungsten above 3500°F (1926°C).

Criscione et al[63] investigated iridium as a protective
coating for graphite and concluded that a number of its desirable characteristics would be equally useful in a coating
for the refractory metals. Some of these considerations are
summarized in Table 5.

Iridium was found to be virtually impervious to oxygen diffusion up to 4000°F (2204°C). Permability experiments were
conducted at this temperature by exposing a 0.015 inch thick
iridium membrane to flowing air on one side for 2½ hours
with a mass spectrometer monitoring the other side. No oxygen was observed and the upper limit for the permeability

TABLE 4

MINIMUM MELTING POINTS, °C

PLATINUM METAL - REFRACTORY METAL SYSTEMS

	Cb	Mo	Ta	W
Ir	1840	2080	1950	2305
Ru	1774	1945	1950	2205
Rh	1500	1940	1740	1966
Pt	1700 (Approx.)	1769	1600 (Approx.)	1769

TABLE 5

IRIDIUM AS A REFRACTORY METAL COATING FOR SERVICE NEAR 2000° C

FAVORABLE	UNFAVORABLE
DUCTILE	DIFFICULT TO APPLY
LOW O_2 PERMABILITY	SUPPLY INADEQUATE FOR LARGE-SCALE APPLICATION
MODERATE RECESSION RATE	HIGH DENSITY
IMPROVES WITH REDUCED O_2 PRESSURE	INTERDIFFUSION WITH Ta, Cb
GOOD EXPANSION MATCH	
MODERATELY HIGH-TEMPERATURE EUTECTICS	

was indicated to be less than 10^{-14} grams per square centimeter per second.

The coefficient of thermal expansion of iridium closely matches that of molybdenum and tantalum, as shown in Figure 17. The degree of mismatch is small relative to columbium and larger with respect to tungsten but still smaller than that of a non-metallic coating. Also, thermal stresses will be reduced by yielding and thermal stresses will be minimized by the high conductivity of iridium.

The behavior of iridium in a high temperature oxidizing environment differs from that of most protective coatings and determines the conditions under which its use is advantageous. The exact mechanisms are not clear, but it appears that a surface oxide is first formed and that this volatilizes at a rate which is dependent upon air pressure and flow rate as well as temperature. The rate-determining factor at the surface under engineering conditions is the ability of oxygen to reach the surface and form IrO_3. This accounts for the very large discrepancy between the high recession rates calculated from equilibrium vapor pressure data[64] and the moderate recession rates or weight losses observed by many investigators[65, 66, 67]. Berkowitz-Mattuck has calculated a recession rate of about 100 mils per hour from the extrapolated data of Alcock and Hooper and 0.005 mils per hour from the measurements of Krier and Jaffee, both for air at 1 atmosphere and 1400°C (2552°F). A second implication of the nature of the oxidation process is that the rate of surface recession is decreased at low partial pressures of oxygen. This is the reverse of what may happen with silicide coatings and suggests the possibility of application under conditions where silicides are unsuitable.

Iridium coatings suffer from several shortcomings, aside from the obvious one of price and availability. Low and non-reproducible emittance is one of these[68]; by reducing the heat-rejection ability of a coating-substrate combination this may result in a sufficient temperature rise to negate the protective ability of the coating. Another is the difficulty of application. Iridium cannot be plated from aqueous solution. Limited success has been obtained by electrodeposition from fused salts in an inert atmosphere[67] but the process is sensitive to small amounts of interstitial contaminants. Roll bonding has been successfully used, but it is shape-limited. Other techniques which have been explored[68] include slurry dipping and sintering[69] and vapor plating. One of the more serious difficulties results from the interdiffusion of iridium with the substrate. This results in a growth of the interfacial zone, which does not have the oxidation resistance of the iridium itself, toward the surface. The condition is particularly troublesome with a tantalum or columbium substrate. Criscione, Rexer and Fenish observed[67]

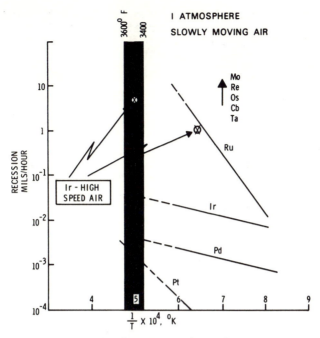

Fig. 16. Recession rate.

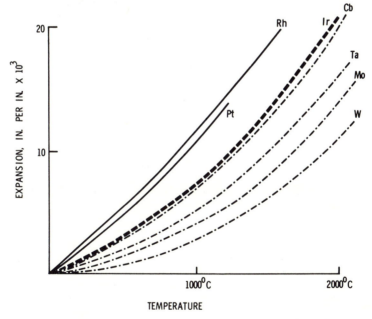

Fig. 17. Thermal expansion – refractory metals – platinum metals.

that the rate at which iridium was consumed by oxidation at
the surface in slowly moving air at 1850°C (3362°F) was only
about 0.02 mils per hour but that the rate of formation of
an interfacial diffusion zone with tantalum was six times as
high.

Several attempts have been made to increase the life of the
iridium coating and change its linear recession characteristics by modification with a variety of additions, including
other platinum metals. Dickson, Wimber and Stetson[70] reported that the addition of 30% rhodium improved the oxidation
resistance of the unalloyed iridium without lowering the melting point significantly. The surface recession rate observed
for this alloy subjected to plasma-arc oxidation at 2120°C
(3850°F) in a simulated air environment was 1 to 4 mils per
hour. Various refractory oxide and metal additions produced
no comparable improvement. This confirmed, in general,
earlier work[71] which indicated that iridium-rhodium alloys
have slightly better oxidation resistance than pure iridium
at 2000°C (3632°F).

This brief summary does not exhaust the list of programs in
progress. Combinations of iridium, rhenium, hafnium, platinum, refractory oxides and other materials have been studied
to provide diffusion barriers, increase emissivity and improve surface recession characteristics. Comparisons of
performance are very difficult at this time. It is hoped
that this situation will be improved with open publication
and more complete disclosure of test results. Lack of
standardization of test conditions is another handicap which
is recognized by active committees of the American Society
for Testing and Materials and the Materials Advisory Board.

An indication of the potentialities of the hafnium-tantalum
and iridium coatings relative to one another and the commercial coatings is in order. Both have a much higher temperature capability for short-time service, particularly at
low partial pressures of oxygen, than the silicides. Under
engineering conditions iridium has a lower recession rate
than hafnium-tantalum by as much as an order of magnitude,
but the rate is dependent upon the travel of oxygen through
the boundry layer to the surface and hence cannot be specified,
even approximately, without reference to the test conditions.
Iridium is also extremely limited in supply and incompatible
with such substates as tantalum alloys for more than a short
period because of interdiffusion and brittle compound formation. Neither is likely to displace the silicides and
aluminides except for limited use -- iridium as a thin coating for very high temperature service at low oxygen mass
flow rates and hafnium-tantalum as a thick cladding for short
time service at very high temperatures or possibly for lower
temperature service involving many thermal cycles of short
duration.

Acknowledgments

No individual can be expected to have a knowledge in depth
of so diverse a subject. The author acknowledges the in-
valuable assistance of R. Kemper of the Battelle Memorial
Institute, Hanford Washington, who reviewed the material on
technetium; Dr. D. R. Mash of the Fansteel Metallurgical
Corp., Compton, California, who supplied the figures on the
processing of the Hf-20% Ta alloy; W. D. Klopp and P. L. Raffo
of the NASA Lewis Research Center, Cleveland, Ohio, who
contributed information on rhenium; and D. J. Maykuth of the
Battelle Memorial Institute, Columbus, Ohio, who provided
data from much of the work of his own laboratory.

REFERENCES

(1) R. I. Jaffee, D. J. Maykuth and R. W. Douglass,
     Rhenium and the Refractory Platinum-Group Metals,
     A.I.M.E. Metallurgical Society Conference Vol. 11,
     Ed. by M. Semchyshen and J. J. Harwood, Interscience
     Publishers, 1961, 383-463

(2) B. W. Gonser, Editor, Rhenium, Elsevier Publishing Co.,
     1962 (225 pages). Proceedings of a symposium sponsored
     by the Electrochemical Society, May 1960

(3) J. G. Booth, R. I. Jaffee and E. I. Salkovitz, The
     Mechanisms of the Rhenium-Alloying Effect in Group
     VI-A Metals, Plansee Proceedings 1964, Edited by F.
     Benesovsky, Metallwerk Plansee AG, 547-564

(4) R. D. Peacock, The Chemistry of Technetium and Rhenium,
     Elsevier Publishing Company, 1966

(5) R. P. Elliot, Constitution of Binary Alloys, First
     Supplement, McGraw-Hill Book Co., 1965

(6) E. M. Savitskii and M. A. Tylkina, Phase Diagrams of
     Rhenium with the Transition Metals, pp. 67-83 of
     reference (2)

(7) J. Niemiec, Arch. Hutnictwa (Poland) 10 (1965) 121-141
     (Eng.)

(8) W. D. Klopp, Rhenium Ductility Effect in Group VIa
     Metals, NASA Technical Note TN D-   , 1968.

(9) N. D. Tarasov, R. A. Ulyanov and Y. D. Mikhaylov, The
     Effect of Alloying on the Physiomechanical Properties
     of Niobium, NASA technical translation from the Russian,
     TT-F-9374 (1964)

(10) G. D. McAdam, Substitutional Niobium Alloys of High
     Creep Strength, J. Inst. Metals 93 (1965), 559

(11) J. H. Port and J. M. Pontenandolfo, Fabrication and
     Properties of Rhenium and Rhenium-Molybdenum Alloy, in
     AIME Metallurgical Society Conferences Vol. 2, Reactive
     Metals, Interscience Publishers, New York/London (1959)
     pp. 555-574, with additional data by Chase Brass and
     Copper Co. dated March, 1965.

(12) C. T. Sims and R. I. Jaffee, Further Studies of the
     Properties of Rhenium Metal, AIME Trans. 206 (1956),
     913-917

(13) C. T. Sims, C. M. Craighead and R. I. Jaffee, Physical and Mechanical Properties of Rhenium, AIME Trans. 203 (1955) 168-179

(14) A. L. Field, Jr., R. L. Ammon, A. I. Lewis and L. S. Richardson, Fabrication and Properties of Tantalum-Base Alloys, in High Temperature Materials II, AIME Metallurgical Society Conferences Vol. 18, Interscience Publishers, 1963, pp. 139-157

(15) W. D. Klopp, W. R. Witzke and P. L. Raffo, Mechanical Properties of Dilute Tungsten-Rhenium Alloys, NASA Technical Note TN D-3483, September 1966

(16) J. W. Pugh, L. H. Amra and D. T. Hurd, Properties of Tungsten-Rhenium Lamp Wire, Trans. ASM 55 (1962), 451-461

(17) J. R. Stephens and R. G. Garlick, Compatibility of Tantalum, Columbium and their Alloys with Hydrogen in the Presence of a Temperature Gradient, NASA Technical Note TN D-3546, August 1966

(18) R. A. Rapp and G. N. Goldberg, The Oxidation of Cb-Zr and Cb-Zr-Re Alloys in Oxygen at 1000°C, Trans. Met. Soc. AIME 236 (1966), 1619-1628

(19) N. D. Tomashov, G. P. Chernova and O. N. Marcova, Effect of Supplementary Alloying Elements on Pitting Corrosion Susceptibility of 18 Cr-14Ni Stainless Steel, Corrosion 20 (1964), 166t-173t

(20) P. N. Flagella and C. O. Tarr, Creep-Rupture Properties of Rhenium and Some Alloys of Rhenium at Elevated Temperature, in Refractory Metals and Alloys IV, AIME Metallurgical Conference series, to be published by Gordon & Breach, 1968

(21) Metals Handbook, 8th Edition Vol. 1, American Society for Metals, 1961

(22) C. J. Smithells, Metals Reference Book, 4th Edition, Plenum Press, 1967

(23) R. W. Powell, R. P. Tye and M. J. Woodman, The Thermal Conductivity and Electrical Resistivity of Rhenium, J. Less-Common Metals 5 (1963), 49-56

(24) Electro-Optical Systems, Inc. NASA Contract NAS 7-514 (July 1967), unpublished data.

(25) J. G. Donaldson, F. W. Hoertel and A. A. Cochran, A Preliminary Study of Vapor Deposition of Rhenium and Rhenium-Tungsten, J. Less-Common Metals 14 (1968), 93-101

(26)   G. E. Boyd and Q. V. Larson, Report of the Occurrence of
       Technetium in the Earths' Crust, J. Phys. Chem 60 (1956)
       707-715

(27)   C. Perrier and E. Segré, Some Chemical Properties of
       Element 43, J. Chem. Phys. 5 (1937), 712-716

(28)   D. E. Baker, The Thermal Conductivity of Technetium,
       J. Less-Common Metals 8 (1965), 435-436

(29)   J. A. Wheeler, Jr., G. R. Love and M. L. Picklesimer,
       Technetium and its Alloys, Oak Ridge National Labor-
       atory ORNL-3870 (annual report for year ending June
       30, 1965)

(30)   R. S. Kemper and D. O. O'Keefe, Quarterly Progress
       Reports, NASA Contract R-48-005-001

(31)   G. H. Cartledge, J. Amer. Chem. Soc. 77 (1955), 2658

(32)   J. Niemiec, X-Ray Analysis of Technetium Binary Alloys
       with Tungsten and Rhenium (in English), Bulletin de
       L'Académie Polonaise des Sciences, Série des Sciences
       Chimiques, XI No. 6 (1963), 311-316

(33)   M. G. Chasanov, I. Johnson and R. V. Schablaske, The
       System Zinc-Technetium-99, J. Less-Common Metals 7
       (1964), 127-132

(34)   J. B. Darby, Jr., J. W. Downey and L. J. Norton, Inter-
       mediate Phases in Technetium-Aluminum and Technetium-
       Silicon Systems, J. Less-Common Metals 8 (1965), 15-19

(35)   J. B. Darby, Jr., A. F. Berndt and J. W. Downey, Some
       Intermediate Phases in the Thorium-Technetium and
       Uranium-Technetium Systems, J. Less-Common Metals
       9 (1965), 466-468

(36)   A. L. Giorgi and E. G. Szklarz, Los Alamos Scientific
       Laboratory, unpublished data, November 1964

(37)   C. C. Koch and G. R. Love, The Electrical Resistivity
       of Technetium from $8.0^\circ K$ to $1700^\circ K$, J. Less-Common
       Metals 12 (1967), 29-35

(38)   S. T. Sekula, R. H. Kernohan and G. R. Love, Super-
       conducting Properties of Technetium, Phys. Rev. 155
       (1967) 364-369

(39)   A. L. Giorgi and E. G. Szklarz, Superconductivity of
       Technetium and Technetium Carbide, J. Less-Common
       Metals 11 (1966), 455-456

(40) A. L. Giorgi and E. G. Szklarz, The Nb-Re and Nb-Tc
      Systems, Los Alamos Scientific Lab., Unpublished
      data, 1964

(41) C. C. Koch, R. H. Kernohan and S. T. Sekula, Super-
      conductivity in the Technetium-Vanadium Alloy System,
      J. Appl. Physics 38 (1967), 4359-4364

(42) C. C. Koch and G. R. Love, Reaction Morphologies and
      Superconducting Properties in Technetium-Vanadium
      Alloys, presented to AIME, February 1968.

(43) S. H. Autler, J. K. Hulm and R. S. Kemper, Super-
      conducting Technetium-Tungsten Alloys, Phys. Rev.
      140 (1965), A1177

(44) Anon., The Specialty Metals Industry, Materials Today
      (American Society for Metals) March 1968, 28-31

(45) D. E. Thomas and E. T. Hayes, the Metallurgy of Haf-
      nium, Atomic Energy Commission, U. S. Government
      Printing Office, 1959 (384 pp.)

(46) P. D. Raffo and W. D. Klopp, Mechanical Properties of
      Solid-Solution and Carbide-Strengthened Arc-Melted
      Tungsten Alloys, Technical Note TN D-3248.   National
      Aeronautics and Space Administration, February 1966.
      U. S. Government Printing Office.

(47) K. Marnoch, High-Temperature Oxidation-Resistant
      Hafnium-Tantalum Alloys, J. of Metals, (Nov. 1965)
      1225-1231

(48) J. J. Rausch, IIT Research Institute, Protective
      Coatings for Tantalum-Base Alloys, Tech. Report No.
      AFML-TR-64354, Air Force Materials Lab, Nov. 1964
      Part II, January 1966 and Part III, Sept. 1966

(49) R. T. Wimber, Solar Division of International Har-
      vester Company, Development of Protective Coatings
      for Tantalum Base Alloys, Tech. Summary Report ML-
      TDR-64-294, Pt II, Air Force Materials Lab., Nov.
      1965

(50) P. Kofstad and S. Espevik, Kinetic Study of High-
      Temperature Oxidation of Hafnium, J. Less-Common
      Metals 12 (May 1967), 382

(51) L. L. Oden, D. K. Deardoff, M. I. Copeland and H.
      Kato, The Hafnium-Tantalum Equilibriam Diagram,
      Bureau of Mines Station, Albany, Oregon, Report of
      Investigation 6521, 1964

(52)   P. A. Ramans, O. G. Paasche and H. Kato, The Trans-
       formation Temperature of Hafnium, J. Less-Common
       Metals 8 (1965), 213

(53)   A. Taylor and N. J. Doyle, The Constitution Diagram
       of the Niobium-Hafnium System, J. Less-Common Metals
       7 (1964), 37.

(54)   V. L. Hill, Development of Oxidation-Resistant Haf-
       nium Alloys, Final Report IITRI-136062-4, July 1967,
       under Naval Air Systems Command Contract NOw 66-0212-
       d

(55)   D. R. Mash, A. L. Donlevy and D. W. Bauer, Tantalum-
       Hafnium Alloys: Metallurgical Processing Technology.
       Interim report under NASA Contract NAS 7-417, 9 Oct.
       1966

(56)   R. W. Douglass and R. I. Jaffee, Elevated Temperature
       Properties of Rhodium, Iridium and Ruthenium, Proc.
       ASTM 62 (1962), 627-637

(57)   C. A. Brookes and B. Harris, Tensile Properties of
       Refractory Metals at High Temperatures, Plansee
       Proceedings 1961, F. Benesovsky, ed., pp. 712-722

(58)   R. D. Berry and J. Hope, High Temperature Mechanical
       Properties and Corrosion Resistance of Iridium and
       its Alloys, Rev. Met. 34 (1966) 339-345

(59)   A. S. Bufferd, K. M. Zwilsky, J. T. Blucher and
       N. J. Grant, Oxide Dispersion Strengthened Platinum,
       Int. J. Powder Met. 3 (1967), 17-26

(60)   A. S. Darling, G. L. Selman and A. A. Bourne, Dis-
       persion Strengthened Platinum, Plat. Metals Rev.
       12 (Jan. 1968)

(61)   International Nickel Co. (Mond), Ltd. British Patent
       974,057

(62)   New England Materials Laboratory, Inc. U. S. Patent
       3,175,904

(63)   J. C. Criscione et al, High Temperature Protective
       Coatings for Graphite, Air Force Reports AFML-TDR-
       64-173 Pt. II (Jan. 1965), AD 608 092; Pt. III (Dec.
       1965), AD 479131; and Pt. IV (Feb. 1967), AD 805
       438.

(64)   D. B. Alcock and G. W. Hooper, Thermodynamics of the
       Gaseous Oxides of the Platinum-Group Metals, Proc.
       Royal Society A. 254 (1960), 551-561

(65)   C. A. Krier and R. I. Jaffee, Oxidation of the Platinum-
        Group Metals, J. Less-Common Metals 5 (1963), 411-431

(66)   General Electric Flight Propulsion Laboratory, High-
        Temperature Materials Program, Prog. Report No. 17
        Part A (USAEC GEMP-17A), Nov. 15, 1962

(67)   J. M. Criscione, J. Rexer and R. G. Fenish, High
        Temperature Protective Coatings for Refractory Metals,
        Yearly Summary Report, NASA Accession Number 66-1022,
        March 1966.

(68)   H. F. Volk, Status of Oxidation Protective Coatings
        for Graphite, to be issued as part of report of
        Materials Advisory Board Panel on Protective Coatings,
        1968

(69)   V. L. Hill and J. J. Rausch, Protective Coatings for
        Tantalum-Base Alloys, Final Summary Technical Report,
        Air Force AFML-TR-64-354 Pt. III, Sept. 1966

(70)   D. T. Dickson, R. T. Wimber and A. R. Stetson, Very
        High Temperature Coatings for Tantalum Alloys, Final
        Summary Technical Report, Air Force AFML-TR-66-317,
        October 1966

(71)   E. D. Zysk, D. A. Toenshoff and J. Penton, Compati-
        bility of Iridium with Other High Temperature Materials
        at Elevated Temperatures, Engelhard Industries Tech.
        Bull. IV, (1963) 52-58

(72)   A. S. Darling, The Elastic and Plastic Properties of
        the Platinum Metals, Plat. Metals Rev. 10 (1966),
        14-18

(73)   W. Hume-Rothery, The Platinum Metals and their Alloys,
        Plat. Metals Rev. 10 (1966), 94-100

(74)   G. Reinacher, Metall, 18 (1964), 731-740, See Pro-
        perties of Iridium at High Temperatures, Platinum
        Metals Rev. 9 (1965), 18-19

(75)   G. Reinacher, Metall. 17 (1963), 699-705, See Pro-
        perties of Rhodium at High Temperatures, Platinum
        Metals Rev. 7 (1963), 144-146

(76)   G. A. Geach, A. G. Knapton and A. A. Woolf, Certain
        Alloys of Ruthenium with Molybdenum, same as ref. (57)
        pp. 750-758.

(77) International Nickel Co. (Mond) Ltd., British Patent
      974,057.

(78) L. Pauling, The Chemical Bond, Cornell Univ. Press
      1967, p. 210

Index

489